Lecture Notes in Mathematics

Edited by J.-M. Morel, F. Takens and B. Teissier

Editorial Policy
for the publication of monographs

1. Lecture Notes aim to report new developments in all areas of mathematics and their applications - quickly, informally and at a high level. Mathematical texts analysing new developments in modelling and numerical simulation are welcome.

   Monograph manuscripts should be reasonably self-contained and rounded off. Thus they may, and often will, present not only results of the author but also related work by other people. They may be based on specialised lecture courses. Furthermore, the manuscripts should provide sufficient motivation, examples and applications. This clearly distinguishes Lecture Notes from journal articles or technical reports which normally are very concise. Articles intended for a journal but too long to be accepted by most journals, usually do not have this „lecture notes" character. For similar reasons it is unusual for doctoral theses to be accepted for the Lecture Notes series, though habilitation theses may be appropriate.

2. Manuscripts should be submitted (preferably in duplicate) either to Springer's mathematics editorial in Heidelberg, or to one of the series editors (with a copy to Springer). In general, manuscripts will be sent out to 2 external referees for evaluation. If a decision cannot yet be reached on the basis of the first 2 reports, further referees may be contacted: The author will be informed of this. A final decision to publish can be made only on the basis of the complete manuscript, however a refereeing process leading to a preliminary decision can be based on a pre-final or incomplete manuscript. The strict minimum amount of material that will be considered should include a detailed outline describing the planned contents of each chapter, a bibliography and several sample chapters.

   Authors should be aware that incomplete or insufficiently close to final manuscripts almost always result in longer refereeing times and nevertheless unclear referees' recommendations, making further refereeing of a final draft necessary.

   Authors should also be aware that parallel submission of their manuscript to another publisher while under consideration for LNM will in general lead to immediate rejection.

3. Manuscripts should in general be submitted in English. Final manuscripts should contain at least 100 pages of mathematical text and should always include

   - a table of contents;
   - an informative introduction, with adequate motivation and perhaps some historical remarks: it should be accessible to a reader not intimately familiar with the topic treated;
   - a subject index: as a rule this is genuinely helpful for the reader.

   For evaluation purposes, manuscripts may be submitted in print or electronic form (print form is still preferred by most referees), in the latter case preferably as pdf- or zipped ps-files. Lecture Notes volumes are, as a rule, printed digitally from the authors' files. To ensure best results, authors are asked to use the LaTeX2e style files available from Springer's web-server at:

   ftp://ftp.springer.de/pub/tex/latex/mathegl/mono/ (for monographs) and

   ftp://ftp.springer.de/pub/tex/latex/mathegl/mult/ (for summer schools/tutorials).

   Additional technical instructions, if necessary, are available on request from lnm@springer-sbm.com.

Continued on inside back-cover

Lecture Notes in Mathematics 1890

M. Bunge · J. Funk

Singular Coverings of Toposes

Springer

Authors

Marta Bunge (Professor Emerita)
Department of Mathematics
McGill University
805 Sherbrooke Street West
Montréal, Québec, H3A 2K6
Canada
e-mail: marta.bunge@mcgill.ca

Jonathon Funk (Lecturer)
Department of Computer Science,
Mathematics, and Physics
The University of the West Indies
Cave Hill Campus
P.O. Box 64, Bridgetown
BB11000, Barbados
e-mail: funk@uwichill.edu.bb

Library of Congress Control Number: 2006928617

Mathematics Subject Classification (2000): Primary: 18B25, 57M12;
Secondary: 18C15, 06E15

ISSN print edition: 0075-8434
ISSN electronic edition: 1617-9692
ISBN-10 3-540-36359-9 Springer Berlin Heidelberg New York
ISBN-13 978-3-540-36359-0 Springer Berlin Heidelberg New York

DOI 10.1007/3-540-36359-9

Springer is a part of Springer Science+Business Media
springer.com

Typesetting by the authors and SPi using a Springer LaTeX package
Cover design: *design & production* GmbH, Heidelberg

Printed on acid-free paper SPIN: 11794899 VA41/3100/SPi 5 4 3 2 1 0

Dedicated to F. W. Lawvere

Preface

This book gives an introduction to a theory of complete spreads, which basically uses the same strategy employed by R.H. Fox [Fox57] for dealing with branched coverings, and which we carry out in close connection with (and parallel to) a theory of distributions in the sense of F. W. Lawvere [Law66, Law83, Law92].

Rather than elucidating the concepts of toposes, distributions, and complete spreads in this preface, we give a preliminary taste of these concepts by including certain quotations which authoritatively describe the original settings and motivations behind them, and let the reader explore these concepts and their interplay in the often new guises in which we present them in this book, itself based on our own work on these topics during the past ten years.

Toposes

"The original notion of a *topos*, as a 'generalized space' suitable for supporting the exotic cohomology theories required in algebraic geometry, sprang from the fertile brain of Alexandre Grothendieck in the early 1960's, and was developed in his Séminaire de Géométrie Algébrique du Bois-Marie particularly during the academic year 1963-64. The duplicated notes of that seminar circulated widely among algebraic geometers and category-theorists over the next decade, until Springer-Verlag did the world a service by publishing a revised and expanded edition in three volumes of *Lecture Notes in Mathematics* in 1972 [AGV72b]. But by then the subject had already been 'reborn' in its second incarnation, as an elementary theory having links with higher-order intuitionistic logic, through the collaboration of Bill Lawvere and Myles Tierney during 1969-70 (which, in turn, built upon Lawvere's work on providing a categorical foundation for mathematics, which had been developing since the early 1960's)."

P. T. Johnstone, *Sketches of an Elephant. A Topos Theory Compendium.* Volumes 1 and 2, Oxford University Press, 2002.

Distributions

"Following up on a 1966 Oberwolfach talk where I had proposed a theory of distributions (not only in but) on presheaf toposes, in 1983 at Aarhus I posed several questions concerning distributions on $\mathscr{S}$-toposes [where $\mathscr{S}$ is an elementary topos, thought of to be *Set*, the category of sets and functions]. The base for the definition and questions is a pair of analogies with known theories (commutative algebra and measure theory) for variable quantities, coupled with the fact that there are many important examples of variable $\mathscr{S}$-'quantities' where the domains of variation are $\mathscr{S}$-toposes. The intensively variable quantities are taken to be the sheaves on the topos, i.e., simply the objects in the category. Of course, the term 'topos' means 'place' or 'situation', but Grothendieck treats the general situation by dealing instead with the category of *Set*-valued quantities which vary continuously over it, as an affine k-scheme is described by dealing with the k-algebra of functions on it. (...) Then we follow the lead of analysis and define a *distribution* or extensively variable quantity on an $\mathscr{S}$-topos to be a continuous linear functional, or generalized point, i.e., a functor to $\mathscr{S}$ which preserves $\mathscr{S}$-colimits, but not necessarily the finite limits."

F. W. Lawvere, Comments on the Development of Topos Theory, in: Jean-Paul Pier (editor), *Development of Mathematics 1950-2000*, Birkhäuser Verlag, Basel Boston Berlin, 2000, 715-734.

Complete Spreads

"The principal object of this note is to formulate as a topological concept the idea of a 'branched covering space'. This topological concept encompasses the combinatorial concept used by Heegard, Tietze, Alexander, Reidemeister and Seifert. This has as a consequence that the knot-invariants defined by Seifert (the linking invariants of the cyclic coverings) are invariants of the topological type of the knot (i.e., unaltered by an orientation-preserving autohomeomorphism of 3-space). Without the developments of this note I am unable to see any simple proof that these invariants are invariants of anything more than the combinatorial type of the knot. It appears that the best way to look at a branched covering is as a *completion* of an unbranched covering. This completion process appears in its simplest form if it is applied to a somewhat wider class of objects. It is for this reason that I introduce the concept of a *spread* (a concept that encompasses, in particular, the 'branched and folded coverings' of Tucker.)"

R. H. Fox, Covering Spaces with Singularities, in R. H. Fox et al. (editors), *Algebraic Geometry and Topology; A Symposium in Honor of S. Lefschetz*, Princeton University Press, 1957, 243-257.

Acknowledgments

We are grateful to Bill Lawvere for suggesting that we write a book on distributions on toposes and complete spread geometric morphisms based on our work, and for his continued support.

In addition, we have profited from interactions with several colleagues (including our valuable collaborators) since we started this project ten years ago, among them, Bill Boshuck, Aurelio Carboni, Marcelo Fiore, Marco Grandis, Mamuka Jibladze, Peter Johnstone, André Joyal, John Kennison, Anders Kock, Steve Lack, Ieke Moerdijk, Susan Niefield, Bob Paré, Andy Pitts, Gonzalo Reyes, Richard Squire, Ross Street, Thomas Streicher, Ed Tymchatyn, Japie Vermeulen, Steve Vickers, and Richard Wood.

The first-named author has been partially supported throughout by an individual grant from the Natural Sciences and Engineering Research Council of Canada (NSERC) which, in particular, has made possible several visits by the second-named author to McGill University. The second-named author has been partially supported by an individual NSERC grant since April, 2005.

Montréal and Bridgetown,
February 2006

Marta Bunge
Jonathon Funk

Contents

Part III Aspects of Distributions and Complete Spreads

Introduction

Our main objective in this book is to develop a fairly self-contained theory of certain singular coverings of toposes which we name complete spreads. For instance, locally constant coverings are complete spreads, and are essential in describing those complete spreads which are branched coverings. Our theory extends the complete spreads in topology due to R. H. Fox (1957) but, unlike the classical theory, it emphasizes an unexpected connection with topos distributions in the sense of F. W. Lawvere (1983): topos distributions and complete spread geometric morphisms with a locally connected domain are opposite sides of the same coin.

We think of complete spreads as the geometry of distributions. The two notions complete spread and distribution come together and reinforce each other. Our constructions, though often motivated by classical theories, are sometimes quite different from them. We study special classes of distributions and of complete spreads, inspired respectively by functional analysis and topology. Among the former are the probability distributions; the branched coverings are singled out amongst the latter.

An étale geometric morphism, or local homeomorphism, is familiar to every topos-theorist. A complete spread is a kind of geometric morphism that is dual to an étale geometric morphism. Informally, if an étale geometric morphism is a generalized open part of a topos frame, then a complete spread is like a generalized closed part. Here we think of "closed" as orthogonal to a naturally prior notion of dense, as in a factorization system, rather than as the formal complement of open.

Just as sheaves and local homeomorphisms form equivalent categories, so do distributions (cosheaves) and complete spread geometric morphisms. In fact, the two equivalences may be established by the same basic construction, but they differ wildly in several respects. For instance, local homeomorphisms are exponentiable, but complete spreads need not be. Complete spreads are as basic and fundamental as local homeomorphisms. We hope that the reader will gain an appreciation of this, and of how we have come to this view.

Throughout we work in, or over, an elementary topos $\mathscr{S}$, of which we usually make no special assumptions. Thus, our methods and results are necessarily constructive and relative to $\mathscr{S}$. This perspective is important for other reasons. For instance, it helps to clarify and understand separately the completeness condition.

The symmetric topos classifies distributions and is part of a Kock-Zöberlein-monad (KZ-monad, for short) M on the 2-category $\mathbf{Top}_{\mathscr{S}}$ of toposes bounded over $\mathscr{S}$, geometric morphisms over $\mathscr{S}$, and 2-cells between them. We develop an axiomatic theory of certain KZ-monads (completion, closed, linear) on a 2-category $\mathscr{K}$, of which M is the motivating example. In particular, this gives a theory of complete spreads as an instance of the theory of discrete fibrations relative to a particular one such KZ-monad, namely the symmetric KZ-monad on $\mathbf{Top}_{\mathscr{S}}$.

We specialize and exemplify our theory in several different contexts: algebraic, logical, topological. In particular, we begin to develop a theory of branched coverings as certain kinds of complete spreads.

The present work stems from (but is not limited to) a number of papers written by the authors, individually, and in collaboration with each other or with others, since 1995. In the process of almost totally reorganizing and revising the existing material in a coherent manner, we have included numerous new concepts and results, new proofs of old results, and new examples from various areas of mathematics.

Although primarily aimed at topos theorists, this book may also be used as a textbook for an advanced one-year course introducing topos theory with an emphasis on geometric applications.

The book is divided into three interrelated, yet distinct parts, both in content as well as in background and approach. They are meant to be read in the order in which they are presented. However, an alternative way of discussing the material in two one-semester courses, with the second as a follow-up of the first, would be (roughly) to devote the first course to chapters 1, 2, 7, and 9, relegating chapters 3, 4, 5, 6, and 8 to a second course. Another possibility would be to cover chapters 1, 2, 3, 4, 5, ancd 6 in the first part, leaving chapters 7, 8, and 9 for the second part of the year, in view of the several open problems stated in the latter chapters.

In *Part 1* we introduce distributions and complete spread geometric morphisms, motivating them by means of examples. We then establish the main fact in this book, namely that there exists an adjoint equivalence between distributions on a topos $\mathscr{E}$ and complete spreads maps over $\mathscr{E}$ with locally connected domain. We emphasize that our investigations of complete spread geometric morphisms benefit from this far-reaching equivalence. Thus, the distribution concept from functional analysis and a concept from topology, namely complete spread, come together in the topos realm to form a coherent theory. The analogies multiply. Just as sheaves and cosheaves are dual notions, so are local homeomorphisms and complete spreads, functions and distributions, and discrete fibrations and discrete opfibrations. The rest of the book is devoted to an exploration of these aspects and to giving interesting examples and applications in various fields of mathematics.

Local connectedness plays a central role. Barr and Paré's (1980) investigation into local connectedness points to the relevant notion of complemented subobject: what they term a definable subobject (more generally, definable morphism). It is easy to see that a locally connected geometric morphism is subopen, and that definable morphisms compose in its domain topos. Such a geometric morphism is therefore what we name a definable dominance. These conditions form part of the Barr-Paré characterization of local connectedness.

The 'comprehensive factorization' is one of our basic tools. It says that every geometric morphism with locally connected domain admits a unique factorization into a pure geometric morphism followed by a complete spread.

We also deal separately with the notions of spread and completeness, and prove that a geometric morphism is a complete spread if and only if it is both a spread and complete.

In *Part 2* we develop an abstract (axiomatic) theory of complete spreads. Distributions on a topos $\mathscr{E}$ have a classifier: the symmetric topos. The study of the symmetric topos as the functor part of a monad is an instance of what we call a completion Kock-Zöberlein monad, or completion KZ-monad, for short. From certain key axioms (completion, closed, linear), which are all satisfied by the basic model, namely, the symmetric monad M, we are able to deduce, by using only 2-category theory, all of the main theorems of the subject, such as the comprehensive factorization and in turn, the equivalence between distributions on a topos $\mathscr{E}$ (or points of the topos $\mathrm{M}(\mathscr{E})$), and complete spreads over $\mathscr{E}$ with locally connected domain (or discrete M-fibrations over $\mathscr{E}$), where M is the symmetric monad. For instance, we show with these methods that complete spreads with locally connected domain are stable under bipullback along an $\mathscr{S}$-essential geometric morphism. It is not clear to us how to prove this directly from the original definition of complete spread.

There is a natural notion of density of a distribution that may be cast within the theory of closed completion KZ-monads. In connection with this, we make a special assumption on an object B of the 2-category $\mathscr{K}$ on which a given completion KZ-monad is given, namely, the existence of a Gleason core of B, or equivalently, the existence of the terminal distribution on B, meaning the terminal point of $\mathrm{M}(B)$. The 'Gleason core axiom' holds for all Grothendieck toposes, but a general criterion for when does it hold in more general situations is not available.

The pullback definition of a complete spread is used to establish an intrinsic characterization of completeness saying that a geometric morphism's pure factor is a surjection iff a certain cover-refinement property holds. The bicomma construction of a complete spread, on its part, shows that this same property (the pure factor is a surjection) holds iff every cogerm comes from a point, or that 'cogerms converge.' Thus, cogerms converge iff the cover-refinement property holds. This statement may be interpreted as a sort of 'Heine-Borel theorem'.

Lawvere has posed the question of finding a suitable 'single universe' in which cosheaves and sheaves exist and interact . For instance, the Kleisli category of the symmetric monad M is a natural candidate for such a single universe. Another approach to single universes uses a glueing construction along the density functor from distributions (respectively complete spreads, cosheaves, closed parts) to functions (respectively local homeomorphisms, sheaves, open parts), when such exists.

In *Part 3* we deal with further aspects of distributions and complete spreads, such as localic, algebraic, lattice-theoretic, and topological.

We begin with the lattice-theoretic aspects. A distribution on an object $\mathscr{E}$ of $\mathbf{Top}_{\mathscr{S}}$ is a pair of adjoint functors $\mu \dashv \mu_*$ in the sense of $\mathscr{S}$-indexed categories. We usually identify the notion of a distribution with just the $\mathscr{S}$-cocontinuous functor $\mu : \mathscr{E} \longrightarrow \mathscr{S}$ since, by the indexed version of the (special) adjoint functor theorem, such a μ must have an $\mathscr{S}$-indexed right adjoint μ_*. We show that the distribution μ is completely determined by the 'distribution algebra' $H = \mu_*(\Omega_{\mathscr{S}})$. Distribution algebras are part of a duality similar to the classical Stone duality. Indeed, the duality between *Set* and complete atomic Boolean algebras is a special case of the distribution algebra duality. We prove for instance that the opposite of the category distributions on a Grothendieck topos is monadic over the given topos.

Pure geometric morphisms are also the first factor of our relative version of Johnstone's (1982) pure, entire factorization theorem, in which we introduce relative versions of Stone locale and entire geometric morphism. A notion of relative Boolean algebra due to Jibladze and investigated further by Kock and Reyes (1999) is relevant in these investigations. Since a distribution algebra H in $\mathscr{E}$ over $\mathscr{S}$ is in particular an $\Omega_{\mathscr{S}}$-Boolean algebra, we may also consider the pure, entire factorization (relative to a base topos $\mathscr{S}$), and compare it with the comprehensive factorization for a geometric morphism with a locally connected domain.

In what we view as localic and algebraic aspects of distributions, we begin with an analysis of two important examples of completion KZ-monads that are related to the symmetric monad M. One is the lower power locale P_L, of interest in theoretical computer science, and the other (also of interest in theoretical computer science, but not just therein) is the lower bagdomain topos B_L. Not surprisingly, we obtain the decomposition $\mathrm{M} = \mathrm{B}_L \cdot \mathrm{T}$, where T classifies probability distributions. The KZ-monad T is thus the difference between distributions and bags of points. In this connection, a notion of discreteness that we call 'discrete complete spread structure' emerges as yet another single universe for local homeomorphisms and complete spreads. We discuss and analyse other algebraic aspects of distributions, such as coschemes. We make a special analysis of distributions on the so-called Jonsson-Tarski topos.

In the final chapter on topological aspects of distributions and complete spreads, we deal with branched coverings and, in particular, with locally constant coverings. We do not attempt to formally deal with folded coverings in this context, although they too ought to be instances of spreads. Our definition of a branched covering is inspired by the classical construction of a branched covering of a space with a given set of singularities as the spread completion of a universal (locally trivial) covering of the non-singular part of the space. Implicit in such a definition is the existence and uniqueness of the completion of a spread. We characterize branched coverings (with a locally connected domain) in an axiomatic way that relies only on the notions of complete spread, locally trivial covering, pure subobject, and on a newly isolated notion of purely skeletal geometric morphism. In effect, in this book we make the latter our formal definition of a branched covering and then prove its equivalence with the classically-inspired notion.

We also explain the difference between the usual covering spaces (locally constant, or locally trivial) and the unramified coverings which, in our context, form precisely the intersection of the two main classes of geometric morphisms that we have been considering: local homeomorphisms and complete spreads. We find something common to the usual coverings and the unramified ones: they both satisfy a general van Kampen theorem with respect to the same class of geometric morphisms of effective descent, namely, locally connected surjections. We discuss the fundamental groupoid as well as the branched fundamental groupoid, but we leave aside questions pertaining to path lifting and homotopy as this would take us too far from the main subject of the book.

Throughout we emphasize open problems whenever it seems appropriate to do so. Several routine proofs are left as exercises, but also as 'exercises' the reader will find open questions for possible future work. We leave it to the reader to judge the difficulty involved in each problem posed, partly because we do not always have a complete appraisal of it ourselves, but also because difficulty is a subjective matter, not inherent to the questions themselves.

Part I

Distributions and Complete Spreads

1

Lawvere Distributions on Toposes

In this chapter we begin, after some preliminaries, by recalling the notion of a topos distribution as understood by F. W. Lawvere (1966) in the spirit of the widely used 'Riesz paradigm', which models extensive quantities as linear functionals on intensive quantities. We refer the reader to the various papers of Lawvere [Law66, Law83, Law92, Law00a, Law00b] for a proper understanding of distributions in his sense, and of how they differ from those of L. Schwartz [Sch66]. A few comments about this interpretation and terminology are also made in this chapter.

In connection with Lawvere distributions, cosheaves were first studied by A. M. Pitts (1985). Cosheaves appear explicitly in the Borel-Moore homology theory (1960) arising by duality from sheaf cohomology. In universal algebra, G. Bergman (1991) is led to cosheaves by studying sheaves with values in the opposite of a category of algebras.

We then present Barr and Paré's (1980) concept of definable morphism in connection with their characterization of the notion of a locally connected topos. We introduce a weaker notion than locally connected geometric morphism, which we call a definable dominance by analogy with domain theory, as in the work of G. Rosolini (1986). All toposes defined over a Boolean base topos $\mathscr{S}$ are definable dominances, which may be a reason why classically it has not been isolated. Definable dominances are used in Chapter 7 to give a constructive version of the pure, entire factorization of a geometric morphism.

1.1 Basic Background in Topos Theory

An *elementary topos* is a category that:

1. has all finite limits (and all finite colimits),
2. is cartesian closed, and
3. has a subobject classifier $\langle \Omega, 1 \xrightarrow{\top} \Omega \rangle$, where 1 denotes the terminal object of the category.

Throughout, the script letters $\mathscr{S}$, $\mathscr{E}$, $\mathscr{Y}$, etc., usually denote an elementary topos. We reserve $\mathscr{S}$ for what we call the base topos - this is like specifying a set theory (possibly not the usual ZFC) in which all constructions take place. $\mathscr{S}(A, B)$ denotes the collection of morphisms $A \longrightarrow B$ in $\mathscr{S}$. We sometimes refer to a morphism $1 \longrightarrow B$ in a topos as a *global section.* When it suits our purpose we shall assume that $\mathscr{S}$ has a *natural numbers object.* The existence of a natural numbers object is a categorical analogue of the axiom of infinity in set theory. We shall think of $\mathscr{S}$ as the category *Set* of sets and functions, but we shall refrain from using any of its classical connotations such as the law of the excluded middle, or the axiom of choice. Different equivalent definitions of topos are known. For instance, we may assume just the existence of finite limits and power objects Ω^F. Lemma 1.1.1 (part of topos theory folklore) indicates how to construct the general exponential G^F. Intuitively, the lemma says that a subobject of $F \times G$ is the graph of a function $F \longrightarrow G$ iff it is everywhere-defined and well-defined.

Lemma 1.1.1 *An exponential* G^F *in a topos may be obtained as the following limit diagram.*

$$\begin{array}{ccccc} G^F & \rightarrowtail & E_1 & & \\ \downarrow & & \downarrow & & \\ E_2 & \rightarrowtail & \Omega^{F\times G} & \underset{\exists_G}{\overset{1_F}{\rightrightarrows}} & \Omega^F \\ & & {\scriptstyle p_1^*\wedge p_2^*}\downarrow\;\downarrow{\scriptstyle \exists_\delta} & & \\ & & \Omega^{F\times G\times G} & & \end{array}$$

G^F *is the pullback of the equalizers* E_1 *and* E_2. *The projections* $F \times G \times G \longrightarrow F \times G$ *are denoted* p_1, p_2.

We shall use the notation $P(\mathbb{C})$ for the category $\mathscr{S}^{\mathbb{C}^{\mathrm{op}}}$ of internal presheaves on a category $\mathbb{C}$ in an elementary topos $\mathscr{S}$. The reader may think of $P(\mathbb{C})$ as the category of contravariant $\mathscr{S}$-valued functors on a small category $\mathbb{C}$, and natural transformations between them. The well-known Yoneda functor

$$h : \mathbb{C} \longrightarrow P(\mathbb{C})$$

associates with an object c of $\mathbb{C}$ the representable presheaf $h(c) = h_c : \mathbb{C}^{\mathrm{op}} \longrightarrow \mathscr{S}$. $P(\mathbb{C})$ is also an elementary topos. For instance, an exponential G^F in $P(\mathbb{C})$ has the value $P(\mathbb{C})(F \times h_c, G)$ at c.

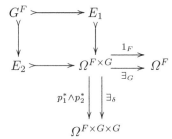

A *sieve on* c is a subobject of the representable h_c, usually denoted $R \rightarrowtail h_c$. The subobject classifier in $P(\mathbb{C})$ is the presheaf $\Omega : \mathbb{C}^{op} \longrightarrow \mathscr{S}$ whose value at an object c is the set $\Omega(c) = \{R \rightarrowtail h_c\}$ of all sieves on c, where $1 \xrightarrow{\top} \Omega$ is the natural transformation whose component at c is given by the identity natural transformation $h_c \longrightarrow h_c$. A *site* in $\mathscr{S}$ (which we continue to regard as if it were *Set*) is a pair $\langle \mathbb{C}, J \rangle$, where $\mathbb{C}$ is a category in $\mathscr{S}$ and where J is a *Grothendieck topology* on $\mathbb{C}$, meaning that for each object c of $\mathbb{C}$ there is assigned a family $J(c)$ of sieves on c such that:

1. for each c, $h_c \in J(c)$,
2. for each $d \xrightarrow{\alpha} c$, if $R \in J(c)$, then
$$\alpha^* R = \{e \xrightarrow{f} d \mid \alpha f \in R\} \in J(d)\ ,$$
3. if a sieve $S \rightarrowtail h_c$ has the property that there is $R \in J(c)$ such that for every $d \xrightarrow{\alpha} c \in R$, we have $\alpha^* S \in J(d)$, then $S \in J(c)$.

We sometimes refer to a member of $J(c)$ as a J-sieve, or as a J-cover. The notion of a site is equivalently (and often more conveniently) given by a pair $\langle \mathbb{C}, J \rangle$ where $\mathbb{C}$ is a category in $\mathscr{S}$ and $J \rightarrowtail \Omega$ is a subobject in $P(\mathbb{C})$ whose characteristic map $j : \Omega \longrightarrow \Omega$ is *a local operator*, meaning that it satisfies three conditions:

1. there exists a global section $1 \xrightarrow{d} J$ such that the triangle
$$\begin{array}{ccc} 1 & \xrightarrow{d} & J \\ & \searrow^{\top} & \downarrow \\ & & \Omega \end{array}$$
commutes;
2. the morphism $\Omega \xrightarrow{j} \Omega$ is idempotent (i.e., $j \cdot j = j$), and
3. the morphism $\Omega \xrightarrow{j} \Omega$ commutes with the characteristic map $\wedge : \Omega \times \Omega \longrightarrow \Omega$, classifier of the subobject $(\top, \top) : 1 \longrightarrow \Omega \times \Omega$.

A monomorphism $U \rightarrowtail Z$ in $P(\mathbb{C})$ is said to be J-*dense* if its characteristic morphism $m : Z \longrightarrow \Omega$ factors through $J \rightarrowtail \Omega$, so that there is a pullback

$$\begin{array}{ccc} U & \rightarrowtail & Z \\ \downarrow & & \downarrow m \\ 1 & \underset{d}{\rightarrowtail} & J \end{array}$$

in $P(\mathbb{C})$. Associated with any site $\langle \mathbb{C}, J \rangle$ is the full subcategory

$$i_* : Sh(\mathbb{C}, J) \rightarrowtail P(\mathbb{C})$$

of sheaves. *A sheaf* is a presheaf $F : \mathbb{C}^{op} \longrightarrow \mathscr{S}$ having the property that for any J-dense monomorphism $\sigma : U \rightarrowtail Z$ in $P(\mathbb{C})$, the induced $F^\sigma : F^Z \longrightarrow F^U$ is an isomorphism. The *associated sheaf functor*

$$i^* : P(\mathbb{C}) \longrightarrow Sh(\mathbb{C}, J)$$

is left adjoint to the inclusion i_*. It follows that $Sh(\mathbb{C}, J)$ is an elementary topos, and that i^* is *left exact, or lex*, meaning that it preserves finite limits. Yoneda followed by sheafification provides a canonical functor

$$i^* \cdot h : \mathbb{C} \longrightarrow Sh(\mathbb{C}, J) .$$

The adjoint pair $i^* \dashv i_*$ is an example of what is known as a geometric morphism. A *geometric morphism* between elementary toposes

$$\varphi : \mathscr{F} \longrightarrow \mathscr{E}$$

is an adjoint pair $\varphi^* \dashv \varphi_*$ such that $\varphi^* : \mathscr{E} \longrightarrow \mathscr{F}$ is left exact.

We say that $\mathscr{E}$ is *bounded over* $\mathscr{S}$ (by means of a geometric morphism $e : \mathscr{E} \longrightarrow \mathscr{S}$) if there is an equivalence $\mathscr{E} \cong Sh(\mathbb{C}, J)$ over $\mathscr{S}$, where $Sh(\mathbb{C}, J)$ is the category of sheaves on a site $\langle \mathbb{C}, J \rangle$ in $\mathscr{S}$. In this case, the inverse image functor $e^* : \mathscr{S} \longrightarrow \mathscr{E}$ associates with any "set" a constant sheaf (associated sheaf of the constant presheaf), and its right adjoint $\mathscr{E} \xrightarrow{e_*} \mathscr{S}$ is given by the global sections functor $e_*(X) = \mathscr{E}(1, X)$. If $\mathscr{S}$ is *Set*, then there is up to unique isomorphism only one possible such e. We refer to a bounded topos $\mathscr{E} \longrightarrow Set$ as a Grothendieck topos.

The objects of the 2-category $\mathbf{Top}_{\mathscr{S}}$ are bounded toposes over $\mathscr{S}$, which we call $\mathscr{S}$-toposes for short. A morphism in $\mathbf{Top}_{\mathscr{S}}$ is a pair (φ, t) consisting of a geometric morphism

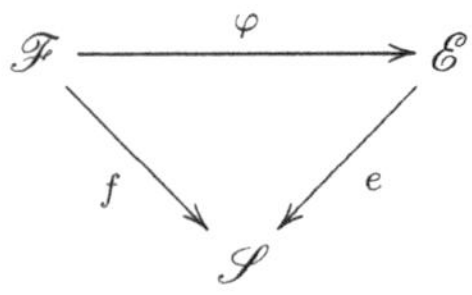

a natural isomorphism $t : \varphi^* \cdot e^* \cong f^*$. Usually we omit t from the notation. We say that φ *is over* $\mathscr{S}$. A 2-cell $(\varphi, t) \Rightarrow (\psi, s)$ in $\mathbf{Top}_{\mathscr{S}}$ is a natural transformation $r : \varphi^* \longrightarrow \psi^*$ such that $s \cdot re^* = t$.

1.2 Essential and Locally Connected Geometric Morphisms

Essential and locally connected geometric morphisms play a crucial role in the sequel.

Associated with a base topos $\mathscr{S}$ are the 2-categories $\mathbf{Top}_{\mathscr{S}}$ (defined above) and its opposite $\mathbf{Frm}_{\mathscr{S}}$ (reverse the 1-cells). The 2-category $\mathbf{Frm}_{\mathscr{S}}$ has the same objects and 2-cells as $\mathbf{Top}_{\mathscr{S}}$, but its 1-cells are the inverse image functors of geometric morphisms. We sometimes refer to the objects of $\mathbf{Frm}_{\mathscr{S}}$ as *topos frames*, since they are the domain of variation, in that 2-category, of lex $\mathscr{S}$-cocontinuous functors.

We may associate with any $\mathscr{S}$-topos $\mathscr{E} \longrightarrow \mathscr{S}$ an $\mathscr{S}$*-fibration* in the sense of Bénabou (1975), or with a presentation of the latter - namely an $\mathscr{S}$*-indexed* *category* in the sense of Paré and Schumacher (1978), in a natural way. The fiber $\mathscr{E}^I$ corresponding to an object I of $\mathscr{S}$ is given by the slice topos $\mathscr{E}/e^*I$, and to a morphism $K \xrightarrow{\alpha} I$ corresponds the induced functor $\alpha^* : \mathscr{E}^I \longrightarrow \mathscr{E}^K$ given by pulling back along $e^*\alpha$. We denote this $\mathscr{S}$-fibration by

$$\mathscr{E}/e^* \longrightarrow \mathscr{S} \,.$$

We shall meet this construction again in § 3.3.

The transition functors α^* for $\mathscr{E}/e^*$ have left and right adjoints Σ_α and Π_α. Moreover, these adjoints $\Sigma_\alpha \dashv \alpha^*$ and $\alpha^* \dashv \Pi_\alpha$ satisfy the Beck-Chevalley condition (BCC).

If $\mathscr{F} \xrightarrow{\varphi} \mathscr{E}$ is any geometric morphism, then both φ^* and φ_* are $\mathscr{E}$-indexed in an obvious sense, as is the adjointness $\varphi^* \dashv \varphi_*$.

Definition 1.2.1 *A geometric morphism $\mathscr{F} \xrightarrow{\varphi} \mathscr{E}$ is said to be* locally connected *if its inverse image functor φ^* has an $\mathscr{E}$-indexed left adjoint, denoted $\varphi_!$. If the structure morphism of a topos $\mathscr{E} \xrightarrow{e} \mathscr{S}$ is locally connected, then we say that $\mathscr{E}$ is* a locally connected topos, *as an object of* $\boldsymbol{Top}_{\mathscr{S}}$.

To say that a left adjoint $\varphi_!$ is $\mathscr{E}$-indexed means that whenever we have a pullback (below left)

$$\begin{array}{ccc} X & \longrightarrow & Y \\ \big\downarrow & & \big\downarrow \\ \varphi^* A & \xrightarrow{\varphi^* m} & \varphi^* B \end{array} \qquad\qquad \begin{array}{ccc} \varphi_! X & \longrightarrow & \varphi_! Y \\ \big\downarrow & & \big\downarrow \\ A & \xrightarrow{m} & B \end{array}$$

then the transposed square in $\mathscr{E}$ (above right) is a pullback.

If the structure map $\mathscr{E} \xrightarrow{e} \mathscr{S}$ is locally connected, then we sometimes denote the $\mathscr{S}$-indexed adjoints $e_! \dashv e^* \dashv e_*$ by the more traditional symbols $\pi_0 \dashv \Delta \dashv \Gamma$. We refer to $e_!$ or to π_0 as *the connected components functor.*

We recall a well-known site description of local connectedness that we sometimes rely on. A sieve $R \rightarrowtail h_c$ is said to be a *connected sieve* if it is non-empty, and if any two morphisms in R have a finite connecting diagram in R over c. We say that a site $\langle \mathbb{C}, J \rangle$ is a *locally connected site* if every J-sieve is connected.

Proposition 1.2.2 *If a site $\langle \mathbb{C}, J \rangle$ is locally connected, then the sheaf topos $Sh(\mathbb{C}, J)$ is locally connected. Any locally connected topos can be presented by a locally connected site.*

An $\mathscr{S}$-essential (or just essential) geometric morphism is weaker than a locally connected one. Only the $\mathscr{S}$-indexing of the geometric morphism (and not the $\mathscr{E}$-indexing) is taken into account.

Definition 1.2.3 *A geometric morphism $\mathscr{F} \xrightarrow{\varphi} \mathscr{E}$ in $\boldsymbol{Top}_{\mathscr{S}}$ is said to be* $\mathscr{S}$-essential *(or* essential*) if its inverse image functor φ^* has an $\mathscr{S}$-indexed left adjoint, denoted $\varphi_!$.*

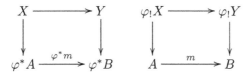

We sometimes omit the prefix "$\mathscr{S}$-" when the base topos is understood.

Remark 1.2.4

1. *A locally connected morphism in* $\mathbf{Top}_{\mathscr{S}}$ *is* $\mathscr{S}$*-essential, and a geometric morphism into the base topos* $\mathscr{S}$ *is locally connected iff it is essential.*
2. *Locally connected geometric morphisms are closed under composition and pullback. In general,* $\mathscr{S}$*-essential geometric morphisms are closed under composition, but are not necessarily stable under pullbacks in* $\mathbf{Top}_{\mathscr{S}}$*.*
3. *A geometric morphism* φ *over Set is Set-essential iff* φ^* *has a left adjoint, because the left adjoint, if it exists, is automatically Set-indexed.*
4. *Explicitly,* $\mathscr{F} \xrightarrow{\varphi} \mathscr{E}$ *is* $\mathscr{S}$*-essential if whenever the left square below is a pullback, then so is the transposed one (below right).*

$$\begin{array}{ccc} X & \longrightarrow & Y \\ \downarrow & & \downarrow \\ f^*I & \xrightarrow{f^*m} & f^*K \end{array} \qquad \begin{array}{ccc} \varphi_!X & \longrightarrow & \varphi_!Y \\ \downarrow & & \downarrow \\ e^*I & \xrightarrow{m} & e^*K \end{array}$$

Exercises 1.2.5

1. *By definition, a topological space is locally connected if its connected open subsets form a base for its topology. Show that a topological space is locally connected iff the connected components of any open subset are open. Show that a space is locally connected iff its topos of sheaves is locally connected.*
2. *(R. Auer) The Zariski spectrum of a commutative ring* R *is the set of prime ideals* P *of* R*, which we regard as a topological space whose topology is generated by the sets* $D(r) = \{P \mid r \notin P\}$*. Show that the Zariski spectrum of* R *is locally connected iff every localization* $R[r^{-1}]$ *has only finitely many idempotents.*
3. *Show that any presheaf topos* $\mathrm{P}(\mathbb{C}) \longrightarrow \mathscr{S}$ *is locally connected. Explicitly describe the connected components functor* π_0*, and show that for every object* c *of* $\mathbb{C}$*, we have* $\pi_0(h_c) = 1$*.*
4. *Show that if a collection of sieves satisfying only the first two requirements of a Grothendieck topology is locally connected, then the topos of sheaves for the generated topology is locally connected.*

1.3 Topos Frames and Distributions

An $\mathscr{S}$-indexed category $\mathscr{A}$ is said to be $\mathscr{S}$*-cocomplete* if:

1. each fiber $\mathscr{A}^I$ has stable finite colimits, and

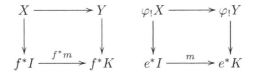

2. for each $K \xrightarrow{\alpha} I$ there is a left adjoint Σ_α to α^*, such that for each pullback square in $\mathscr{S}$ of the form

$$\begin{array}{ccc} K \times_I J & \xrightarrow{\pi_0} & K \\ \downarrow{\scriptstyle \pi_1} & & \downarrow{\scriptstyle \alpha} \\ J & \xrightarrow{\beta} & I \end{array}$$

the canonical arrow $\Sigma_{\pi_1} {\pi_0}^* \longrightarrow \beta^* \Sigma_\alpha$ is an isomorphism.

It follows readily from the above remarks that $\mathscr{E}/e^*$ is $\mathscr{S}$-cocomplete and complete.

Definition 1.3.1 *An* $\mathscr{S}$-cocontinuous *functor* $F : \mathscr{A} \longrightarrow \mathscr{B}$ *between* $\mathscr{S}$*-cocomplete* $\mathscr{S}$*-indexed categories is an* $\mathscr{S}$*-indexed functor* F *such that:*

1. *for each* I*, the functor* $F^I : \mathscr{A}^I \longrightarrow \mathscr{B}^I$ *preserves finite colimits, and*
2. *for any morphism* $J \xrightarrow{\alpha} I$*, the fiber functors commute with the* Σ_α*.*

An $\mathscr{S}$-additive *functor is defined by requiring that only coproducts (finite and* Σ*) are preserved.*

If $\psi : \mathscr{F} \longrightarrow \mathscr{E}$ is a geometric morphism over $\mathscr{S}$, then the $\mathscr{S}$-indexed functor $\psi^* : \mathscr{E} \longrightarrow \mathscr{F}$ is $\mathscr{S}$-cocontinuous.

Let $\mathbf{Coc}_{\mathscr{S}}$ denote the 2-category of $\mathscr{S}$-cocomplete categories, $\mathscr{S}$-cocontinuous functors, and natural transformations. There is a 2-functor

$$U : \mathbf{Frm}_{\mathscr{S}} \longrightarrow \mathbf{Coc}_{\mathscr{S}} \tag{1.1}$$

which simply forgets that a topos frame $\mathscr{E}$ is anything more than an $\mathscr{S}$-cocomplete category and that a topos frame morphism $\psi^* : \mathscr{E} \longrightarrow \mathscr{F}$ is anything more than cocontinuous.

The 2-category $\mathbf{Top}_{\mathscr{S}}$ may be regarded as "a category of space". This idea is given substance by the following result. Recall that if $\mathscr{K}$ is a 2-category with finite coproducts, then we have a pseudo-functor $\Phi : \mathscr{K} \times \mathscr{K} \longrightarrow \mathscr{K}$ given by coproduct; if A and B are objects of $\mathscr{K}$, we may then form

$$\Phi_{A,B} : \mathscr{K}/A \times \mathscr{K}/B \simeq (\mathscr{K} \times \mathscr{K})/(A,B) \longrightarrow \mathscr{K}/(A+B) \, .$$

A 2-category $\mathscr{K}$ with finite coproducts is *extensive* if for all pairs of objects A, B of $\mathscr{K}$, the functor $\Phi_{A,B}$ is a biequivalence. Equivalently, $\mathscr{K}$ is extensive if the contravariant pseudo-functor that assigns a 0-cell A to the 2-category $\mathscr{K}/A$ preserves binary products. If $\mathscr{K}$ has finite coproducts and a terminal object T, then $\mathscr{K}$ is extensive iff Φ is a biequivalence for $A = B = T$.

Proposition 1.3.2 *$\mathbf{Top}_{\mathscr{S}}$ is extensive.*

Proof. $\mathbf{Top}_{\mathscr{S}}$ has (bi)coproducts and $\mathscr{S} \xrightarrow{id} \mathscr{S}$ is the terminal object. The topos-frame of the topos coproduct $\mathscr{F} + \mathscr{G}$ is the product category $\mathscr{F} \times \mathscr{G}$ (not to be confused with the topos product $\mathscr{F} \times_{\mathscr{S}} \mathscr{G}$). We must show that

$$\Phi : \mathbf{Top}_{\mathscr{S}} \times \mathbf{Top}_{\mathscr{S}} \longrightarrow \mathbf{Top}_{\mathscr{S}}/(\mathscr{S} + \mathscr{S})$$

is a biequivalence. $\mathscr{S} + \mathscr{S}$ classifies binary coproduct decompositions of the terminal object of a topos. Indeed, if we have a geometric morphism

$$\begin{array}{ccc} \mathscr{E} & \xrightarrow{\ h\ } & \mathscr{S} + \mathscr{S} \\ & {\scriptstyle e}\searrow \qquad \swarrow & \\ & \mathscr{S} & \end{array}$$

then $h^*(1,1) \cong h^*(1,0) + h^*(0,1)$ is a coproduct decomposition of $1_{\mathscr{E}}$.

On the other hand, given a coproduct decomposition $1_{\mathscr{E}} \cong X + Y$, define $h^*(A, B) = X \times e^*A + Y \times e^*B$. We now define a pseudo-inverse

$$\Psi(h) = (\mathscr{E}/X \longrightarrow \mathscr{S}, \mathscr{E}/Y \longrightarrow \mathscr{S}) ,$$

where the decomposition $1_{\mathscr{E}} \cong X+Y$ corresponds to h. $\mathscr{E}$ is itself an extensive category, so the topos coproduct $\mathscr{E}/X + \mathscr{E}/Y$ is equivalent to $\mathscr{E}$. It follows easily that Φ and Ψ are mutually pseudo-inverse. □

Corollary 1.3.3 *The full sub-2-category* $\boldsymbol{LTop}_{\mathscr{S}}$ *of* $\boldsymbol{Top}_{\mathscr{S}}$ *whose objects are the locally connected toposes is extensive.*

Proof. $\mathbf{LTop}_{\mathscr{S}}$ is closed in $\mathbf{Top}_{\mathscr{S}}$ under finite coproducts. □

We now come to one of the central notions in this book.

Definition 1.3.4 *(Lawvere) Let* $\mathscr{E}$ *be an* $\mathscr{S}$*-topos.* A distribution *on* $\mathscr{E}$ *with values in an* $\mathscr{S}$*-topos* $\mathscr{X}$ *is an* $\mathscr{S}$*-indexed adjoint pair* $\mu \dashv \mu_*$*:*

$$\mathscr{E} \underset{\mu}{\overset{\mu_*}{\rightleftarrows}} \mathscr{X} .$$

We denote by $\boldsymbol{Dist}_{\mathscr{S}}(\mathscr{X}, \mathscr{E})$ *the category of* $\mathscr{X}$*-valued distributions on* $\mathscr{E}$*, and natural transformations between them. This gives rise to a 2-category* $\boldsymbol{Dist}_{\mathscr{S}}$*.*

Equating with an adjoint pair $\mu \dashv \mu_*$ the $\mathscr{S}$-cocontinuous functor μ is an equivalence $\mathbf{Dist}_{\mathscr{S}}(\mathscr{X}, \mathscr{E}) \simeq \mathbf{Coc}_{\mathscr{S}}(\mathscr{E}, \mathscr{X})$. This is justified by the $\mathscr{S}$-indexed version of the (Special) Adjoint Functor Theorem due to Paré and Schumacher.

Remark 1.3.5 *We usually identify the term '*$\mathscr{S}$*-valued distribution on* $\mathscr{E}$*' with the* $\mathscr{S}$*-cocontinuous left adjoint* $\mathscr{E} \xrightarrow{\mu} \mathscr{S}$*. We sometimes denote* μ *by* $\int_{\mathscr{E}}(-)d\mu$*. If* $\mathscr{E}$ *is locally connected* ($e_! \dashv e^*$)*, then we denote* $\int_{\mathscr{E}}(-)de_!$ *by* $\int_{\mathscr{E}}(-)dx$*.*

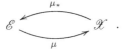

Assigning to each $\mathscr{S}$-topos $\mathscr{E}$ the category

$$\mathbf{E}(\mathscr{E}) = \mathbf{Dist}_{\mathscr{S}}(\mathscr{S},\mathscr{E}) \simeq \mathbf{Coc}_{\mathscr{S}}(\mathscr{E},\mathscr{S})$$

of $\mathscr{S}$-valued distributions on $\mathscr{E}$ extends to a *covariant* 2-functor

$$\mathbf{E} : \mathbf{Top}_{\mathscr{S}} \longrightarrow \mathbf{Coc}_{\mathscr{S}}$$

by letting $\mathbf{E}(\psi)$ be composition with ψ^*. We think of $\mathbf{E}$ as an *extensive quantity type* of $\mathscr{S}$-toposes, and of an 'element' of $\mathbf{E}(\mathscr{E}) = \mathbf{Coc}(\mathscr{E},\mathscr{S})$ as an extensive quantity of type $\mathscr{S}$ and domain $\mathscr{E}$. Notice that $\mathbf{E}(\mathscr{S}) \simeq \mathscr{S}$, and that for any $\mathscr{E} \xrightarrow{e} \mathscr{S}$ there is induced a 'total' $\mathbf{E}(\mathscr{E}) \longrightarrow \mathbf{E}(\mathscr{S}) \simeq \mathscr{S}$ sending $\mu \mapsto \mu(1)$, whereas any point $p : \mathscr{S} \longrightarrow \mathscr{E}$ in $\mathbf{Top}_{\mathscr{S}}$ induces a cocontinuous functor $\mathbf{E}(p)$, which is a member of the category

$$\mathbf{Coc}_{\mathscr{S}}(\mathbf{E}(\mathscr{S}),\mathbf{E}(\mathscr{E})\,) \simeq \mathbf{E}(\mathscr{E})\,.$$

Under this equivalence, we see that $\mathbf{E}$ assigns to a point p of $\mathscr{E}$ the 'Dirac measure' p^* on $\mathscr{E}$ concentrated on the point p.

The topos $\mathscr{R} = \mathscr{S}^{\mathscr{S}_f}$, where $\mathscr{S}_f$ is the topos of finite sets, is the *object classifier* in $\mathbf{Top}_{\mathscr{S}}$. Its category of points is the topos $\mathscr{S}$. The existence of the object classifier $\mathscr{R}$, whose category of points is $\mathscr{S}$, shows that the intuition of sheaves as intensive quantities is correct. Indeed, as a topos frame, an object $\mathscr{E}$ of $\mathbf{Top}_{\mathscr{S}}$ is equivalent to the category $\mathbf{Top}_{\mathscr{S}}(\mathscr{E},\mathscr{R})$, by an equivalence that assigns to a geometric morphism $\varphi : \mathscr{E} \longrightarrow \mathscr{R}$ the object $\varphi^*(U)$, where U is the generic object in $\mathscr{R}$: U is the inclusion functor $\mathscr{S}_f \rightarrowtail \mathscr{S}$. In turn, this suggests the corresponding notion of distribution as an extensive quantity, namely as a linear functional on the intensive quantities.

Denote $\mathbf{Top}_{\mathscr{S}}(\mathscr{E},\mathscr{R})$ by $\mathbf{A}(\mathscr{E})$. This extends to a *contravariant* 2-functor

$$\mathbf{A} : \mathbf{Top}_{\mathscr{S}}^{\mathrm{op}} \longrightarrow \mathbf{Coc}_{\mathscr{S}}\,.$$

We think of $\mathbf{A}$ as an *intensive quantity type* of $\mathscr{S}$-toposes, and of an element of $\mathbf{A}(\mathscr{E}) = \mathbf{Top}_{\mathscr{S}}(\mathscr{E},\mathscr{R})$ as an intensive quantity of type $\mathscr{R}$ and domain $\mathscr{E}$. $\mathbf{A}$ and the previous U (1.1) are one and the same functor.

Remark 1.3.6 *The* Riesz paradigm *in the topos context can be explained more clearly by analogy with measure theory. This analogy is based on replacing the real line $\mathbb{R}$ by 'the space of sets,' which we have denoted suggestively by $\mathscr{R}$. In* measure theory *one has*

1. *The Riesz algebra $A(X)$ consists of all the (Borel) measurable functions $X \longrightarrow \mathbb{R}$.*
2. *A measure on X is a linear map $A(X) \longrightarrow \mathbb{R}$.*

In Lawvere distributions theory *one has*

1. *The topos frame* $\mathbf{A}(\mathscr{E})$ *is the category of geometric morphisms from* $\mathscr{E}$ *to* $\mathscr{R}$ *(the space of sets).*
2. *A distribution on* $\mathscr{E}$ *is an* $\mathscr{S}$*-cocontinuous functor* $\mathbf{A}(\mathscr{E}) \longrightarrow \mathscr{S}$.

The Riesz paradigm affords (in general) a universal class of extensive examples, namely linear combinations of points. *Of couse, in the topos case, the coefficients are sets. Lawvere had chosen the term "distribution" rather than "measure" because the connotation of the latter as a functional on subsets is inappropriate in the topos context.*

Remark 1.3.7 *It is natural to regard distributions on* $\mathscr{E}$ *as* generalized points *of* $\mathscr{E}$. *We have an inclusion*

$$\boldsymbol{Top}_{\mathscr{S}}(\mathscr{X}, \mathscr{E}) \hookrightarrow \boldsymbol{Dist}_{\mathscr{S}}(\mathscr{X}, \mathscr{E})$$

since a geometric morphism $\mathscr{X} \longrightarrow \mathscr{E}$ *is equivalently given by a left exact* $\mathscr{S}$*-cocontinuous functor* $\mathscr{E} \longrightarrow \mathscr{X}$, *whereas an* $\mathscr{X}$*-valued distribution on* $\mathscr{E}$ *is only* $\mathscr{S}$*-cocontinuous. The choice of the 1-cells in the 2-category* $\boldsymbol{Dist}_{\mathscr{S}}$ *is appropriately given in order to render notationally evident this inclusion of points into distributions.*

There is an *action of functions on distributions*. For any object X of $\mathscr{E}$ (i.e., a sheaf regarded as a function) and any distribution μ on $\mathscr{E}$, we have a new distribution $X.\mu : \mathscr{E} \longrightarrow \mathscr{S}$ on $\mathscr{E}$, given by the formula

$$X.\mu\,(E) = \mu(X \times E)\,.$$

In integral notation, this is the familiar formula $\int_{\mathscr{E}} g\, d(f.\mu) = \int_{\mathscr{E}} (gf)\, d\mu$.

Exercises 1.3.8

1. $\mathscr{G} = \mathscr{E}/e^*$ *is the glueing category along* e^* *of* $\mathscr{S}$ *and* $\mathscr{E}$. $\mathscr{G} \xrightarrow{g} \mathscr{S}$ *is a (totally connected) topos that includes* $\mathscr{E}$ *as a closed subtopos, and has an open point* $p : \mathscr{S} \longrightarrow \mathscr{G}$ *such that* $g \dashv p$. *Moreover, if* $\mathscr{E}$ *is bounded over* $\mathscr{S}$, *then so is* $\mathscr{G}$. *Exercise 2 explains* $\mathscr{G}$ *when* $\mathscr{E}$ *is the sheaf topos of a space.*
2. *Let* X *be any space. Let* $\widehat{X}$ *denote the space obtained when a point* p *is added to* X *such that the open sets of* $\widehat{X}$ *are sets* $U \cup \{p\}$, *such that* $U \subseteq X$ *is open. Show that* $Sh(\widehat{X})$ *is equivalent to the glueing topos* $Sh(X)/\Delta$.
3. *Show that there is a canonical equivalence* $\mathscr{S}/I \simeq \mathbf{E}(\mathscr{S}/I)$, *for any object* I *of* $\mathscr{S}$.

1.4 Cosheaves and the Bergman Functor

The connection between distributions and cosheaves was first made explicit by A. M. Pitts.

Definition 1.4.1 *Let* $\langle \mathbb{C}, J \rangle$ *be a site in* $\mathscr{S}$*. An* $\mathscr{S}$*-valued* J-cosheaf *on* $\mathbb{C}$ *is a functor* $\mathbb{C} \longrightarrow \mathscr{S}$ *that takes* J*-covers to colimiting cones. Denote by* $\mathrm{Cosh}(\mathbb{C}, J)$ *the full subcategory of* $\mathscr{S}^{\mathbb{C}}$ *whose objects are the* $\mathscr{S}$*-valued* J*-cosheaves on* $\mathbb{C}$*. Similarly, we may define* $\mathrm{Cosh}_{\mathscr{F}}(\mathbb{C}, J)$*, replacing* $\mathscr{S}$ *by an arbitrary* $\mathscr{S}$*-topos* $\mathscr{F}$*.*

An arbitrary functor $F : \mathbb{C} \longrightarrow \mathscr{F}$ may be lifted in a canonical way to an $\mathscr{S}$-cocontinuous functor $P(\mathbb{C}) \longrightarrow \mathscr{F}$, called the left Kan extension of F along h.

We leave the proof of the following lemma as an exercise.

Lemma 1.4.2 *Let* $\langle \mathbb{C}, J \rangle$ *be a site in* $\mathscr{S}$*. Then a functor* $\mathbb{C} \longrightarrow \mathscr{F}$ *is a* J*-cosheaf iff its left Kan extension* $P(\mathbb{C}) \longrightarrow \mathscr{F}$ *along the Yoneda functor* h *inverts* J*-dense monomorphisms.*

Proposition 1.4.3 *Let* $\langle \mathbb{C}, J \rangle$ *be a site. Let* $\mathscr{F}$ *be a bounded* $\mathscr{S}$*-topos. Then composition with the canonical functor* $\mathbb{C} \longrightarrow Sh(\mathbb{C}, J)$ *is an equivalence*

$$\mathbf{Coc}_{\mathscr{S}}(Sh(\mathbb{C}, J), \mathscr{F}) \simeq \mathrm{Cosh}_{\mathscr{F}}(\mathbb{C}, J)\,.$$

Proof. Composition with the Yoneda functor h is an equivalence

$$\mathbf{Coc}_{\mathscr{S}}(P(\mathbb{C}), \mathscr{F}) \simeq \mathscr{F}^{\mathbb{C}}\,. \tag{1.2}$$

If $\mathbb{C}$ is equipped with a Grothendieck topology J, then composition with the ($\mathscr{S}$-cocontinuous) left adjoint i^* of the inclusion

$$i^* \dashv i_* : Sh(\mathbb{C}, J) \rightarrowtail P(\mathbb{C})$$

provides a fully faithful functor

$$\mathbf{Coc}_{\mathscr{S}}(i^*, \mathscr{F}) : \mathbf{Coc}_{\mathscr{S}}(Sh(\mathbb{C}, J), \mathscr{F}) \longrightarrow \mathbf{Coc}_{\mathscr{S}}(P(\mathbb{C}), \mathscr{F})\,.$$

The equivalence (1.2) restricts along this functor to the equivalence stated in the Proposition. In other words, a functor $F : \mathbb{C} \longrightarrow \mathscr{F}$ is a J-cosheaf iff its left Kan extension $P(\mathbb{C}) \longrightarrow \mathscr{F}$ factors through i^*. Indeed, let F^* denote the right adjoint of the left Kan extension of F, so that $F^*(Y)(c) = \mathscr{F}(F(c), Y)$. Then the left Kan extension of F factors through i^* iff F^* factors through i_* iff F is a J-cosheaf. The last step requires Lemma 1.4.2. □

A crucial fact about a topos frame is that it may be presented as a coinverter. This holds because cosheaves are an inverter.

Definition 1.4.4 *A* coinverter *in a 2-category* $\mathscr{K}$ *is a weighted colimit of a diagram in* $\mathscr{K}$ *of the following kind.*

$$A \underset{g}{\overset{f}{\underset{t\Downarrow}{\rightrightarrows}}} B$$

A coinverter is a 1-cell $k : B \longrightarrow C$ *such that* kt *is an isomorphism, which is universal with this property. The above diagram together with its coinverter is called* a coinverter diagram. *The dual notion is that of an* inverter diagram.

A coinverter (inverter) is unique up to equivalence, so that any two equivalences between coinverters (inverters) of the same diagram are isomorphic up to unique isomorphism.

Proposition 1.4.5 *Let* $\langle \mathbb{C}, J \rangle$ *be a site. Then the category of* J*-cosheaves makes the diagram*

$$\mathrm{Cosh}_{\mathscr{F}}(\mathbb{C}, J) \rightarrowtail \mathscr{F}^{\mathbb{C}} \underset{\Phi}{\overset{\Psi}{\underset{\Downarrow}{\rightrightarrows}}} \mathscr{F}^{\mathbb{J}}$$

an inverter (Def. 1.4.4) *in* $\mathbf{Coc}_{\mathscr{S}}$, *where* $\mathbb{J} \longrightarrow \mathbb{C}$ *is the discrete fibration corresponding to the presheaf* J. *The functors* Φ *and* Ψ *are:*

$$\Phi(F)(R,c) = F(c) \text{ and } \Psi(F)(R,c) = \varinjlim_{R} F ,$$

where $R \rightarrowtail h_c$ *is a* J*-covering sieve, and* F *is a functor* $\mathbb{C} \longrightarrow \mathscr{F}$.

Proof. Just consider the definition of cosheaf, and the fact that inverters in $\mathbf{Coc}_{\mathscr{S}}$ are created in categories. □

The extensive quantity $\mathbf{E}(\mathscr{E})$ of type $\mathscr{S}$ and domain $\mathscr{E}$ has the structure of an $\mathscr{E}$-indexed category. For an object X, the fiber $\mathbf{E}(\mathscr{E})^X$ is the category $\mathbf{E}(\mathscr{E}/X)$, and the transition functor $\mathbf{E}(\mathscr{E}/Y) \longrightarrow \mathbf{E}(\mathscr{E}/X)$ along a morphism $X \xrightarrow{\alpha} Y$ of $\mathscr{E}$ is given by composition with the ($\mathscr{S}$-cocontinuous) functor $\mathscr{E}/X \xrightarrow{\Sigma} \mathscr{E}/Y$. The action of sheaves on distributions

$$X.\mu\,(E) = \mu(X \times E)$$

is derived from the $\mathscr{E}$-indexed structure of $\mathbf{E}(\mathscr{E})$.

Remark 1.4.6 *If* $\mathbb{C}$ *is a small category, then* $\mathbf{E}(P(\mathbb{C})) \simeq \mathscr{S}^{\mathbb{C}}$ *is a* $P(\mathbb{C})$*-indexed category, whose* $P(\mathbb{C})$*-fibers are toposes.* $\mathbf{E}(P(\mathbb{C}))$ *is a topos over* $\mathscr{S}$*; however, in general* $\mathbf{E}(P(\mathbb{C}))$ *is not a topos over* $P(\mathbb{C})$ *in the sense that the corresponding* $P(\mathbb{C})$*-fibration is engendered by a geometric morphism* $\mathscr{X} \longrightarrow P(\mathbb{C})$.

Let us now investigate the existence of certain coinverters in $\mathbf{Coc}_{\mathscr{S}}$. Suppose that $\lambda : \mathscr{E} \longrightarrow \mathscr{F}$ is a distribution and $X \xrightarrow{f} Y$ is a morphism of $\mathscr{E}$. Then we have a diagram

$$\mathscr{E} \underset{Y.\lambda}{\overset{X.\lambda}{\rightrightarrows}} \mathscr{F} \quad (f.\lambda \Downarrow)$$

in $\mathbf{Coc}_{\mathscr{S}}$. By definition, $f.\lambda$ is the natural transformation such that $(f.\lambda)_E = \lambda(f \times E)$.

Proposition 1.4.7 *Let* $1 \xrightarrow{d} J$ *denote a topology in a presheaf topos* $P(\mathbb{C})$, *with subtopos* $i^* \dashv i_* : \mathscr{E} \rightarrowtail P(\mathbb{C})$. (We regard d as a morphism $d \xrightarrow{d} 1_J$ of $P(\mathbb{C})/J$.) *Then*

$$P(\mathbb{C})/J \underset{\Sigma_J}{\overset{d.\Sigma_J}{\rightrightarrows}} P(\mathbb{C}) \xrightarrow{i^*} \mathscr{E} \quad (d.\Sigma_J \Downarrow) \tag{1.3}$$

is a coinverter diagram in $\mathbf{Coc}_{\mathscr{S}}$. *Moreover, the unique extension to* $\mathscr{E}$ *of a distribution* $\lambda : P(\mathbb{C}) \longrightarrow \mathscr{X}$ *that inverts* $d.\Sigma_J$ *is isomorphic to* $\lambda \cdot i_*$. *Hence the correspondence restricts to left exact distributions.* (This diagram is not a coinverter in $\mathbf{Frm}_{\mathscr{S}}$, but only because the functors Σ_J and $d.\Sigma_J$ do not preserve the terminal.)

Proof. If $Z \xrightarrow{m} J$ is any morphism and

$$\begin{array}{ccc} U & \rightarrowtail & Z \\ \downarrow & & \downarrow m \\ 1 & \underset{d}{\rightarrowtail} & J \end{array}$$

is a pullback in $P(\mathbb{C})$, then $(d.\Sigma_J)_m : d.\Sigma_J(m) \longrightarrow \Sigma_J(m)$ is equal to $U \rightarrowtail Z$. This is a dense monomorphism, so i^* inverts it. Thus, i^* inverts the natural transformation $d.\Sigma_J$. Now let $f : \mathscr{F} \longrightarrow \mathscr{S}$ be any topos. We must show that $\mathbf{Coc}_{\mathscr{S}}(\_, \mathscr{F})$ turns (1.3) into an inverter diagram.

$$\mathbf{Coc}_{\mathscr{S}}(\mathscr{E}, \mathscr{F}) \longrightarrow \mathbf{Coc}_{\mathscr{S}}(P(\mathbb{C}), \mathscr{F}) \underset{}{\overset{\Downarrow}{\rightrightarrows}} \mathbf{Coc}_{\mathscr{S}}(P(\mathbb{C})/J), \mathscr{F})$$

If $J : \mathbb{J} \longrightarrow \mathbb{C}$ denotes J as a discrete fibration, then $P(\mathbb{C})/J \simeq P(\mathbb{J})$, and the above diagram is equivalent to the following one.

$$\mathbf{Coc}_{\mathscr{S}}(\mathscr{E}, \mathscr{F}) \longrightarrow \mathscr{F}^{\mathbb{C}} \overset{\Downarrow}{\rightrightarrows} \mathscr{F}^{\mathbb{J}}$$

By Propositions 1.4.3 and 1.4.5 we are done. □

We will use the following generalized form of Proposition 1.4.7, which may be proved in a similar manner.

Proposition 1.4.8 *Let K denote any subobject of Ω in a presheaf topos $P(\mathbb{D})$. Suppose that $1 \xrightarrow{t} \Omega$ factors through K, denoted $1 \xrightarrow{d} K$. Let $i^* \dashv i_* : \mathscr{F} \rightarrowtail P(\mathbb{D})$ be the subtopos of sheaves for the topology generated by K. Then*

$$P(\mathbb{D})/K \underset{\Sigma_K}{\overset{d.\Sigma_K}{\underset{\Downarrow d.\Sigma_K}{\rightrightarrows}}} P(\mathbb{D}) \xrightarrow{i^*} \mathscr{F}$$

is a coinverter diagram in $\mathbf{Coc}_{\mathscr{S}}$.

The *tensor* $\mathscr{A} \otimes_{\mathscr{S}} \mathscr{B}$ (or binary coproduct) of two $\mathscr{S}$-cocomplete categories $\mathscr{A}$ and $\mathscr{B}$ classifies bi-$\mathscr{S}$-cocontinuous functors on $\mathscr{A} \times \mathscr{B}$.

We have mentioned that $\mathbf{A}$ may be thought of as an intensive quantity type of toposes, in the sense that it carries products to coproducts (tensors). On objects this says that

$$\mathbf{A}(\mathscr{E}) \otimes_{\mathscr{S}} \mathbf{A}(\mathscr{F}) \cong \mathbf{A}(\mathscr{E} \times_{\mathscr{S}} \mathscr{F}) .$$

By analogy with a property of vector spaces, we say that the *nuclearity property* holds. We now show that this is the case. Indeed, we have just seen that for any topos $\mathscr{E}$, the cocomplete category $\mathbf{A}(\mathscr{E})$ (which we denote just by $\mathscr{E}$) forms part of a coinverter

$$P(\mathbb{D}) \underset{}{\overset{}{\underset{\theta \Downarrow}{\rightrightarrows}}} P(\mathbb{C}) \longrightarrow\!\!\!\!\!\to \mathscr{E} .$$

If $\mathscr{F}$ is any other object of $\mathbf{Top}_{\mathscr{S}}$, regarded as a cocomplete category, then the topos product $\mathscr{E} \times_{\mathscr{S}} \mathscr{F}$, as a cocomplete category, forms part of another coinverter

$$\mathscr{F}^{\mathbb{D}^{\mathrm{op}}} \underset{}{\overset{}{\underset{\theta \Downarrow}{\rightrightarrows}}} \mathscr{F}^{\mathbb{C}^{\mathrm{op}}} \longrightarrow\!\!\!\!\!\to \mathscr{E} \times_{\mathscr{S}} \mathscr{F} , \tag{1.4}$$

and since one can argue that

$$P(\mathbb{C}) \otimes_{\mathscr{S}} \mathscr{F} \simeq \mathscr{F}^{\mathbb{C}^{\mathrm{op}}} ,$$

the nuclearity property follows easily.

Example 1.4.9 *Consider the following preliminary examples of distributions.*

1. *If X is a topological space, then the category of cosheaves on X is equivalent to the category of* Set*-valued distributions on* $\mathrm{Sh}(X)$*, denoted* $\mathbf{E}(X)$*. It contains distributions that are defined by a point of X: if $x \in X$, then the inverse image functor*

   $$x^*(F) = \text{ the stalk of the sheaf } F \text{ at } x$$

   is a (left exact) distribution.

2. *More generally, we have seen that a point* $\mathscr{S} \xrightarrow{p} \mathscr{E}$ *gives the distribution* $\mathbf{E}(p) = p^*$, *which is a sort of* Dirac measure *on* $\mathscr{E}$ *concentrated at the point* p.
3. *If* $\mathscr{E} \xrightarrow{e} \mathscr{S}$ *is locally connected, then the connected components functor* $e_!$ *is a distribution. Its evaluation on an object* X *of* $\mathscr{E}$, *identified with a function* f, *may be thought of as the integration* $\int_{\mathscr{E}} f dx$.
4. *If* $\mathscr{E} \xrightarrow{e} \mathscr{S}$ *is a local topos, then the right adjoint* e_* *is a left exact distribution.*
5. *Any object* A *of* $\mathscr{E}$ *for which the endofunctor* $(\_)^A$ *of* $\mathscr{E}$ *has a right adjoint gives rise to an* infinitesimal distribution $e_!((\_)^A)$.

We leave the following assertion as an exercise.

Proposition 1.4.10 *Constant functors on a locally connected site are cosheaves* (Def. 1.4.1).

Corollary 1.4.11 *If* $\mathscr{E}$ *is locally connected, then* $e_!$ *is the terminal distribution.*

Proof. Consider a locally connected site for $\mathscr{E}$. By Prop. 1.4.10, the constant functor 1 on this site is a cosheaf, which is clearly the terminal cosheaf. This cosheaf corresponds to the distribution $e_!$. In particular, any locally connected topos $\mathscr{E}$ has a terminal distribution. □

Remark 1.4.12 *The existence of a terminal distribution on a topos* $\mathscr{E}$ *in* $\mathbf{Top}_{\mathscr{S}}$ *is not confined to locally connected toposes. For instance, any Grothendieck topos* $\mathscr{E}$ *has a terminal distribution because* $\mathbf{E}(\mathscr{E})$ *is a locally presentable category, hence it is complete (as well as cocomplete). We are led to make assumptions of the sort that the terminal distribution exists in order to carry out certain special developments such as the density of a distribution, or the free distribution algebra. Making such an assumption in general may be too restrictive; however, it is worth pointing out that we know of no pair of toposes* $\mathscr{E}$ *and* $\mathscr{S}$ *for which the terminal distribution* $\mathscr{E} \longrightarrow \mathscr{S}$ *does not exist. There exist Grothendieck toposes with no non-zero distributions, in which case the terminal distribution equals the zero distribution.*

Recall that a *Stone space* is a compact, Hausdorff space in which the clopens generate the topology.

Proposition 1.4.13 (Bergman) *If* X *is a Stone space, then* $\mathbf{E}(X) \simeq \mathit{Set}/|X|$, *where* $|X|$ *denotes the underlying set of* X.

Proof. Associate with a function $A \xrightarrow{f} |X|$ the cosheaf $G_f(U) = f^{-1}(U)$. On the other hand, associate with a cosheaf G on X the function $G(X) \xrightarrow{f_G} |X|$ such that

$$f_G(\xi) = \bigcap \{\text{clopen } V \mid \xi \in G(V)\}\,.$$

Note: "$\xi \in G(V)$" makes sense because for clopen V we have $G(X) \cong G(V) + G(X - V)$. This family of closed sets satisfies the finite intersection property.

By compactness, the intersection is non-empty. Since X is Hausdorff, the intersection is a singleton. It follows directly that a cosheaf G is isomorphic to the cosheaf G_{f_G}, and that a function f is equal to f_{G_f}. □

We know that $\mathbf{E}(P(\mathbb{C})) \simeq \mathscr{S}^{\mathbb{C}}$ is a topos over $\mathscr{S}$, but Remark 1.4.6 takes a different perspective. Proposition 1.4.13 shows that for a space X, $\mathbf{E}(X)$ may be a topos; however, we cannot jump to the conclusion that $\mathbf{E}(\mathscr{E})$ is always a topos. The following is a counterexample.

Example 1.4.14 *If X is a complete metric space, then*

$$\mathbf{E}(X) \simeq \mathrm{Cosh}(X)$$

is equivalent to the category of complete spread spaces over X. (It turns out that over a complete metric space, a localic complete spread must be spatial.) It follows that in this case the point-cosheaves $\{p_x | x \in X\}$ are a generating set for Cosh(X), *where for an open subset $U \subseteq X$, $p_x(U) = 1$ if $x \in U$, and $p_x(U) = \emptyset$ if $x \notin U$. Moreover, this generating set is a discrete category, as a full subcategory of* Cosh(X). *Therefore, we have a faithful functor*

$$\Phi : \mathrm{Cosh}(X) \longrightarrow \mathit{Set}^{|X|}$$

such that $\Phi(G)(x) = \mathrm{Cosh}(X)(p_x, G)$. Associated with any space is a cosheaf $D(U) = |U|$ that carries an open set U to its underlying set. If X is also locally connected, but not discrete (such as the reals), then the unique morphism $D \longrightarrow \pi_0$ to the terminal cosheaf (connected components) is not an isomorphism. But Φ carries $D \longrightarrow \pi_0$ to an isomorphism, so that $D \longrightarrow \pi_0$ is both a monomorphism and an epimorphism in Cosh(X). *Thus,* Cosh(X) *is not a* balanced *category in this case. In particular, it cannot even be an elementary topos.*

Remark 1.4.15 *The following comments may be of interest.*

1. *(A. Zouboff) Proposition 1.4.13 can also be obtained by weakening the hypotheses to X sober and T_1, but restricting to* flabby *cosheaves, i.e., cosheaves G on X such that for each $U \rightarrowtail V$, the induced $G(U) \longrightarrow G(V)$ is injective.*
2. *Cosheaves of* vector spaces *on a locally compact space X often arise by* dualization *from sheaves of vector spaces, as in the example of the cosheaf G on a topological space X that associates to an open subset U of X the vector space $G(U)$ of measures with compact support on U (which are linear functionals on an appropriate space of functions on U).*
3. *G. Bergman has observed that we may assign to a continuous map $l : L \longrightarrow X$, with L locally connected, a cosheaf*

$$\lambda_l : \mathscr{O}(X) \longrightarrow \mathit{Set}$$

such that

$$\lambda_l(U) = \pi_0(l^{-1}(U)) \; ; \; U \in \mathcal{O}(X) \; .$$

The assignment λ *is a functor, which we shall call* the Bergman functor. *Proposition 1.4.13 may be misleading, but then Stone spaces are very special. Announced without either proof or follow-up is the statement by Bergman that if* X *is a complete metric space, then the general cosheaf (of sets) is such a* λ_l. *But for what sort of map* l *is not described. We now know that these maps occur naturally in the theory of Fox. The corresponding topos version is the subject of Chapter 2.*

We generalize the Bergman functor λ (Remark 1.4.15 (3)). For any topos $\mathscr{E} \longrightarrow \mathscr{S}$, consider the 2-category $\mathbf{LTop}_{\mathscr{S}}/\mathscr{E}$ of toposes over $\mathscr{E}$ with locally connected domain: an object of $\mathbf{LTop}_{\mathscr{S}}/\mathscr{E}$ is a geometric morphism $\mathscr{F} \longrightarrow \mathscr{E}$ for which $\mathscr{F} \longrightarrow \mathscr{S}$ is locally connected. The 1-cells and 2-cells are geometric morphisms and natural transformations over $\mathscr{E}$. We may associate with such an object

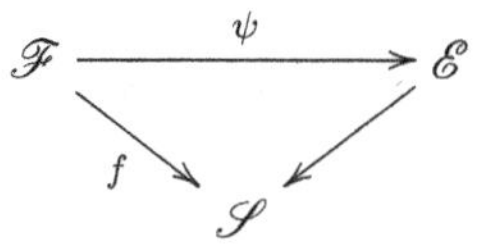

the distribution $f_! \cdot \psi^*$. Indeed, this defines a 2-functor into the distribution category $\mathbf{E}(\mathscr{E})$:

$$\Lambda : \mathbf{LTop}_{\mathscr{S}}/\mathscr{E} \longrightarrow \mathbf{E}(\mathscr{E}) \; ; \; \Lambda(\psi) = f_! \cdot \psi^* \; .$$

This 2-functor carries a diagram such as

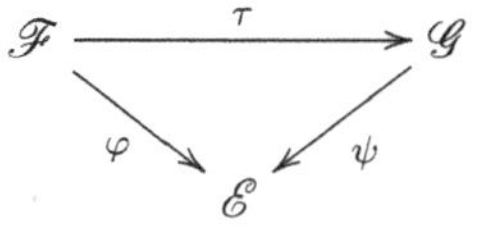

to a natural transformation of distributions $f_! \cdot \varphi^* \longrightarrow g_! \cdot \psi^*$. This natural transformation comes from first applying τ^* and ψ^* to the unit of adjointness $g_! \dashv g^*$, yielding a natural transformation

$$\tau^* \psi^* \longrightarrow \tau^* g^* g_! \psi^* \; .$$

This is naturally equivalent to $\varphi^* \longrightarrow f^* g_! \psi^*$, which transposes to the desired $f_! \cdot \varphi^* \longrightarrow g_! \cdot \psi^*$.

In Chapter 2 we shall see how to pass from distributions back to geometric morphisms by a right adjoint to Λ.

Exercises 1.4.16

1. *(A. M. Pitts) Use the nuclearity property to show that locally connected geometric morphisms are stable under pullback, such that the BCC is satisfied.*
2. *Show that Λ is indeed a 2-functor from a 2-category to a category: a 2-cell $t : \tau \Rightarrow \rho$ between two geometric morphisms over $\mathscr{E}$ passes to an identity 2-cell. Hint: if $\mathscr{G} \xrightarrow{\psi} \mathscr{E}$ is the codomain of τ and ρ, then $t\psi^*$ is an isomorphism.*
3. *Prove Lemma 1.4.2.*
4. *Prove Proposition 1.4.10.*

1.5 Definable Dominances

M. Barr and R. Paré had introduced the concept of a *definable morphism* in order to study local connectedness. We shall review the basic facts surrounding definable morphisms. As always, our base topos is $\mathscr{S}$. We usually do not refer to $\mathscr{S}$ unless it is necessary to do so.

Definition 1.5.1 *A morphism $X \xrightarrow{m} Y$ in a topos $\mathscr{F}$ is* definable *if it can be put in a pullback square as follows.*

$$\begin{array}{ccc} X & \xrightarrow{m} & Y \\ \downarrow & & \downarrow \\ f^*A & \xrightarrow[f^*n]{} & f^*B \end{array}$$

A definable subobject *is a monomorphism that is definable.*

Definable morphisms do not compose in general, not even over a Boolean topos. We ask the reader in the exercises to show that definables compose in a locally connected topos.

We denote the characteristic map of $f^*(\top) : f^*1 \longrightarrow f^*(\Omega_{\mathscr{S}})$ by $\tau : f^*(\Omega_{\mathscr{S}}) \longrightarrow \Omega_{\mathscr{F}}$.

Definition 1.5.2 *f is said to be* subopen *if τ is a monomorphism.*

For instance, an open geometric morphism is subopen.

Theorem 1.5.3 is a characterization of locally connected toposes that is due to M. Barr and R. Paré. We shall not use this result and we do not include a proof, but the idea of their proof provides the key for our proof of Proposition 2.2.12 having to do with pure geometric morphisms.

Let $\mathscr{F} \xrightarrow{f} \mathscr{S}$ be a topos. Consider the following three properties:

1. f is subopen,
2. definable morphisms compose in $\mathscr{F}$,
3. there is a functor $F : \mathscr{F} \longrightarrow \mathscr{S}$ such that for every object Y of $\mathscr{F}$, we have a lattice isomorphism ${\Omega_{\mathscr{S}}}^{FY} \cong f_*(f^*(\Omega_{\mathscr{S}})^Y)$, naturally in Y.

Theorem 1.5.3 (Barr and Paré) *A locally connected topos has the above three properties, and conversely, if the conditions hold, then $\mathscr{F}$ is locally connected and the functor F is the $\mathscr{S}$-indexed left adjoint of f^*.*

Definition 1.5.4 *A topos $f : \mathscr{F} \longrightarrow \mathscr{S}$ is a* definable dominance *if f is subopen, and definable subobjects compose in $\mathscr{F}$.*

We have adopted the term definable dominance because if these two conditions hold, then the class of definable monomorphisms form a *dominance* in the sense of G. Rosolini. This means that the members of this class are stable under pullback and composition, and are classifiable.

Any topos over a Boolean topos is a definable dominance because in this case a subobject is definable iff it is complemented. In particular, every Grothendieck topos (over *Set*) is a definable dominance.

A locally connected topos is a definable dominance. The work of M. Barr and R. Paré shows that a definable dominance is locally connected iff in it suprema of $\mathscr{S}$-indexed families of definable subobjects are again definable in the following sense, and has property (3) above. They use Lemma 1.5.6 to prove Theorem 1.5.3.

Definition 1.5.5 *We shall say that* definable suprema of (families of) definable subobjects are definable *(in a topos over $\mathscr{S}$) if for every square*

$$
\begin{array}{ccc}
U & \twoheadrightarrow & V \\
\downarrow & & \downarrow \\
X & \xrightarrow[m]{} & Y
\end{array}
$$

in which m and $U \rightarrowtail X$ are definable, $V \rightarrowtail Y$ is definable.

Lemma 1.5.6 *Let $\mathscr{F}$ be a definable dominance in which definable suprema of definable subobjects are definable* ($\mathscr{F}$ may not be locally connected because we do not assume property (3)). *Then an exponential $f^*(A)^X$ may be constructed as a limit diagram*

$$
\begin{array}{ccccc}
f^*(A)^X & \rightarrowtail & E_1 & & \\
\downarrow & & \downarrow & & \\
E_2 & \rightarrowtail & f^*(\Omega_{\mathscr{S}})^{X\times f^*(A)} & \underset{\exists_{f^*A}}{\overset{1_X}{\rightrightarrows}} & f^*(\Omega_{\mathscr{S}})^X \\
 & & {\scriptstyle p_1^*\wedge p_2^*}\downarrow\downarrow{\scriptstyle \exists_\delta} & & \\
 & & f^*(\Omega_{\mathscr{S}})^{X\times f^*(A\times A)} & &
\end{array}
$$

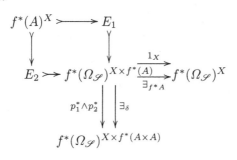

similar to Lemma 1.1.1. $f^*(A)^X$ *is the pullback of the equalizers* E_1 *and* E_2*. The projections* $X \times f^*(A \times A) \longrightarrow X \times f^*A$ *are denoted* p_1, p_2*.*

Proof. If f is subopen, then

$$f^*(\Omega_{\mathscr{S}})^{X \times f^*(A)} \rightarrowtail \Omega_{\mathscr{F}}^{X \times f^*(A)}$$

is a subobject. If definable monomorphisms in $\mathscr{F}$ compose, then $\exists_\delta$ suitably restricts. The 'definable suprema' hypothesis implies that $\exists_{f^*A}$ restricts. The infimum of two definable subobjects is again definable. We may now use Lemma 1.1.1. We also need Exercise 8. □

Exercises 1.5.7

1. *Show that if a monomorphism is definable, then it is definable by a monomorphism. Consequently, show that the classifying map of a definable subobject factors through the canonical map* $\tau : f^*(\Omega_{\mathscr{S}}) \longrightarrow \Omega_{\mathscr{F}}$*, although this factorization may not be unique.*
2. *Show that the inverse image functor of a geometric morphism preserves definable morphisms. Show that the direct image functor of a pure geometric morphism (Def. 2.2.6) preserves definable subobjects.*
3. *Show that in a locally connected topos* $\mathscr{F} \xrightarrow{f} \mathscr{S}$ *a morphism is definable iff the following adjunction square is a pullback.*

$$\begin{array}{ccc} X & \xrightarrow{m} & Y \\ \downarrow & & \downarrow \\ f^*f_!X & \xrightarrow[f^*f_!m]{} & f^*f_!B \end{array}$$

This says that in a locally connected topos, definable morphisms are (canonically) definable by the connected components. Hence, show that definable morphisms compose in a locally connected topos.

4. *Show that if* f *is subopen, then the pair* $\langle f^*(\Omega_{\mathscr{S}}), f^*(\top)\rangle$ *classifies definable subobjects in* $\mathscr{F}$*.*
5. *Show that a locally connected topos has the three properties in the Barr-Paré characterization.*
6. *Show that the inverse image functor of a locally connected geometric morphism preserves exponentials.*
7. *Show that a subobject in a topos over a Boolean base topos is definable iff it is complemented. Thus, show that definable subobjects compose in this case.*
8. *Show that for any morphism* $X \xrightarrow{m} e^*A$*, we have a pullback*

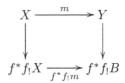

$$
\begin{array}{ccc}
X & \rightarrowtail & X \times e^*A \\
{\scriptstyle m}\downarrow & & \downarrow{\scriptstyle m \times 1_A} \\
e^*A & \rightarrowtail & e^*(A \times A)
\end{array}
$$

so that $X \rightarrowtail X \times e^*A$ *is a definable subobject.*

9. *Show that for any object* Z*, any morphism* $X \xrightarrow{m} Y$ *over* Z*,* m *is definable in* $\mathscr{E}/Z$ *iff* $\Sigma_Z(m)$ *is definable in* $\mathscr{E}$*, where* $\Sigma_Z \dashv Z^*$*.*
10. *Show that the left adjoint* $\varphi_!$ *of an essential geometric morphism* $\mathscr{F} \xrightarrow{\varphi} \mathscr{E}$ *preserves definable morphisms. (Hint: Remark 1.2.4, 4)*

Further reading: Artin & Grothendieck & Verdier [AGV72a], Barr & Paré [BP80], Bénabou [Ben75], Bird [Bir84], Bergman [Ber91], Borel and Moore [BM60], Bredon [Bre67], Bunge [Bun95], Bunge & Funk [BF96b], Bunge & Lack [BL03], Grothendieck [Gro55], Johnstone [Joh77, Joh02], Joyal & Tierney [JT84], Kelly [Kel82], Kock [Koc83], Lawvere [Law66, Law83, Law92, Law00a, Law00b], Mac Lane & Moerdijk [MM92], Moens [Moe82], Paré & Schumacher [PS78], Pitts [Pit85], Rosolini [Ros86], Schwartz [Sch66].

2

Complete Spread Maps of Toposes

In this chapter we introduce a notion of complete spread geometric morphism over a base topos $\mathscr{S}$, inspired by the topological notion introduced by R. H. Fox (1957) [Fox57] as a tool to understand the combinatorial invariants of knots from a topological viewpoint.

In § 2.1 we illustrate the notion of a complete spread by means of two topological examples. The first is the example of (the path of) the shadow cast over the earth (in a variable fashion) of a floating balloon. We shall also describe Nielsen cosheaves, which give rise to the complete spreads associated with universal branched coverings of a marked sphere.

In § 2.2 we introduce and study a notion of pure geometric morphism relative to the base topos $\mathscr{S}$. This notion is relevant to complete spreads when the domain is locally connected, but also in other more general contexts.

Unlike the topological or localic situations, we formally define in § 2.3 and § 2.4 a complete spread geometric morphism as a certain topos bipullback. Distributions and complete spreads with locally connected domain over a base topos $\mathscr{S}$ form equivalent categories. The equivalence, shown by Bunge and Funk (1996) [BF96b], reveals that every geometric morphism with a locally connected domain admits a unique factorization (said to be comprehensive) into a pure geometric morphism followed by a complete spread with a locally connected domain.

We prove in § 2.5 that complete spreads with locally connected domain are stable under composition, and under pullback along a locally connected geometric morphism, such that the Beck-Chevalley condition holds (BCC). In fact, pullback stability along an essential geometric morphism holds, but we must leave this significant and deeper result for later when we examine the symmetric monad.

2.1 Fox Complete Spreads

We illustrate the notion of a complete spread in topology with two examples: a floating balloon and Nielsen cosheaves.

We begin with the simple example of a floating balloon. In this connection and for future developments in this chapter, we provide the definition of complete spread in the sense of Fox. We shall see how the symmetric topos can inform us about how to describe the temporal component of the moving balloon.

A floating balloon may at any instant in time cast a shadow on the earth, which we may describe with a map $S^2 \longrightarrow S^2$, so that the image of the map is the balloon's shadow. This map is a complete spread, whose singular set is a circle in both the domain and codomain spheres. (The singular sets are where the map is not a local homeomorphism.) We are ignoring the balloon's motion for now, but we shall bring that into the picture later. Consider the following definitions.

Suppose that $\varphi : Z \longrightarrow X$ is any map of spaces, such that Z is locally connected. We construct another map $\psi : Y \longrightarrow X$, whose fiber, or costalk, at a point $x \in X$ is the set of cogerms of the cosheaf $G = \pi_0 \cdot \varphi^{-1} = \lambda(\varphi)$ (considered in § 1.4) at x:

$$\psi^{-1}(x) = \varprojlim_{x \in U} G(U) \,.$$

A *cogerm* ξ of G at x is thus a consistent selection $\xi_U \in G(U)$ as U runs over the neighbourhood system of x. Then Y is the space of all cogerms of G, and ψ is the evident projection. The topology for Y is generated by the basic sets

$$(V, \alpha) = \{\text{cogerms } \xi \mid \psi(\xi) \in V \text{ and } \xi_V = \alpha\}$$

where V is an open set of X and $\alpha \in G(V)$.

The map ψ is the best complete spread over X associated with φ. We define a map $Z \xrightarrow{\rho} Y$ over X, such that $\rho(z)$ is the cogerm:

$$\rho(z)_U = \text{ component of } \varphi^{-1}(U) \text{ that contains } z \,.$$

The passage from φ to ψ is the reflection (left adjoint with unit ρ) of maps over X with locally connected domain into complete spreads over X.

Definition 2.1.1 *A map $Z \xrightarrow{\varphi} X$ for which Z is locally connected is* a complete spread *if the unit ρ from Z into the space of cogerms of the cosheaf*

$$G(U) = \pi_0(\varphi^{-1}(U))$$

is a homeomorphism.

R. H. Fox's topological notions of spread and complete spread are the following.

Definition 2.1.2 *A* spread *is a map* $Y \xrightarrow{\psi} X$ *with locally connected domain such that the components of inverse image sets* $\psi^{-1}(U)$, *for* U *open, generate the topology in* Y. *A spread is* complete *if for every* $x \in X$ *and every cogerm* ξ *of* G *at* x, *the set* $\bigcap_{x \in U} \xi_U$ *is non-empty.*

We wish to compare Definitions 2.1.1 and 2.1.2.

Proposition 2.1.3 *For any map* $Z \xrightarrow{\psi} X$ *with locally connected domain, the following are equivalent:*

1. ψ *is a complete spread in the sense of Definition 2.1.1.*
2. ψ *is a spread and* ρ *is a bijection, in which case the inverse of* ρ *is given by:*
$$\rho^{-1}(x, \xi) = (\bigcap_{x \in U} \xi_U) \cap \psi^{-1}(x) .$$
3. ψ *is a spread, and for every* $x \in X$ *and every cogerm* ξ *of* G_ψ *at* x, *the set*
$$(\bigcap_{x \in U} \xi_U) \cap \psi^{-1}(x)$$
is equal to a singleton.

Proof. 1 implies 2 because a complete spread is a spread with locally connected domain, and ρ is a bijection, being a homeomorphism. Conversely, if ρ is a bijection, then the image set $\rho(\beta)$ is equal to (U, β), where β is a component of $\psi^{-1}(U)$. If ψ is a spread, then the β are a base for Y, so that ρ is open, hence a homeomorphism. The formula given for ρ^{-1} is easily seen to hold. The equivalence of 2 and 3 is also straightforward to establish. □

Remark 2.1.4 *Proposition 2.1.3 reveals that Definition 2.1.1 does not coincide exactly with Fox's slightly wider topologically defined notion. There is just one meaning of the term 'spread' for topological spaces (2.1.2), but for us 'complete spread' will always refer to Definition 2.1.1 for topological spaces.*

Of course, the balloon moves in time. Since the symmetric topos $\mathrm{M}(S^2)$ (meaning $\mathrm{M}(\mathit{Sh}(S^2))$) is 'the space' of complete spreads over S^2, we may interpret the shadow cast over the earth by the moving balloon as a path in $\mathrm{M}(S^2)$, by which we mean a geometric morphism
$$B : \mathit{Sh}(I) \longrightarrow \mathrm{M}(S^2) ,$$
where I denotes the unit interval $[0, 1]$. We know that such a geometric morphism amounts to a distribution $\mathit{Sh}(S^2) \longrightarrow \mathit{Sh}(I)$, or equivalently to a cosheaf
$$B : \mathcal{O}(S^2) \longrightarrow \mathit{Sh}(I)$$
from the frame of S^2 into sheaves on I. For any open $U \subseteq S^2$, the stalk of the sheaf space $B(U)$ at time t, $0 \leq t \leq 1$, is equal to the set of components of the part of the balloon that is above U at time t.

Proposition 2.1.5 *The balloon cosheaf B preserves the terminal.*

Proof. The sheaf $B(S^2)$ is the terminal sheaf on I because the balloon's shadow never eclipses the entire surface of the earth. Thus, the stalk $B(S^2)_t$ is always a single component, namely the whole balloon. This is the terminal sheaf on I. □

In this example, sheaves (on I) are indispensable. For instance, the following calculation shows that B does not preserve monomorphisms. (In any case, if B were left exact, then B would amount to an ordinary path $I \longrightarrow S^2$.) Suppose that $U \subset S^2$ is a fixed open disk that lies in the path of the moving balloon's shadow (a closed disk), which is eventually totally eclipsed by the balloon, and then later again falls entirely outside its shadow. Then the stalks $B(U)_t$ of the sheaf space $B(U) : X_U \longrightarrow I$ have cardinality:

$$\begin{cases} 0\ , \text{ at time } t \text{ the balloon's shadow does not meet } U \\ 1\ , \text{ at time } t \text{ the balloon's shadow meets, but does not cover } U \\ 2\ , \text{ at time } t \text{ the balloon's shadow covers } U\ . \end{cases}$$

The total space X_U is not Hausdorff, and the set $\{t \mid B(U)_t = 1\}$ is an open subset of I (not connected).

We shall now illustrate how the notion of a complete spread in topology is connected with cosheaves through what we call a Nielsen cosheaf from the theory of branched coverings. These complete spreads are distinctly different from the floating balloon because they arise as the completion of a local homeomorphism.

A finitely punctured 2-sphere has a universal covering space. In this covering space we may canonically provide fibers for the points deleted from the sphere. We refer to the deleted points as marked points on the sphere. The resulting map $\psi : Y \longrightarrow S^2$ is an instance of a complete spread. Every Riemann surface is a topological quotient of such a ψ. In § 9.3 we shall define branched covering as a special kind of complete spread. The map ψ is a branched covering in this sense.

Let us examine ψ and its construction more closely. The upstairs space Y is homeomorphic to a subspace of the 'topological' Poincaré disk with the circle at infinity included. Figure 2.1 depicts ψ when the sphere has three marked points a, b, c. The topology of the interior of the disk is Euclidean (for our purposes we ignore the hyperbolic metric, although the leaves of the covering space may be patched along geodesics); however, the topology on the rim is finer than Euclidean. Indeed, the open tangent disk at c plus c is also an open neighbourhood of c, a so-called horoball. Then for a sufficiently small open ball B containing c on the sphere, $\psi^{-1}(B)$ is equal to a union of such horoballs. The points of the fiber of a marked point lie on the circle at infinity. ψ sends the indicated leaf, which in the diagram is depicted as a curved diamond shaped region, to S^2 in an evident way. This leaf is adjacent to four other leaves of the same shape, and indeed every leaf is adjacent to four leaves.

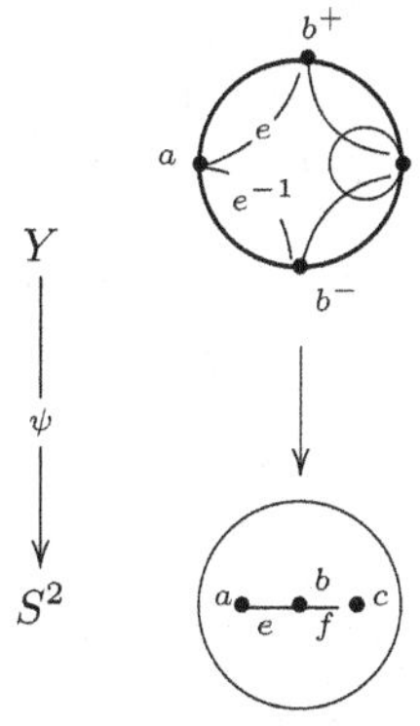

Fig. 2.1. A complete spread

Consider the following construction of ψ. Let $Z \xrightarrow{p} P$ denote the universal cover of the 3-punctured sphere, where $P = S^2 - \{a, b, c\}$. The functor

$$G(U) = \pi_0(p^{-1}(P \cap U))$$

is a cosheaf on S^2, where U is an open subset of S^2. We call G a Nielsen cosheaf. The map ψ is the complete spread associated with the cosheaf G, and ψ is the spread completion of the map $Z \xrightarrow{p} P \hookrightarrow S^2$.

Remark 2.1.6 *The space Y is not locally compact. Indeed, because the open tangent disks plus their tangency points are open sets, a point on the circle at infinity does not have an open neighbourhood whose closure is compact. In particular, ψ is not an exponentiable object in the category of spaces over S^2. By contrast, a local homeomorphism is always an exponentiable object.*

Remark 2.1.7 *The Nielsen cosheaves and their complete spreads may be used to provide a geometric explanation of the ordering of a braid group discovered by Patrick Dehornoy. The Dehornoy ordering of a braid group is a canonical linear ordering that is invariant under left multiplication in the braid group.*

Exercises 2.1.8

1. *Sketch the non-Hausdorff sheaf space $B(U) : X_U \longrightarrow I$.*
2. *The floating balloon can also change shape in time. For instance, a depression, or concavity, may appear in the balloon. How is the presence of such a depression (depending on the shadow projection) reflected in a change in the singular sets?*
3. *The traditional notion of a homotopy $I \times S^2 \longrightarrow S^2$ also applies to the floating balloon. How does this compare with the idea of a path in $\mathrm{M}(S^2)$? Suppose that the situation is more complicated, such as when there are two balloons (possibly colliding).*

4. *Provide a combinatorial description of the Nielsen cosheaf for three marked points, including its transition maps. One way to do this is by introducing a naming system for the components of* $p^{-1}(P \cap U)$. *Let* e, f *denote the two line segments connecting the three marked points* a, b, c, *as depicted in Figure 2.1. Let* $e \cup f$ *denote the union of the line segments and the marked points. We may restrict* G *to open disks* V *that contain at most one marked point, so that* $\pi_0(V - (e \cup f)) \leq 2$.
5. V *contains no marked points: let* $G(V)$ *be the (underlying set of the) free group on* $\{e, f\}$. *An element of the free group names a component of* $p^{-1}(P \cap V)$ *by recording the edges a path crosses in order to reach the component from a fixed starting leaf. If* V *lies on* e *or* f, *then we must arbitrarily decide at the outset which way a component of* $p^{-1}(P \cap V)$ *'falls' in order to name it.*
6. V *contains a single marked point: let* $G(V)$ *consist of certain infinite reduced strings in* e, f, *i.e., certain functions*

$$N \longrightarrow \{e, e^{-1}, f, f^{-1}\}$$

emanating from the natural numbers; only those infinite reduced strings that eventually 'spiral' into a branch point are included.
7. *Transition maps: if* $V \subset V'$, *where* V' *contains a single marked point that is not in* V, *then the transition map* $G(V) \longrightarrow G(V')$ *makes a finite word into an infinite string by adding an infinite spiraling sequence (either consistently left or consistently right). The spiraling sequence depends on which marked point is contained in* V', *and in which component of* $V' - (e \cup f)$ *the smaller disk* V *lies. Again, we must take care when* V *lies on* e *or* f.

2.2 Pure Geometric Morphisms

Pure geometric morphisms are the 'epimorphism' class of the comprehensive factorization system in $\mathbf{Top}_{\mathscr{S}}$.

A continuous map of spaces $X \stackrel{\rho}{\longrightarrow} Y$ will be called *pure* if for every connected open set $U \hookrightarrow Y$, $\rho^{-1}(U)$ is connected. By convention the empty set is not connected, so that implicitly a pure map is a dense map.

Example 2.2.1 *The complement of a knot in* $\mathbb{R}^3$ *(Euclidean 3-space) is a pure open subset of* $\mathbb{R}^3$. *But the complement in* $\mathbb{R}^2$ *of a circle is not a pure open subset.*

We shall seek an appropriate notion of pure geometric morphism. This notion simultaneously generalizes pure maps and initial functors (Exercises 1, 3).

Proposition 2.2.2 *Let* $\mathscr{F} \stackrel{\rho}{\longrightarrow} \mathscr{E}$ *denote a geometric morphism over* $\mathscr{S}$. *Then the following conditions are equivalent:*

1. *For every object A of the base topos $\mathscr{S}$, the unit*

$$e^*A \longrightarrow \rho_*\rho^*(e^*A) \cong \rho_*(f^*A)$$

is a monomorphism.

2. *The unit $e^*\Omega_{\mathscr{S}} \longrightarrow \rho_*\rho^*(e^*\Omega_{\mathscr{S}})$ is a monomorphism.*

Proof. If 2 holds, then for any X in $\mathscr{E}$ the map

$$\mathscr{E}(X, e^*\Omega_{\mathscr{S}}) \xrightarrow{\rho^*} \mathscr{F}(\rho^*X, f^*\Omega_{\mathscr{S}})$$

is injective. Suppose for any A in $\mathscr{S}$ that the unit in 1 equalizes two morphisms $X \xrightarrow{x,y} e^*A$. The equalizer $E \rightarrowtail X$ of x and y fits in the pullback

$$\begin{array}{ccc} E & \rightarrowtail & X \\ \downarrow & & \downarrow{\scriptstyle (x,y)} \\ e^*A & \rightarrowtail & e^*(A \times A) \end{array}$$

where the bottom horizontal is the diagonal. The equalizer is therefore a definable subobject. Since the unit equalizes x and y, we have $\rho^*E \cong \rho^*X$. Therefore, if χ denotes a factorization of the characteristic map of $E \rightarrowtail X$ through $e^*\Omega_{\mathscr{S}}$, then $\rho^*(\chi)$ is equal to $\rho^*X \longrightarrow 1 \xrightarrow{\top} f^*\Omega_{\mathscr{S}}$. Therefore, χ is equal to $X \longrightarrow 1 \xrightarrow{\top} e^*\Omega_{\mathscr{S}}$. Thus, $E \rightarrowtail X$ is an isomorphism, whence $x = y$. This shows that the two conditions are equivalent. □

Definition 2.2.3 *We shall say that a geometric morphism is $\mathscr{S}$-*dense*, or just* dense*, if it satisfies either equivalent condition of Proposition 2.2.2.*

Lemma 2.2.4 *If a dense geometric morphism has subopen domain, then its codomain topos is also subopen.*

Proof. Let $\mathscr{F} \xrightarrow{\psi} \mathscr{E}$ be dense with $\mathscr{F}$ subopen. Consider the following diagram in $\mathscr{E}$.

$$\begin{array}{ccc} e^*(\Omega_{\mathscr{S}}) & \rightarrowtail & \psi_*f^*(\Omega_{\mathscr{S}}) \\ \downarrow{\scriptstyle \tau_{\mathscr{E}}} & & \downarrow{\scriptstyle \psi_*(\tau_{\mathscr{F}})} \\ \Omega_{\mathscr{E}} & \longrightarrow & \psi_*(\Omega_{\mathscr{F}}) \end{array}$$

The τ's are the transposes of the canonical frame morphisms. We conclude that $\tau_{\mathscr{E}}$ is a monomorphism, so $\mathscr{E}$ is subopen. □

Proposition 2.2.5 *If a geometric morphism $\mathscr{F} \xrightarrow{\rho} \mathscr{E}$ is dense, then for any definable subobject $U \rightarrowtail Y$ in $\mathscr{E}$, the adjunction square*

$$\begin{array}{ccc} U & \longrightarrow & \rho_*\rho^*U \\ \downarrow & & \downarrow \\ Y & \longrightarrow & \rho_*\rho^*Y \end{array}$$

is a pullback. The converse holds if $\mathscr{E}$ is subopen.

Proof. Suppose that ρ is dense. Let $U \rightarrowtail Y$ be definable, 'classified' by a morphism $Y \longrightarrow e^*\Omega_{\mathscr{S}}$ (possibly not unique). Both squares in the following diagram are pullbacks.

$$\begin{array}{ccccc} U & \longrightarrow & 1 & \longrightarrow & 1 \\ \downarrow & & \downarrow & & \downarrow \\ Y & \longrightarrow & e^*\Omega_{\mathscr{S}} & \rightarrowtail & \rho_*\rho^*(e^*\Omega_{\mathscr{S}}) \end{array}$$

The outer square above is equal to the outer square of

$$\begin{array}{ccccc} U & \longrightarrow & \rho_*\rho^*U & \longrightarrow & 1 \\ \downarrow & & \downarrow & & \downarrow \\ Y & \longrightarrow & \rho_*\rho^*Y & \longrightarrow & \rho_*\rho^*(e^*\Omega_{\mathscr{S}}) \end{array}$$

which is therefore a pullback. The right hand square is also a pullback so we are done.

For the converse, we assume $\mathscr{E}$ is subopen. If the stated pullback condition holds, then sending a definable subobject $U \rightarrowtail Y$ to $\rho^*U \rightarrowtail \rho^*Y$ is injective. It follows that the unit $e^*\Omega_{\mathscr{S}} \rightarrowtail \rho_*\rho^*(e^*\Omega_{\mathscr{S}})$ is a monomorphism. □

Definition 2.2.6 *We shall say that a geometric morphism ρ is* pure *if the unit*

$$e^*\Omega_{\mathscr{S}} \longrightarrow \rho_*\rho^*(e^*\Omega_{\mathscr{S}})$$

is an isomorphism. We say that ρ_* preserves $\mathscr{S}$-indexed coproducts *if the unit $e^* \longrightarrow \rho_*\rho^*e^*$ is an isomorphism.*

Proposition 2.2.7 *If the domain topos of a pure geometric morphism over $\mathscr{S}$ is a definable dominance, then so is the codomain topos.*

Proof. Let $\psi : \mathscr{F} \longrightarrow \mathscr{E}$ be pure, such that $\mathscr{F}$ is a definable dominance. By Lemma 2.2.4, $\mathscr{E}$ is subopen so we have only to show that definable subobjects compose in $\mathscr{E}$, assumming that they do in $\mathscr{F}$. Definable subobjects are classified in $\mathscr{E}$ and in $\mathscr{F}$ by $e^*\Omega_{\mathscr{S}}$ and $f^*\Omega_{\mathscr{S}}$ respectively. Thus, if ψ is pure, then application of $\psi^* : \mathrm{Sub}(E) \longrightarrow \mathrm{Sub}(\psi^*E)$ restricts to an isomorphism

$$\mathrm{Sub}_{\mathrm{def}}(E) \cong \mathrm{Sub}_{\mathrm{def}}(\psi^*E)\ .$$

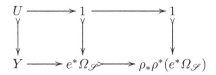

In particular, if the inverse image of a definable subobject $A \rightarrowtail E$ in $\mathscr{E}$ is the top subobject in $\mathscr{F}$, then $A = E$. Now suppose that $A \rightarrowtail B \rightarrowtail E$ are definable subobjects in $\mathscr{E}$. Then $\psi^* A \rightarrowtail \psi^* E$ is definable in $\mathscr{F}$. Therefore, there is a unique definable subobject $S \rightarrowtail E$ such that $\psi^* S = \psi^* A$. We must show that $S = A$, so that $A \rightarrowtail E$ is definable. The pullback $S \wedge A \rightarrowtail A$ is definable, and $\psi^*(S \wedge A) = \psi^* A$. Therefore $S \wedge A = A$. But also $S \wedge B \rightarrowtail S$ and $S \wedge A \rightarrowtail S \wedge B$ are definable. We have $\psi^*(S \wedge A) = \psi^* S$, so $\psi^*(S \wedge A) = \psi^*(S \wedge B)$ and $\psi^*(S \wedge B) = \psi^* S$. Therefore $S \wedge A = S \wedge B = S$. This shows $S = A$. □

A geometric morphism ρ is *connected* if the unit $id \longrightarrow \rho_* \rho^*$ is an isomorphism. Thus, connected geometric morphisms are pure. A connected geometric morphism is a surjection, but surjections are not always pure. Both pure geometric morphisms and surjections are dense. $\mathscr{S}$-dense geometric morphisms, and geometric morphisms whose lower star functor preserves $\mathscr{S}$-coproducts generalize surjection, and respectively connected: a geometric morphism into the base topos $\mathscr{S}$ is $\mathscr{S}$-dense (lower star preserves $\mathscr{S}$-coproducts) iff it a surjection (connected).

Proposition 2.2.8 *Consider a commutative diagram of geometric morphisms.*

$$\begin{array}{ccc} \mathscr{F} & & \\ {\scriptstyle\rho}\downarrow & \searrow^{\tau} & \\ \mathscr{G} & \xrightarrow[\xi]{} & \mathscr{E} \end{array}$$

Then we have the following:

1. *If ξ and ρ are pure, then τ is pure* (closure under composition).
2. *If τ and ρ are pure, then ξ is pure.*
3. *If τ is pure, and ξ is an inclusion, then ρ is pure* (and hence ξ is also pure). *Consequently, the surjection and inclusion factors of a pure geometric morphism are both pure.*

Proof. 1 and 2 follow by considering how the units of the three geometric morphisms combine.

$$\begin{array}{ccc} e^* & \longrightarrow & \tau_* \tau^* e^* \\ \downarrow & & \downarrow {\scriptstyle 1} \\ \xi_* \xi^* e^* & \longrightarrow & \tau_* \tau^* e^* \end{array}$$

The bottom horizontal morphism involves the unit of ρ.

For 3, we transpose the above square under $\xi^* \dashv \xi_*$, using $\xi^* e^* \cong g^*$ and $\tau_* \tau^* e^* \cong \xi_* \rho_* \rho^* g^*$.

$$\begin{array}{ccc} \xi^* e^* & \longrightarrow & \xi^* \xi_* \rho_* \rho^* g^* \\ \big\downarrow & & \big\downarrow \\ g^* & \longrightarrow & \rho_* \rho^* g^* \end{array}$$

The right vertical morphism must be the counit for ξ, which is an isomorphism if ξ is an inclusion. The left vertical is an isomorphism, and so is the top horizontal if τ is pure. Thus, the bottom horizontal morphism is an isomorphism, which says that ρ is pure. □

Remark 2.2.9 *Every assertion of Proposition 2.2.8 holds with 'pure' replaced by 'direct image functor preserves $\mathscr{S}$-coproducts.'*

Clearly a pure geometric morphism $\mathscr{F} \xrightarrow{\rho} \mathscr{E}$ provides a bijection between the lattices of definable subobjects of an object E of $\mathscr{E}$ and the object $\rho^* E$; however, we would like to make this bijection explicit. The bijection is given in one direction by applying ρ^*, and in the other by the pullback in Lemma 2.2.10, 2, below. We leave a proof of Lemma 2.2.10 as an exercise.

Lemma 2.2.10 *If $\mathscr{F} \xrightarrow{\rho} \mathscr{E}$ is pure, then:*

1. *for any definable subobject $V \rightarrowtail E$ in $\mathscr{E}$, the naturality square*

$$\begin{array}{ccc} V & \longrightarrow & \rho_* \rho^* V \\ \big\downarrow & & \big\downarrow \\ E & \xrightarrow{\eta} & \rho_* \rho^* E \end{array}$$

   is a pullback (ρ is dense),
2. *for any definable subobject $U \rightarrowtail \rho^* E$ in $\mathscr{F}$, the pullback in $\mathscr{E}$*

$$\begin{array}{ccc} V & \longrightarrow & \rho_* U \\ \big\downarrow & & \big\downarrow \\ E & \xrightarrow{\eta} & \rho_* \rho^* E \end{array}$$

   is a definable suboject, and
3. *for any definable subobject $U \rightarrowtail \rho^* E$ in $\mathscr{F}$, the counit $\rho^* \rho_* U \xrightarrow{\epsilon} U$ has a section $U \xrightarrow{s} \rho^* \rho_* U$ $(\epsilon \cdot s = 1_U)$ such the square*

$$\begin{array}{ccc} U & \xrightarrow{s} & \rho^* \rho_* U \\ \big\downarrow & & \big\downarrow \\ \rho^* E & \xrightarrow{\rho^* \eta} & \rho^* \rho_* \rho^* E \end{array}$$

   is a pullback. (The section s is necessarily unique with this property.)

Conversely, if $\mathscr{F}$ is subopen, then the above three conditions imply that ρ is pure and that $\mathscr{E}$ is also subopen.

Proposition 2.2.11 *Pure geometric morphisms are stable under pullback along a locally connected geometric morphism. Pullback along a locally connected surjection reflects pure geometric morphisms.*

Proof. We use the well-known fact that the pullback of a locally connected is again locally connected, and that the BCC holds: the connected components functors commute with the inverse image functors. The BCC may be put equivalently in terms of the direct image functors, from which we deduce that the pullback is pure. The reflection property for locally connected surjections follows in a similar fashion because a surjection reflects isomorphisms. □

We would like to refine the notion of pure geometric morphism (Definition 2.2.6) when the domain (and codomain) toposes are locally connected.

Proposition 2.2.12 *Let $\mathscr{F} \xrightarrow{\rho} \mathscr{E}$ denote a geometric morphism with locally connected domain. Suppose that in $\mathscr{E}$ definable suprema of definable subobjects are again definable* (Def. 1.5.5). *Then the following conditions are equivalent:*

1. *$\mathscr{E}$ is locally connected, and the canonical natural transformation $f_!\rho^* \longrightarrow e_!$ is an isomorphism;*
2. *The unit $e^* \longrightarrow \rho_*\rho^*e^*$ is an isomorphism;*
3. *The unit $e^*\Omega_{\mathscr{S}} \longrightarrow \rho_*\rho^*(e^*\Omega_{\mathscr{S}})$ is an isomorphism.*

Proof. If 1 holds, then the adjoint transpose $e^* \longrightarrow \rho_* f^*$ is an isomorphism, hence 2 holds. Clearly 2 implies 3.

Assume that 3 holds, and consider the $\mathscr{S}$-indexed functor $f_!\rho^* : \mathscr{E} \longrightarrow \mathscr{S}$. By Lemma 1.1.1, the exponential $A^{f_!\rho^*X}$ in $\mathscr{S}$ is the limit diagram

$$
\begin{array}{ccccc}
A^{f_!\rho^*X} & \rightarrowtail & E_1 & & \\
\downarrow & & \downarrow & & \\
E_2 & \rightarrowtail & \Omega_{\mathscr{S}}^{f_!\rho^*(X)\times A} & \underset{\exists_A}{\overset{1_{f_!\rho^*X}}{\rightrightarrows}} & \Omega_{\mathscr{S}}^{f_!\rho^*X} \\
 & & p_1^*\wedge p_2^* \downarrow\;\downarrow \exists_\delta & & \\
 & & \Omega_{\mathscr{S}}^{f_!\rho^*(X)\times A\times A} & &
\end{array}
$$

By Lemma 2.2.7, $\mathscr{E}$ is a definable dominance, and we have carried forward the hypothesis of Lemma 1.5.6. Global sections of the diagram in Lemma 1.5.6 gives precisely the above diagram. We have

$$A^{f_!\rho^*X} \cong e_*((e^*A)^X)\ ,$$

naturally in A and X. Therefore, $f_!\rho^*$ is an $\mathscr{S}$-indexed left adjoint to e^*. I.e., $\mathscr{E}$ is locally connected and $f_!\rho^* \cong e_!$, which is 1. □

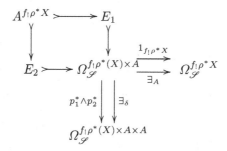

Example 2.2.13 *The inclusion* $x \mapsto \sin(\frac{1}{x})$ *of the (locally connected) positive reals into the topologist's sine curve*

$$Y = \{(0, y) \mid -1 \leq y \leq 1\} \cup \{(x, \sin(\frac{1}{x})) \mid x > 0\}$$

is a pure map, but Y *is not locally connected.* Y *is 'defective' because suprema of definable subobjects are not always again definable.*

Remark 2.2.14 *We emphasize that when the domain and codomain toposes are locally connected, Proposition 2.2.12 holds, and in this case, we often refer to any one of the three conditions in it as pure.*

Proposition 2.2.15 *If a site* $\langle \mathbb{C}, J \rangle$ *is locally connected, then the inclusion geometric morphism* $Sh(\mathbb{C}, J) \rightarrowtail P(\mathbb{C})$ *is pure.*

Proof. Recall that a geometric morphism between locally connected toposes is pure iff its direct image functor preserves $\mathscr{S}$-indexed coproducts. A constant presheaf on a locally connected site is a sheaf. Constant presheaves are sheaves iff the inclusion of sheaves into presheaves preserves $\mathscr{S}$-indexed coproducts. □

The proof of the following helpful fact is related to the proof of Lemma 7.2.13.

Lemma 2.2.16 *Let* $i : \mathscr{F} \rightarrowtail \mathscr{E}$ *be a pure inclusion of toposes, with* $\mathscr{E}$ *locally connected. Then* $\mathscr{F}$ *is locally connected, with* $f_! \cong e_! \cdot i_*$.

Proof. We will first show that i_* preserves $\mathscr{S}$-coproducts. If i is pure, i.e., if the unit $e^*\Omega_{\mathscr{S}} \longrightarrow i_*i^*(e^*\Omega_{\mathscr{S}})$ is an isomorphism, then $e^*\Omega_{\mathscr{S}}$ is an $\mathscr{F}$-sheaf. For any $I \in \mathscr{S}$, we consider (as in 7.2.13) the following pullback in $\mathscr{S}$.

$$\begin{array}{ccc} I & \longrightarrow & 1 \\ {\scriptstyle \{\cdot\}} \big\downarrow & & \big\downarrow {\scriptstyle \top} \\ {\Omega_{\mathscr{S}}}^I & \xrightarrow[\exists!_I]{} & \Omega_{\mathscr{S}} \end{array}$$

We may lift this pullback to $\mathscr{E}$ under e^*. Since by assumption $\mathscr{E}$ is locally connected, e^* preserves exponentials. Therefore $e^*({\Omega_{\mathscr{S}}}^I) \cong (e^*\Omega_{\mathscr{S}})^{e^*I}$ is an $\mathscr{F}$-sheaf, and hence the pullback e^*I is an $\mathscr{F}$-sheaf. Thus, i_* preserves $\mathscr{S}$-coproducts. We have natural bijections:

$$\begin{array}{c} e_!i_*(X) \longrightarrow I \\ \hline i_*X \longrightarrow e^*I \\ \hline i_*X \longrightarrow i_*i^*(e^*I) \\ \hline X \longrightarrow f^*I\,. \end{array}$$

Therefore, $\mathscr{F}$ is locally connected with $f_! \cong e_! \cdot i_*$. □

Exercises 2.2.17

1. *A* pure map *of topological spaces is a map whose inverse image function carries non-empty connected open subsets to the same. Show that the geometric morphism associated with a pure map of locally connected spaces is pure.*
2. *Characterize those maps* $f : X \longrightarrow Y$ *of spaces whose geometric morphism* $f^* \dashv f_* : Sh(X) \longrightarrow Sh(Y)$ *has the property that the unit* $\Delta\Omega_{\mathscr{S}} \longrightarrow f_*(\Delta\Omega_{\mathscr{S}})$ *is an epimorphism.*
3. *A functor* $\mathbb{B} \xrightarrow{F} \mathbb{C}$ *is* initial *if for every object c of* $\mathbb{C}$*, the pullback of* $\mathbb{C}/c \longrightarrow \mathbb{C}$ *along F is a non-empty, connected category. Show that the essential geometric morphism associated with an initial functor is pure.*
4. *Fill in the details of the argument for Proposition 2.2.11.*
5. *Show that a topos is locally connected iff it is a pure subtopos of a presheaf topos.*

2.3 Model Amalgamation

We shall present a construction that appears in several arguments to come. Here we explain this construction as something we call model amalgamation.

Let $\mathbb{C}$ denote a small category. If $\mathscr{F}$ is bounded over $\mathscr{S}$ (as usual), then a geometric morphism $\mathscr{F} \longrightarrow P(\mathbb{C})$ may be equivalently regarded as a flat functor $\mathbb{C} \longrightarrow \mathscr{F}$, or as we shall say, *a model of* $\mathbb{C}$ *in* $\mathscr{F}$. The equivalence is given by composition with the Yoneda model $\mathbb{C} \longrightarrow P(\mathbb{C})$. If $\mathbb{C}$ comes equipped with a Grothendieck topology J, then a model of the site $\langle \mathbb{C}, J \rangle$ is a flat functor $\mathbb{C} \longrightarrow \mathscr{F}$ that in addition carries J-sieves to colimiting cones in $\mathscr{F}$ (this is the cosheaf property). Models of $\langle \mathbb{C}, J \rangle$ correspond to geometric morphisms $\mathscr{F} \longrightarrow Sh(\mathbb{C}, J)$. The Yoneda model combined with sheafification

$$\mathbb{C} \xrightarrow{h} P(\mathbb{C}) \longrightarrow Sh(\mathbb{C}, J)$$

is *the canonical model of* $\langle \mathbb{C}, J \rangle$.

Example 2.3.1 *Let* $P(\mathbb{B}) \xrightarrow{f} P(\mathbb{C})$ *denote the essential geometric morphism associated with a functor* $F : \mathbb{B} \longrightarrow \mathbb{C}$*. Then we may compose Yoneda and* f^**, thereby obtaining a model*

$$\mathbb{C} \xrightarrow{h} P(\mathbb{C}) \xrightarrow{f^*} P(\mathbb{B})$$

such that

$$c \mapsto f^*(h_c) = h_c \cdot F \; .$$

We call this functor the model of $\mathbb{C}$ *associated with F.*

It is always possible to *amalgamate* any two models $\mathbb{C} \longrightarrow \mathscr{E}$ and $\mathbb{C} \longrightarrow \mathscr{F}$ into one model $\mathbb{C} \longrightarrow \mathscr{G}$ in the best possible way. This amalgamation is the topos pullback.

$$\begin{array}{ccc} \mathscr{G} & \longrightarrow & \mathscr{F} \\ \downarrow & & \downarrow \\ \mathscr{E} & \longrightarrow & P(\mathbb{C}) \end{array}$$

We may wish to amalgamate the canonical model of a site $\langle \mathbb{C}, J \rangle$ with the model associated with a functor $\mathbb{B} \xrightarrow{F} \mathbb{C}$. This amalgamation is the topos pullback

$$\begin{array}{ccc} \mathscr{G} & \rightarrowtail & P(\mathbb{B}) \\ {\scriptstyle \psi}\downarrow & & \downarrow{\scriptstyle f} \\ Sh(\mathbb{C}, J) & \rightarrowtail & P(\mathbb{C}) \end{array}$$

where f denotes the essential geometric morphism associated with F. We may think of this amalgamation as the best way to make the model of $\mathbb{C}$ associated with F into a model of the site $\langle \mathbb{C}, J \rangle$. This new model corresponds to the geometric morphism ψ.

The amalgamation of a site $\langle \mathbb{C}, J \rangle$ with a functor F as above may therefore be described in terms of a topology in $\mathbb{B}$, which we shall call the *amalgamation, or pullback, topology*. This topology is the smallest topology K in $\mathbb{B}$ for which $\mathbb{C} \longrightarrow Sh(\mathbb{B}, K) = \mathscr{G}$ is a J-cosheaf. K is generated by the sieves $f^*(R) \rightarrowtail f^*(h_c)$ such that $R \rightarrowtail h_c$ is a member of J. In terms of $\mathbb{B}$, K is generated by all pullbacks

$$\begin{array}{ccc} \hat{m}^* R & \longrightarrow & f^*(R) \\ \downarrow & & \downarrow \\ h_b & \xrightarrow[\hat{m}]{} & f^*(h_c) \end{array}$$

in $P(\mathbb{B})$, where $\hat{m}$ is the transpose under $f_! \dashv f^*$ of a morphism $F(b) \xrightarrow{m} c$ of $\mathbb{C}$. We have

$$\hat{m}^* R = \{ b' \xrightarrow{n} b \mid m \cdot F(n) \in R \} \, .$$

By replacing R with $m^* R$,

$$\begin{array}{ccc} m^* R & \longrightarrow & R \\ \downarrow & & \downarrow \\ h_{Fb} & \xrightarrow[m]{} & h_c \end{array}$$

we may suppose that $F(b) = c$ and $m = 1_c$. In effect, we may suppose that $\hat{m}$ is the unit $h_b \xrightarrow{u} f^* f_! (h_b) = f^*(h_{Fb})$. Then

$$u^*R = \{b' \xrightarrow{n} b \mid F(n) \in R\} .$$

Definition 2.3.2 *We shall denote the sieves u^*R by*

$$F^*R = \{b' \xrightarrow{n} b \mid F(n) \in R\} ,$$

where $R \rightarrowtail h_{Fb}$ is a J-sieve. The sieves $F^*R \rightarrowtail h_b$ include the top sieves, and they are pullback stable. They define a subobject $\widehat{J}$ of $\Omega_{P(\mathbb{B})}$ that generates the amalgamation topology K.

Consider the following result, due to Johnstone, and its application Corollary 2.3.4.

Theorem 2.3.3 *The pullback topology $K \rightarrowtail \Omega_{\mathscr{Y}}$ along a geometric morphism $\gamma : \mathscr{Y} \longrightarrow \mathscr{X}$ of a topology J in $\mathscr{X}$ is given by the upclosure of the image-object $\widehat{J}$ in the diagram:*

$$\begin{array}{ccc} \gamma^* J & \twoheadrightarrow & \widehat{J} \\ \downarrow & & \downarrow \\ \gamma^* \Omega_{\mathscr{X}} & \xrightarrow{\chi} & \Omega_{\mathscr{Y}} \end{array}$$

where χ classifies $\gamma^* \top$.

Corollary 2.3.4 *A sieve $Q \rightarrowtail h_b$ is a covering sieve for the amalgamation topology of a site $\langle \mathbb{C}, J \rangle$ and a functor $\mathbb{B} \xrightarrow{F} \mathbb{C}$ iff there is a J-sieve $R \rightarrowtail h_{Fb}$ such that $F^*R \rightarrowtail Q$.*

The amalgamation of the canonical model of a site with the model associated with a functor is a tool that we shall frequently employ. When the functor is a discrete *op*fibration, this amalgamation is closely related to complete spread geometric morphisms. Our argument for Theorem 3.5.3 depends on Corollary 2.3.4.

Exercises 2.3.5

1. *Let $P(\mathbb{Y}) \xrightarrow{\gamma} P(\mathbb{C})$ denote the essential geometric morphism associated with a discrete opfibration $\mathbb{Y} \xrightarrow{G} \mathbb{C}$. Show that we have*

$$\gamma^*(h_c) = \coprod_{G(y)=c} h_y .$$

Check that the above description of $\gamma^*(h_c)$ agrees with the general one given in 2.3.1.

2. *Show that the amalgamation of the canonical model of a site $\langle \mathbb{C}, J \rangle$ with a discrete fibration $\mathbb{X} \xrightarrow{F} \mathbb{C}$ is the topos $\mathbf{Sh}(\mathbb{C})/aF$, where aF denotes the associated sheaf of F.*
3. *Read the proof of Johnstone's result (herein Thm. 2.3.3) in the Elephant (C3.3.14).*

2.4 Discrete Opfibrations and Complete Spreads

Let μ denote a distribution on a topos $\mathscr{E}$. If $\langle \mathbb{C}, J \rangle$ is any site for $\mathscr{E}$, then μ may be equivalently regarded as a cosheaf $G : \mathbb{C} \longrightarrow \mathscr{S}$ (Definition 1.4.1). Whenever it suits us we may regard a functor $\mathbb{C} \longrightarrow \mathscr{S}$ as a discrete opfibration $\mathbb{Y} \longrightarrow \mathbb{C}$. The objects of $\mathbb{Y}$ are pairs (c, x), such that $x \in G(c)$.

We begin with the following central fact about the amalgamation of the canonical model of a site $\langle \mathbb{C}, J \rangle$ with the model of $\mathbb{C}$ associated with a cosheaf.

Proposition 2.4.1 *Let G denote a cosheaf on a site $\langle \mathbb{C}, J \rangle$. Then the amalgamation site $\langle \mathbb{Y}, K \rangle$ of the canonical model of $\langle \mathbb{C}, J \rangle$ with the model of $\mathbb{C}$ associated with G is locally connected* (Proposition 1.2.2).

Proof. Definition 2.3.2 describes the covering sieves for the amalgamation site $\langle \mathbb{Y}, K \rangle$: $G^*R \rightarrowtail h_y$, where $y = (c, x)$ for $x \in G(c)$, and $R \rightarrowtail h_c$ is a J-sieve. We have

$$G^*R = \{(c', x') \xrightarrow{m} (c, x)(G(m)(x') = x) \mid m \in R\} .$$

Every such sieve G^*R is non-empty and connected because of the cosheaf property $\varinjlim_R G \cong G(c)$. □

We form the model amalgamation of the canonical model of $\langle \mathbb{C}, J \rangle$ with the model of $\mathbb{C}$ associated with G.

$$\begin{array}{ccc} \mathscr{Y} & \overset{\eta}{\rightarrowtail} & P(\mathbb{Y}) \\ {\scriptstyle \psi}\downarrow & & \downarrow{\scriptstyle \gamma} \\ \mathscr{E} & \rightarrowtail & P(\mathbb{C}) \end{array}$$

By Proposition 2.4.1, $\mathscr{Y}$ is locally connected. By Proposition 2.2.15, the inclusion η is pure. The amalgamated model $\mathbb{C} \longrightarrow \mathscr{Y}$ sends c to the associated sheaf of the coproduct $\coprod_{G(y)=c} h_y$ (Exercise 2.3.5, 1). We may retrieve the cosheaf G from the model amalgamation by composing with the connected components functor $y_!$ for $\mathscr{Y}$. Thus, we retrieve the distribution with which we started as

$$\mu \cong y_! \cdot \psi^* . \tag{2.1}$$

Consider again the Bergman 2-functor Λ, defined on objects by $\Lambda(\varphi) = f_!\varphi^*$.

Proposition 2.4.2 *Amalgamation $\mu \mapsto \psi_\mu$ is pseudo right adjoint to Λ: we have category equivalences*

$$\boldsymbol{LTop}_{\mathscr{S}}/\mathscr{E}(\varphi, \psi_\mu) \simeq \mathbf{E}(\mathscr{E})(f_!\varphi^*, \mu) \tag{2.2}$$

natural in φ and μ (although $\mathbf{E}(\mathscr{E})(f_!\varphi^*, \mu)$ is a discrete category, or a set). *Amalgamation is full and faithful.*

Proof. Let $\mathscr{F} \xrightarrow{\varphi} \mathscr{E}$ be a topos over $\mathscr{E}$ with locally connected domain. A natural transformation $f_!\varphi^* \xrightarrow{t} \mu$ defines a functor

$$F : \mathbb{Y} \longrightarrow \mathscr{F}$$

as follows. Let G denote the cosheaf associated with μ. If $y = (c, x)$ is an object of $\mathbb{Y}$, then we define $F(y)$ as the pullback

$$\begin{array}{ccc} F(y) & \longrightarrow & 1 \\ \downarrow & & \downarrow{\scriptstyle f^*x} \\ \varphi^*(h_c) & \underset{\hat{t}_c}{\longrightarrow} & f^*G(c) \end{array}$$

in $\mathscr{F}$, where $\hat{t}$ is the transpose of t under $f_! \dashv f^*$. Then F is a flat functor, which corresponds to a geometric morphism $\mathscr{F} \xrightarrow{\rho} P(\mathbb{Y})$, such that the following diagram of models commutes.

$$\begin{array}{ccc} \mathscr{F} & \overset{\rho^*}{\longleftarrow} & P(\mathbb{Y}) \\ {\scriptstyle \varphi^*}\uparrow & & \uparrow \\ \mathscr{E} & \longleftarrow & \mathbb{C} \end{array}$$

Thus, t is ultimately translated into a geometric morphism $\mathscr{F} \xrightarrow{\tau} \mathscr{Y}$ over $\mathscr{E}$. On the other hand, suppose we start with such a geometric morphism τ over $\mathscr{E}$. We have a unique natural transformation to the terminal distribution on $\mathscr{Y}$: $f_!\tau^* \longrightarrow y_!$ (Proposition 1.4.11). This gives a natural transformation $f_!\tau^*\psi^* \longrightarrow y_!\psi^*$, which is isomorphic to one $f_!\varphi^* \xrightarrow{t} \mu$. By (2.1), amalgamation is full and faithful. □

Remark 2.4.3 *The construction of ψ does not depend on the particular site chosen for $\mathscr{E}$. In fact, the adjointness described in Proposition 2.4.2 shows that ψ depends only on μ.*

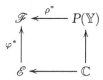

Remark 2.4.4 *An artifact of the proof of Proposition 2.4.2 is the formula: if* $\mathscr{F} \xrightarrow{\tau} \mathscr{Y}$ *corresponds to* t *and* $y = (c, x)$*, then* $f_! \tau^*(h_y) = {t_c}^{-1}(x)$*. Indeed, we have*

$$f_! \tau^*(h_y) \cong f_! \rho^*(h_y) \cong f_! F(y) \cong {t_c}^{-1}(x) \ ,$$

where the last isomorphism comes from transposing the defining pullback square for $F(y)$*.*

In this section we have developed a working definition of complete spread geometric morphism that we can paraphrase as follows: a geometric morphism $\mathscr{Y} \xrightarrow{\psi} \mathscr{E}$ for which $\mathscr{Y}$ is locally connected is a complete spread if ψ is retrieved, by amalgamation, from its distribution $y_! \cdot \psi^*$. Put in terms of models of a site $\langle \mathbb{C}, J \rangle$ for $\mathscr{E}$: ψ is a complete spread if $\mathbb{C} \longrightarrow \mathscr{E} \xrightarrow{\psi^*} \mathscr{Y}$ is the optimal $\langle \mathbb{C}, J \rangle$-model that can be made from the $\mathbb{C}$-model $\mathbb{C} \longrightarrow P(\mathbb{Y})$ associated with the distribution $y_! \cdot \psi^*$.

Corollary 2.4.5 *For any topos* $\mathscr{E}$, $\mathbf{E}(\mathscr{E})$ *is equivalent to a 2-category of geometric morphisms over* $\mathscr{E}$*. These geometric morphisms have locally connected domain.*

Definition 2.4.6 *We shall call the geometric morphisms in Corollary 2.4.5* complete spreads. *We call the geometric morphism constructed from a distribution* the complete spread geometric morphism associated with the distribution.

Example 2.4.7 *A geometric morphism* $\mathscr{F} \longrightarrow P(\mathbb{C})$ *is a complete spread (with locally connected domain) iff it is equivalent to the essential geometric morphism* $P(\mathbb{Y}) \longrightarrow P(\mathbb{C})$ *associated with a discrete opfibration* $\mathbb{Y} \longrightarrow \mathbb{C}$*.*

We have gathered sufficient information about complete spreads and pure geometric morphisms in order to establish the *comprehensive factorization for geometric morphisms*: the unique factorization of a geometric morphism whose domain is locally connected into its pure and complete spread parts.

Theorem 2.4.8 *A geometric morphism with locally connected domain may be factored into a pure geometric morphism followed by a complete spread. This factorization is essentially unique in the sense that any two such factorizations are equivalent by an equivalence that is unique up to unique isomorphism. Moreover, the middle topos in this factorization is locally connected.*

Proof. Existence: let $\mathscr{F} \xrightarrow{\varphi} \mathscr{E}$ be any geometric morphism with locally connected domain. We have only to show that the induced geometric morphism η in the defining pullback diagram is pure.

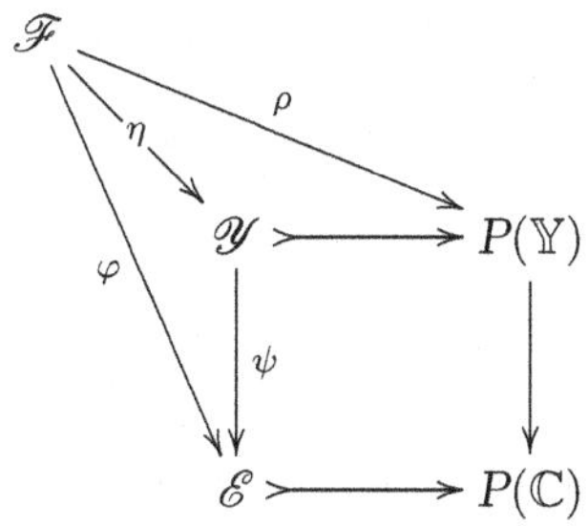

By Proposition 2.2.8, it suffices to show that ρ is pure. Let $F : \mathbb{Y} \longrightarrow \mathscr{F}$ denote the flat functor defined in the proof of Proposition 2.4.2 that corresponds to ρ. If $y = (c, x)$ is any object of $\mathbb{Y}$, then $F(y)$ is given by the following pullback.

$$\begin{array}{ccc} F(y) & \longrightarrow & 1 \\ \downarrow & & \downarrow f^*x \\ \varphi^*(h_c) & \xrightarrow[\hat{t}_c]{} & f^*G(c) \end{array}$$

G is the cosheaf of the distribution $f_!\varphi^*$. We claim that for any object y of $\mathbb{Y}$, we have $f_!F(y) \cong 1$. Indeed, since $\mathscr{F}$ is locally connected, the above square transposes under $f_! \dashv f^*$ to the following pullback.

$$\begin{array}{ccc} f_!F(y) & \longrightarrow & 1 \\ \downarrow & & \downarrow x \\ f_!\varphi^*(h_c) & \xrightarrow[t_c]{} & G(c) \end{array}$$

The morphism t_c is an isomorphism, so that $f_!F(y) \cong 1$. Thus, $f_!F$ is the terminal cosheaf on $\mathbb{Y}$, whence $f_!\rho^* \cong \pi_0^{\mathbb{Y}}$, so that ρ is pure (Prop. 2.2.12). Hence η is pure.

The uniqueness of the factorization is ultimately a consequence of Proposition 2.4.2. □

Remark 2.4.9 *We have defined complete spread geometric morphisms, but we would also like to define separate and independent notions of spread and completeness, such that a geometric morphism with locally connected domain is a complete spread iff it is a spread and complete. We shall accomplish this later.*

Example 2.4.10 *The action of sheaves on distributions can be described in terms of complete spreads. If X is an object of a topos $\mathscr{E}$, and μ is a distribution on $\mathscr{E}$ with complete spread $\mathscr{Y} \xrightarrow{\psi} \mathscr{E}$, then the complete spread corresponding to the distribution $X.\mu$ is the complete spread factor of the composite*

$$\mathscr{Y}/\psi^*X \longrightarrow \mathscr{Y} \xrightarrow{\psi} \mathscr{E}\,.$$

This is the same as the composite of the complete spread factor of

$$\mathscr{Y}/\psi^* X \longrightarrow \mathscr{Y}$$

and ψ.

Example 2.4.11 Model theory. *A model of a site* $\langle \mathbb{C}, J \rangle$ *in a topos* $\mathscr{S}$ *is a flat functor* $\mathbb{C} \longrightarrow \mathscr{S}$ *that carries* J*-sieves to colimit diagrams (the cosheaf property). Let* Mod(T) *denote the category of models of the theory* $T = \langle \mathbb{C}, J \rangle$ *in* $\mathscr{S}$. *Then the topos* $\mathscr{S}[T] = \mathrm{Sh}(\mathbb{C}, J)$ *is the classifying topos of the theory* T *in the sense that* Mod(T) *is canonically equivalent to the category of points of* $\mathscr{S}[T]$. *(A point of a topos* $\mathscr{E}$ *is a geometric morphism* $\mathscr{S} \longrightarrow \mathscr{E}$.*)*

Looking ahead, the equivalence of Corollary 2.5.4 restricts to lex distributions. Thus, if M *is a model of* T, *then the classifying topos* $\mathscr{S}[T_M]$ *for the slice category of models* Mod(T)/M *is structured by a complete spread* $\mathscr{S}[T_M] \xrightarrow{\psi} \mathscr{S}[T]$, *that is, by a complete spread over* $\mathscr{E} = \mathscr{S}[T]$. *The topos* $\mathscr{S}[T_M]$ *is the middle topos in the comprehensive factorization of the point* $\mathscr{S} \longrightarrow \mathscr{E}$ *associated with* M. $\mathscr{S}[T_M]$ *is locally connected, but also totally connected in the sense that its connected components functor is lex. (A space is said to be totally connected if every non-empty open set is connected.) We shall return to totally connected toposes in Part 3.*

In § 7.3 we develop a theory of Stone locales relative to a base topos $\mathscr{S}$. We are interested in this program partly because the pure, entire factorization is related to the problem of obtaining the comprehensive factorization directly from the corresponding distribution algebra, rather than from the distribution as in § 2.4.

Exercises 2.4.12

1. *Complete the proof of Proposition 2.4.2: show that* F *is indeed flat, and that (2.2) is indeed an equivalence.*
2. *Show that if* $\mathscr{E}$ *is locally connected, then the complete spread corresponding to* $e_!$ *is* $\mathscr{E} \xrightarrow{\mathrm{id}} \mathscr{E}$.
3. *Show that* $\mathscr{F} \longrightarrow \mathscr{S}$ *is a complete spread (with locally connected domain) iff* $\mathscr{F}$ *is equivalent to* $\mathscr{S}/f_!(1)$.
4. *A geometric morphism into a presheaf topos* P($\mathbb{C}$) *is a complete spread iff it is equivalent to the essential geometric morphism associated with a discrete opfibration over* $\mathbb{C}$.
5. *Show that a topos* $\mathscr{E}$ *is locally connected iff it is a pure subtopos of a presheaf topos.*
6. *Give a detailed proof of the uniqueness of the comprehensive factorization.*
7. *Show directly without appealing to the comprehensive factorization that a pure complete spread geometric morphism between locally connected toposes is an equivalence.*

8. *Use the theory of locally presentable categories to show that* $\mathbf{E}(\mathscr{E})$ *is complete, for any Grothendieck topos* $\mathscr{E}$*. In particular, a terminal distribution always exists.*

2.5 Properties of Complete Spreads

Remark 2.5.1 *The essential geometric morphism* $P(\mathbb{Y}) \longrightarrow P(\mathbb{C})$ *induced by a discrete opfibration* $\mathbb{Y} \longrightarrow \mathbb{C}$ *is a complete spread (Eg. 2.4.7). S. Niefield has shown, using a general result of P. T. Johnstone and A. Joyal, that there is one such essential geometric morphism which cannot be exponentiable (Remark 2.1.6). Thus, the 2-functor P need not preserve exponentiable morphisms. By contrast, the essential geometric morphism* $P(\mathbb{Y}) \longrightarrow P(\mathbb{C})$ *induced by a discrete fibration is a local homeomorphism, hence exponentiable in* $\boldsymbol{Top}_{\mathscr{S}}$.

Composition of spreads is a relatively easy matter (3.1.9), but it is not entirely obvious that the composition of two complete spreads is again a complete spread. We shall prove this fact by working from the definition of complete spread (Definition 2.4.6), and then again in § 5.2 using the KZ-monad machinery. In § 9.1, we show that a locally trivial covering is unramified (Definition 9.1.7). Therefore, a composite of two locally trivial coverings is unramified, although it may not be locally trivial.

The following lemma will prepare us for Proposition 2.5.3.

Lemma 2.5.2 *Let* $\mathscr{G} \xrightarrow{\psi} \mathscr{Y}$ *be a geometric morphism between locally connected toposes. Suppose that* $\alpha \rightarrowtail X$ *is a component of an object in* $\mathscr{Y}$*, and that* β *is a component of* $\psi^*\alpha$ *in* $\mathscr{G}$*. Then* β *is a component of* $\psi^* X$*.*

Proof. Let γ denote the component determined by β under the composite inclusion

$$\beta \rightarrowtail \psi^*\alpha \rightarrowtail \psi^* X .$$

We want to show that $\beta = \gamma$. There is a unique natural transformation $g_!\psi^* \longrightarrow y_!$ since $y_!$ is the terminal distribution on $\mathscr{Y}$, which we shall denote $\gamma \mapsto \bar{\gamma}$. Intuitively, this map sends a component γ of $\psi^* X$ to the component $\bar{\gamma}$ of X that contains the image of γ under ψ. Then $\bar{\gamma}$ and α have the image of β under ψ in common, so that $\alpha = \bar{\gamma}$. Therefore, $\gamma \rightarrowtail \psi^*\bar{\gamma} = \psi^*\alpha$. Thus, γ must be a component also of $\psi^*\alpha$. Since γ contains β, we must have $\beta = \gamma$. □

Proposition 2.5.3 *The composite of two complete spreads is a complete spread. If geometric morphisms* φ *and* $\varphi \cdot \psi$ *are complete spreads, then so is* ψ.

Proof. Suppose that $\mathscr{Y} \xrightarrow{\varphi} \mathscr{E}$ and $\mathscr{G} \xrightarrow{\psi} \mathscr{Y}$ are two complete spreads over a base topos $\mathscr{S}$, with defining pullbacks below left. The toposes $\mathscr{Y}$ and $\mathscr{G}$ are locally connected. We must show that the far right square is also a pullback.

$$\begin{array}{ccc} \mathscr{Y} & \overset{i}{\rightarrowtail} & P(\mathbb{Y}) \\ {\scriptstyle\varphi}\downarrow & & \downarrow \\ \mathscr{E} & \rightarrowtail & P(\mathbb{C}) \end{array} \qquad \begin{array}{ccc} \mathscr{G} & \rightarrowtail & P(\mathbb{G}) \\ {\scriptstyle\psi}\downarrow & & \downarrow \\ \mathscr{Y} & \underset{i}{\rightarrowtail} & P(\mathbb{Y}) \end{array} \qquad \begin{array}{ccc} \mathscr{G} & \rightarrowtail & P(\mathbb{X}) \\ {\scriptstyle\varphi\cdot\psi}\downarrow & & \downarrow \\ \mathscr{E} & \rightarrowtail & P(\mathbb{C}) \end{array}$$

The objects of the small categories $\mathbb{Y}$, $\mathbb{G}$, and $\mathbb{X}$ are respectively:

$\mathbb{Y}$ (c, α); $c \in \mathbb{C}$ and $\alpha \in y_! \varphi^* c$,
$\mathbb{G}$ (c, α, β); $(c, \alpha) \in \mathbb{Y}$ and $\beta \in g_! \psi^* \alpha$ (i satisfies $i^*(c, \alpha) = \alpha$),
$\mathbb{X}$ (c, β); $c \in \mathbb{C}$ and $\beta \in g_! \psi^* \varphi^* c$.

We will show that the square involving the composite $\varphi \cdot \psi$ is a pullback by showing that $\mathbb{G}$ and $\mathbb{X}$ are equivalent categories over $\mathbb{C}$. We compose the unique natural transformation $g_! \psi^* \longrightarrow y_!$ with φ^* obtaining a natural transformation $g_! \psi^* \varphi^* \longrightarrow y_! \varphi^*$, denoted $\beta \mapsto \bar{\beta}$ as in Lemma 2.5.2. Intuitively, this map sends a component β of $\psi^* \varphi^* X$ to the component $\bar{\beta}$ of $\varphi^* X$ that contains the image of β under ψ. Using this map we may define a functor from $\mathbb{X}$ to $\mathbb{G}$:

$$(c, \beta) \;\mapsto\; (c, \bar{\beta}, \beta)\ .$$

Now suppose we have a component $\alpha \rightarrowtail \varphi^* c$ in $\mathscr{Y}$, and a component β of $\psi^* \alpha$: $\beta \rightarrowtail \psi^* \alpha \rightarrowtail \psi^* \varphi^* c$. By Lemma 2.5.2, β is a component of $\psi^* \varphi^* c$. Thus,

$$(c, \alpha, \beta) \;\mapsto\; (c, \beta)$$

is a functor from $\mathbb{G}$ to $\mathbb{X}$. It is almost clear that these two functors are mutual inverses. What remains to be seen is that for a given (c, α, β), we have $\alpha = \bar{\beta}$. This is so because the two components α and $\bar{\beta}$ of $\varphi^* c$ have the image of β under ψ in common.

The second statement of the theorem follows in a similar way: if the left and right squares above are pullbacks, then so is the middle one. □

We now may combine the canonical correspondence between distributions and complete spreads and Proposition 2.5.3 to obtain the following.

Corollary 2.5.4 *Let ν be a distribution on $\mathscr{E}$, with corresponding complete spread $\mathscr{Y} \xrightarrow{\psi} \mathscr{E}$. Then composition with ψ^* yields an equivalence $\mathbf{E}(\mathscr{Y}) \simeq \mathbf{E}(\mathscr{E})/\nu$.*

Complete spreads are reasonably well-behaved under topos pullback. Propositions 2.5.5 and 2.5.8 are two such pullback results. Proposition 2.5.5 is easy to prove directly from universal properties. Proposition 2.5.8, although less obvious than 2.5.5, has a natural explanation once the appropriate KZ-machinery is in place: we defer its proof to § 5.2.

Proposition 2.5.5 *Consider a topos pullback in which ψ is a complete spread.*

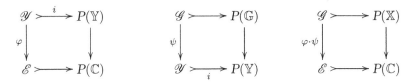

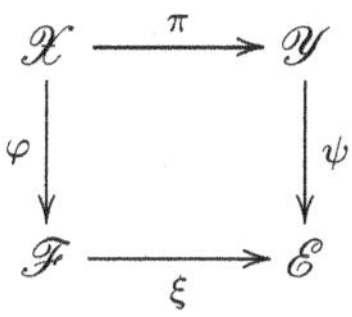

If $\mathscr{X}$ is locally connected, then φ is a complete spread.

Proof. If $\mathscr{X}$ is locally connected, then a diagram-chasing argument is available to us. First factor the pullback φ into its pure and complete spread parts: $\varphi \cong \tau \cdot \rho$. Then factor $\xi \cdot \tau$ into its pure and complete spread parts. We now have a diagram as follows.

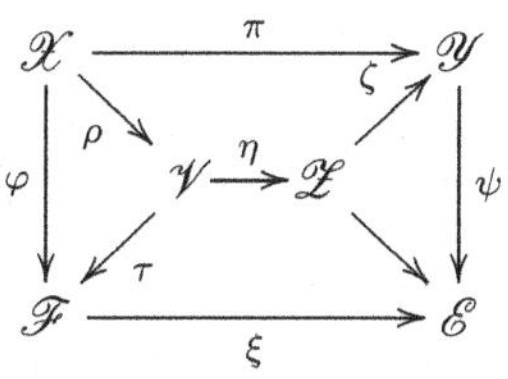

There is such a ζ because ψ is a complete spread, and since $\eta \cdot \rho$ is pure. Since the outer square is a pullback, there is $\mathscr{V} \xrightarrow{\theta} \mathscr{X}$ such that $\pi \cdot \theta \cong \zeta \cdot \eta$ and $\varphi \cdot \theta \cong \tau$. The universal property of the pullback implies that $\theta \cdot \rho \cong id_{\mathscr{X}}$. Hence, $\rho \cdot \theta \cdot \rho \cong \rho$. Therefore, the two geometric morphisms $\rho \cdot \theta$ and $id_{\mathscr{V}}$ from the complete spread τ to itself agree when precomposed with the pure ρ. They must therefore agree: $\rho \cdot \theta \cong id_{\mathscr{V}}$. We have shown that ρ is an equivalence, so that φ is a complete spread. □

Corollary 2.5.6 *Complete spreads are stable under pullback along a locally connected geometric morphism. Furthermore, if μ is the distribution corresponding to a complete spread ψ, then the distribution corresponding to the pullback of ψ along a locally connected ξ is $\mu \cdot \xi_!$.*

Proof. This follows from Proposition 2.5.5 because locally connected geometric morphisms are closed under pullback and composition. Furthermore, the pullback along a locally connected geometric morphism satisfies the BCC condition. □

Corollary 2.5.6 and Proposition 2.2.11 give us the following.

Theorem 2.5.7 *The comprehensive factorization is stable under pullback along a locally connected geometric morphism.*

Corollary 2.5.6 covers the important case of pullback along a local homeomorphism. On the other hand, the proof of Corollary 2.5.6 does not cover the following result because essential geometric morphisms are not generally pullback stable. We shall prove Proposition 2.5.8 in § 5.2 using a result from § 4.3.

Proposition 2.5.8 *The topos pullback of a complete spread along an essential geometric morphism is a complete spread.*

Exercises 2.5.9

1. *After reading about how complete spreads are classified by the symmetric topos, show that the point of the symmetric topos that corresponds to the composite of two complete spreads*

$$\mathscr{Y} \xrightarrow{\varphi} \mathscr{E}\,,\quad \mathscr{G} \xrightarrow{\psi} \mathscr{Y}$$

is $M(\varphi) \cdot q$*, where* $\mathscr{S} \xrightarrow{q} M(\mathscr{Y})$ *is the point that corresponds to* ψ*.*

Further reading: Arnold [Arn00], Bunge & Funk [BF96b, BF98, BF96a], Dehornoy [Deh94], Fox [Fox57], Funk [Fun95, Fun99, Fun01], Johnstone [Joh02], Johnstone & Joyal [JJ82], MacLane & Moerdijk [MM92], Niefield [Nie82], Nielsen [Nie27], Prasolov & Sossinski [PS96], Short & Wiest [SW00], Stillwell [Sti86].

3

The Spread and Completeness Conditions

In this chapter, which has much profited from contributions by our collaborators T. Streicher and M. Jibladze [BFJS00], we define separate notions of spread and completeness, independently of complete spreads as they arise through the equivalence with distributions. Ultimately we shall show that a geometric morphism is a complete spread iff it is a spread and has the completeness property.

The notion of a spread is a natural way in which to capture for geometric morphisms the idea of clopen-generated, or what is sometimes called zero-dimensional. Definable morphisms are sufficient for defining spread. It turns out that spreads are necessarily localic. In the case of a spread with locally connected domain, spreads are generated by families of components, rather than just by the definable families, i.e., by components rather than just by clopens. The coreflection of definable families into families of components is a key ingredient for explicitly describing the topology of the display site in fibrational terms, as well as for isolating a notion of completeness.

3.1 Spreads

A *generating family* for a geometric morphism $\mathscr{F} \xrightarrow{\psi} \mathscr{E}$ is a morphism

$$G \xrightarrow{m} \psi^* X$$

with the property that every $F \xrightarrow{a} \psi^* Y$ (regarded as an object of $\mathscr{F}/\psi^* Y$) can be covered by a pullback of m: this means that a can be put in a diagram of the following form, where the right hand square is a pullback.

$$\begin{array}{ccccc} F & \twoheadleftarrow & P & \longrightarrow & G \\ {\scriptstyle a}\downarrow & & \downarrow & & \downarrow{\scriptstyle m} \\ \psi^* Y & \xleftarrow[\psi^* x]{} & \psi^* Z & \xrightarrow[\psi^* z]{} & \psi^* X \end{array}$$

R. Paré and D. Schumacher prove the following simplifying fact about generating families.

Proposition 3.1.1 *The following are equivalent for* $\mathscr{F} \xrightarrow{\psi} \mathscr{E}$:

1. ψ *has a generating family,*
2. ψ *has a generating family at* 1*: there is* $G \xrightarrow{m} \psi^* X$ *such that every object* F *of* $\mathscr{F}$ *can be put in a diagram of the following form, where the square is a pullback.*

$$\begin{array}{ccccc} F & \twoheadleftarrow & P & \longrightarrow & G \\ & & \big\downarrow & & \big\downarrow{\scriptstyle m} \\ & & \psi^* Z & \underset{\psi^* z}{\longrightarrow} & \psi^* X \end{array}$$

3. *there is a object* G (which we call a Diaconescu object) *with the property that every* F *of* $\mathscr{F}$ *can be put in a diagram of the form*

$$\begin{array}{ccc} S & \twoheadrightarrow & F \\ \big\downarrow & & \\ \psi^* Z \times G & & \end{array}$$

in $\mathscr{F}$,

4. *there is an object* G *of* $\mathscr{F}$ *such that the left vertical of the following pullback composed with the projection is a generating family for* ψ.

$$\begin{array}{ccc} P & \longrightarrow & \{(S, s) \mid s \in S\} \\ \big\downarrow & & \big\downarrow \\ \psi^*\psi_*(\Omega_{\mathscr{F}}{}^G) \times G & \longrightarrow & \Omega_{\mathscr{F}}{}^G \times G \\ \big\downarrow & & \\ \psi^*\psi_*(\Omega_{\mathscr{F}}{}^G) & & \end{array}$$

Proof. The implications $1 \Rightarrow 2 \Rightarrow 3$ and $4 \Rightarrow 1$ are trivial.

$3 \Rightarrow 4$: Paré and Schumacher have shown that for any Diaconescu object G, the family defined in 4 is a generating family at 1. But property 3 localizes: for any X of $\mathscr{E}$, X^*G has the same property in $\mathscr{F}/\psi^*X$. The result follows because the definition of the family $P \longrightarrow \psi^*\psi_*(\Omega_{\mathscr{F}}{}^G)$ is stable under localization. □

Definition 3.1.2 *Let* $\mathscr{F} \xrightarrow{\psi} \mathscr{E}$ *denote a geometric morphism over* $\mathscr{S}$. A definable generating family for ψ *is a generating family* $G \longrightarrow \psi^* X$ *for* ψ *that is a definable morphism. (We usually omit reference to the base topos* $\mathscr{S}$.)

Definition 3.1.3 *A geometric morphism over a base topos $\mathcal{S}$ is said to be a* spread *if it has a definable generating family.*

It is natural to speculate that Proposition 3.1.1 will remain valid if "definable" is appropriately inserted throughout. However, the following example shows that "spread at 1" is much weaker than spread.

Example 3.1.4 *T. Streicher has discovered the following example of a geometric morphism over Set with locally connected domain that has a definable generating family at 1, but is not a spread.*

Let M be the monoid with underlying set $\{1, a, b\}$ where 1 is the neutral element and $xa = a$ and $xb = b$ for all $x \in \{1, a, b\}$. Let $P(M)$ denote the topos of right M-sets. Let $M_0 \rightarrowtail M$ denote the submonoid $\{1, a\}$. Let $\psi : P(M_0) \longrightarrow P(M)$ denote the essential geometric morphism coming from the inclusion of M_0 in M: ψ^* restricts right actions. Evidently, ψ^* is surjective on objects, and therefore, the geometric morphism ψ is not only localic but also a spread at 1 just because identities are Set-definable subobjects. Nevertheless ψ is not a spread over Set. Indeed, let A be the representable M-set, i.e., A is M with right multiplication. Let $R \rightarrowtail \psi^* A$ be the single representable M_0-set ($R = M_0$). Suppose that there is a morphism $B \xrightarrow{u} A$ and Set-definable (i.e., complemented) subobject $Q \rightarrowtail \psi^* B$, such that the restriction of $\psi^* u$ to Q factors through R by an epimorphism $e : Q \twoheadrightarrow R$.

$$\begin{array}{ccc} Q & \stackrel{e}{\twoheadrightarrow} & R \\ \downarrow & & \downarrow \\ \psi^* B & \xrightarrow[\psi^* u]{} & \psi^* A \end{array}$$

(Proposition 3.1.7 characterizes spreads this way.) Then there exists an $x \in Q$ with $u(x) = e(x) = a$. But where does xb lie? We have $(xb)a = xa \in Q$, and since Q is complemented, we must have $xb \in Q$. But then $b = ab = u(x)b = u(xb) = e(xb) \in R$, which is not so.

Proposition 3.1.5 *We have the following:*

1. *An identity geometric morphism is a spread.*
2. *A subtopos of a spread is a spread: in the following diagram if ψ is a spread, then so is φ.*

$$\begin{array}{ccc} \mathcal{G} & \rightarrowtail & \mathcal{F} \\ & {}_{\varphi}\searrow \quad \swarrow_{\psi} & \\ & \mathcal{E} & \end{array}$$

In particular, by 1 an inclusion geometric morphism is a spread.

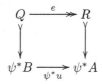

We further investigate the properties of spreads. A geometric morphism $\mathscr{F} \xrightarrow{\psi} \mathscr{E}$ is said to be *localic* if every object A of $\mathscr{F}$ can be put in a diagram of the form

$$\begin{array}{ccc} P & \overset{e}{\twoheadrightarrow} & A \\ {\scriptstyle m}\downarrow & & \\ \psi^* X & & \end{array}$$

for some X, some epimorphism e and monomorphism m.

Proposition 3.1.6 *A geometric morphism $\mathscr{F} \xrightarrow{\psi} \mathscr{E}$ is localic iff every object $A \xrightarrow{a} \psi^* Y$ of $\mathscr{F}/\psi^* Y$ can be put in a diagram of the form*

$$\begin{array}{ccc} P & \overset{e}{\twoheadrightarrow} & A \\ {\scriptstyle r}\downarrow & & \downarrow{\scriptstyle a} \\ \psi^* X & \underset{\psi^* n}{\longrightarrow} & \psi^* Y \end{array}$$

for some epimorphism e and monomorphism r. (Thus, $1_{\mathscr{F}}$ is a Diaconescu object.)

Proof. The condition is clearly sufficient (put $Y = 1$). For the necessity, assume that ψ is localic. Let $A \xrightarrow{a} \psi^* Y$ be given. We have

$$\begin{array}{ccc} P & \overset{e}{\twoheadrightarrow} & A \\ {\scriptstyle m}\downarrow & & \\ \psi^* X & & \end{array}$$

for some X and monomorphism $m : P \rightarrowtail \psi^* X$. Therefore we also have

$$\begin{array}{ccc} P & \overset{e}{\twoheadrightarrow} & A \\ {\scriptstyle (ae,m)}\downarrow & & \downarrow{\scriptstyle a} \\ \psi^*(Y \times X) & \underset{\psi^*(\pi_1)}{\longrightarrow} & \psi^* Y \end{array}$$

□

We have the following result for definable generating families, which is analogous to portions of Proposition 3.1.1. In particular, this result shows that spreads are localic.

Proposition 3.1.7 *The following are equivalent for $\mathscr{F} \xrightarrow{\psi} \mathscr{E}$:*

1. *ψ is a spread* (i.e., ψ has a definable generating family)*;*
2. *every object $F \longrightarrow \psi^*Y$ of $\mathscr{F}/\psi^*Y$ can be put in a diagram of the form*

$$\begin{array}{ccc} S & \twoheadrightarrow & F \\ \downarrow & & \downarrow \\ \psi^*Z & \longrightarrow & \psi^*Y \end{array}$$

for some object $Z \longrightarrow Y$ in $\mathscr{E}/Y$, where $S \longrightarrow \psi^*Z$ is a definable morphism;

3. *every object $F \longrightarrow \psi^*Y$ of $\mathscr{F}/\psi^*Y$ can be put in a diagram of the form*

$$\begin{array}{ccc} S & \twoheadrightarrow & F \\ \downarrow & & \downarrow \\ \psi^*Z & \longrightarrow & \psi^*Y \end{array}$$

for some object $Z \longrightarrow Y$, where $S \rightarrowtail \psi^*Z$ is a definable subobject;

4. *the left vertical of the following pullback is a generating family for ψ.*

$$\begin{array}{ccc} P & \longrightarrow & 1 \\ \downarrow & & \downarrow{\scriptstyle f^*\top} \\ \psi^*\psi_*(f^*\Omega_{\mathscr{S}}) & \longrightarrow & f^*\Omega_{\mathscr{S}} \end{array}$$

Proof. 1 $\Rightarrow$ 2 is immediate.

2 $\Rightarrow$ 3: Let $F \longrightarrow \psi^*Y$ be arbitrary. Assuming 2, we have pullbacks

$$\begin{array}{ccccc} F & \twoheadleftarrow & P & \xrightarrow{a} & f^*A \\ \downarrow & & \downarrow{\scriptstyle p} & & \downarrow{\scriptstyle f^*n} \\ \psi^*Y & \xleftarrow[\psi^*m]{} & \psi^*Z & \xrightarrow[b]{} & f^*B \end{array} \qquad \begin{array}{ccc} P & \xrightarrow{a} & f^*A \\ {\scriptstyle (a,p)}\downarrow & & \downarrow{\scriptstyle f^*(1,n)} \\ f^*A \times \psi^*Z & \xrightarrow[1\times b]{} & f^*(A \times B) \end{array}$$

which shows that $F \longrightarrow \psi^*Y$ can be covered by a definable subobject of a $\psi^*(Z')$, $Z' = e^*A \times Z$. Note: $f^*A \times \psi^*Z \cong \psi^*(e^*A \times Z)$.

3 $\Rightarrow$ 4: $P \rightarrowtail \psi^*\psi_*(f^*\Omega_{\mathscr{S}})$ generates at 1 because any definable subobject $S \rightarrowtail \psi^*Z$ must appear in a pullback diagram as follows.

$$\begin{array}{ccccc} S & \longrightarrow & P & \longrightarrow & 1 \\ \downarrow & & \downarrow & & \downarrow{\scriptstyle f^*\top} \\ \psi^*Z & \longrightarrow & \psi^*\psi_*(f^*\Omega_{\mathscr{S}}) & \longrightarrow & f^*\Omega_{\mathscr{S}} \end{array}$$

The definition of $P \rightarrowtail \psi^*\psi_*(f^*\Omega_{\mathscr{S}})$ is stable under localization at any Y of $\mathscr{E}$, so we may repeat the above argument at Y.

4 $\Rightarrow$ 1 is trivial. □

Corollary 3.1.8 *Spreads are localic.*

Proof. This follows from Propositions 3.1.6 and 3.1.7. □

Proposition 3.1.9 *We have the following:*

1. *If definable subobjects compose in the domain topos of the composite geometric morphism of two spreads, then the composite is a spread.*
2. *If $\rho \cdot \psi$ is a spread, then ψ is a spread.*

The following lemma plays a role in the van Kampen theorem later. In any case, it may be of independent interest.

Lemma 3.1.10 *Pullback along a locally connected surjection reflects spreads.*

Proof. Use the equivalent condition 2 in Prop. 3.1.7 for spread, and the fact that the left adjoint of a locally connected geometric morphism preserves definable morphisms (Exercise 1.5.7 (10)). Notice also that the transpose $\psi_! Y \longrightarrow E$ of an epimorphism $Y \twoheadrightarrow \psi^* E$ is an epimorphism if the locally connected ψ is a surjection. The fact that locally connected geometric morphisms are pullback stable, such that the BCC is satisfied, also comes into play here. □

The following comments lead to an alternative view of spreads. If M is a complete join-semilattice in a topos $\mathscr{E}$, then $K \longrightarrow M$ sup-generates M when $\Omega^K \longrightarrow \Omega^M \xrightarrow{\vee} M$ is an epimorphism. This holds precisely when $K \longrightarrow M$ is a generating family for the $\mathscr{E}$-indexed category M. Suppose M is a frame $\mathscr{O}(X)$ in $\mathscr{E}$, with sheaves $\psi : Sh_{\mathscr{E}}(X) \longrightarrow \mathscr{E}$. Then $K \longrightarrow \mathscr{O}(X) = \psi_* \Omega_X$ sup-generates iff the subobject $S \rightarrowtail \psi^* K$ classified by the transpose $\psi^* K \longrightarrow \Omega_X$ is a generating family for ψ. By Proposition 3.1.7 3, we have the following.

Proposition 3.1.11 $\mathscr{F} \xrightarrow{\psi} \mathscr{E}$ *is a spread iff ψ is localic and the morphism*

$$\psi_*(\tau) : \psi_*(f^* \Omega_{\mathscr{S}}) \longrightarrow \psi_* \Omega_{\mathscr{F}}$$

sup-generates the frame $\psi_* \Omega_{\mathscr{F}}$, where

$$f^* \Omega_{\mathscr{S}} \xrightarrow{\tau} \Omega_{\mathscr{F}}$$

classifies $f^* \top : 1 \rightarrowtail f^* \Omega_{\mathscr{S}}$.

Lemma 3.1.12 *A pure spread is an inclusion. A spread that is also a pure surjection is an equivalence.*

Proof. If $\mathscr{F} \xrightarrow{\psi} \mathscr{E}$ is pure, then the top horizontal in the following commutative square is an isomorphism.

$$\begin{array}{ccc} e^*\Omega_{\mathscr{S}} & \longrightarrow & \psi_*\psi^*e^*\Omega_{\mathscr{S}} \\ \downarrow & & \downarrow{\scriptstyle \psi_*\tau} \\ \Omega_{\mathscr{E}} & \longrightarrow & \psi_*\Omega_{\mathscr{F}} \end{array}$$

If in addition ψ is a spread, then ψ is localic and $\psi_*\tau$ sup-generates $\psi_*\Omega_{\mathscr{F}}$. Therefore, the unique frame morphism $\Omega_{\mathscr{E}} \longrightarrow \psi_*\Omega_{\mathscr{F}}$ sup-generates (bottom horizontal above). This implies that ψ is an inclusion. □

Among other things, we wish to know that the terminology complete spread is justified, i.e., that a complete spread (with locally connected domain) is a spread, at least when the codomain topos is bounded.

Lemma 3.1.13 *The essential geometric morphism associated with a discrete opfibration is a spread, hence localic.*

Proof. Let $\mathbb{Y} \longrightarrow \mathbb{C}$ be a discrete opfibration, with essential geometric morphism k. The unit $h_y \rightarrowtail k^*k_!(h_y)$ of any representable is a definable subobject. More generally, any $\mathscr{S}$-coproduct of such units is again a unit $\coprod_I h_{y_i} \rightarrowtail k^*k_!(\coprod_I h_{y_i})$, and also a definable subobject. Every object of $P(\mathbb{Y})$ is a quotient of such a coproduct $\coprod_I h_{y_i}$, so k is a spread at 1. In order to show that k is a spread at $E \in P(\mathbb{C})$, we may repeat this argument for the pullback of $\mathbb{Y} \longrightarrow \mathbb{C}$ along the discrete fibration $\mathbb{E} \longrightarrow \mathbb{C}$, which is a discrete opfibration. □

Proposition 3.1.14 *A complete spread is a spread.*

Proof. Consider the defining pullback square of a complete spread ψ.

$$\begin{array}{ccc} \mathscr{Y} & \rightarrowtail & P(\mathbb{Y}) \\ {\scriptstyle \psi}\downarrow & & \downarrow{\scriptstyle k} \\ \mathscr{E} & \rightarrowtail & P(\mathbb{C}) \end{array}$$

The result follows by Propositions 3.1.5, 3.1.9, and Lemma 3.1.13. □

Remark 3.1.15 *The fact that complete spreads are localic has two obvious consequences. First, a complete spread over a topos can therefore be regarded as a locale (of a special kind) internal to the topos. Second, because localic geometric morphisms compose, the domain topos of a complete spread over a localic topos is localic. It is also worth mentioning that the comprehensive factorization for geometric morphisms is therefore* sublocalic *in the sense that the second factor of the factorization is localic. Of course, the orthogonal partners of localic morphisms, the so-called hyperconnected geometric morphisms, are pure.*

We may now obtain the following factorization theorem for geometric morphisms as an application of the comprehensive factorization.

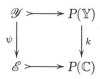

Theorem 3.1.16 *A geometric morphism with locally connected domain may be factored into a pure surjection followed by a spread, where the middle topos is locally connected. This factorization is essentially unique in the sense that any two such factorizations are equivalent by an equivalence that is unique up to unique isomorphism.*

Proof. Existence: factor the pure factor of the comprehensive factorization into a surjection and an inclusion. The surjection and inclusion factors are both pure, and the topos between them is locally connected (Lemma 2.2.16). The inclusion composed with the complete spread is a spread.

Uniqueness: complete the spread factor of a given pure surjection, spread factorization to a complete spread. The geometric morphism from the spread to its completion is a pure spread, hence an inclusion (Lemma 3.1.12). The uniqueness of the comprehensive and surjection, inclusion factorizations can now be applied. □

Exercises 3.1.17

1. *Let X be a locale in a topos $\mathscr{S}$. Show that $\Delta \dashv \Gamma : Sh(X) \longrightarrow \mathscr{S}$ is a spread iff the pullback along the counit*

$$\begin{array}{ccc} P & \rightarrowtail & \Delta\Gamma(\Delta\Omega_{\mathscr{S}}) \\ \downarrow & & \downarrow \\ 1 & \xrightarrow{\Delta\top} & \Delta\Omega_{\mathscr{S}} \end{array}$$

in $Sh(X)$ is a generating family.

2. *Suppose that $\mathscr{E}$ is locally connected. Show that a local homeomorphism $\mathscr{E}/X \longrightarrow \mathscr{E}$ is a spread iff the definable subobject*

$$\{(x,U) \mid x \in U\} \rightarrowtail X \times e^*(\Omega_{\mathscr{S}})^X$$

is a generating family for $\mathscr{E}/X \longrightarrow \mathscr{E}$.

3. *Prove Propositions 3.1.5 and 3.1.9.*
4. *Fill in the details for the proof Lemma 3.1.13.*
5. *Show that localic geometric morphisms are pullback stable and closed under composition.*
6. *Working from the definition of complete spread, show that complete spreads are localic. Use Lemma 3.1.13 and the fact that localic geometric morphisms are pullback stable.*
7. *Investigate the pullback stability of the pure surjection, spread factorization.*
8. *Let $\mathscr{F} \xrightarrow{\varphi} \mathscr{E}$ be any geometric morphism with locally connected domain, and let $I = f_!1$. Then in the topos pullback diagram*

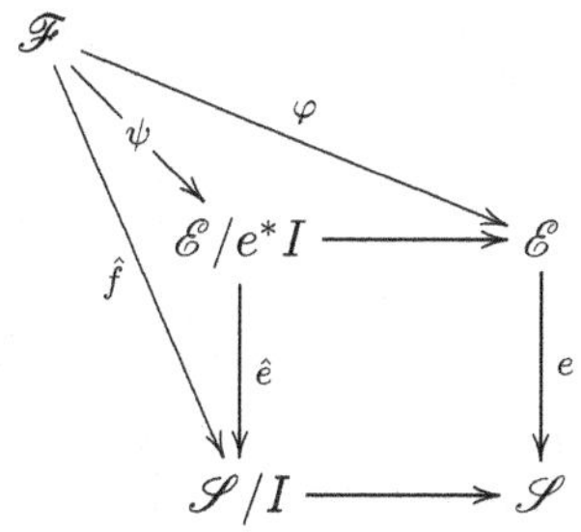

the given φ *is a spread over* $\mathscr{S}$ *iff* ψ *is a spread over* $\mathscr{S}/I$. *The factoring inverse image functor* ψ^* *is given by pullback of* φ^* *along the unit* $1 \longrightarrow f^* f_! 1$.

3.2 Zero-dimensionality

Traditionally, the term zero-dimensional usually means that the clopens generate. Our view is that spread is a natural generalization of this idea (Exercise 1). In any case, it may be worthwhile and interesting to compare spreads with other notions of "clopens generate." We make such a comparison in this section.

We begin with a fact we shall meet again in 7.3 (Lemma 7.3.13). The reader may wish to consult Chapter 7 before reading this section. In fact, our work here depends on some results from 7.3.

Lemma 3.2.1 *Under the identification made in Proposition 7.3.11, the morphism*

$$\psi_*(\tau_{\mathscr{F}}) : \psi_*(f^* \Omega_{\mathscr{S}}) \longrightarrow \psi_* \Omega_{\mathscr{F}}$$

sends a flat function $e^* \Omega_{\mathscr{S}} \xrightarrow{p} \psi_* \Omega_{\mathscr{F}}$ *to* $p(\top)$.

Proof. A flat function p determines an *atlas* for a global section of $e^* \Omega_{\mathscr{S}}$ regarded as a sheaf on the frame $\psi_* \Omega_{\mathscr{F}}$, namely the family

$$\{ (p(u), u) \mid u \in e^* \Omega_{\mathscr{S}} \} .$$

This family is an atlas by Proposition 7.3.12. Then $\psi_* \tau_{\mathscr{F}}$ associates with this atlas the element

$$\bigvee \{ \omega \wedge p(u) \mid u \in e^* \Omega_{\mathscr{S}} \} ,$$

which is equal to $p(\top)$. □

Consider a localic geometric morphism $\psi : \mathscr{F} \longrightarrow \mathscr{E}$ over $\mathscr{S}$, so that $\mathscr{F}$ is equivalent to the topos of sheaves over $\mathscr{E}$ on the frame $\psi_* \Omega_{\mathscr{F}}$. We know there is the composite morphism

$$e^* \Omega_{\mathscr{S}} \xrightarrow{\tau_{\mathscr{E}}} \Omega_{\mathscr{E}} \longrightarrow \psi_* \Omega_{\mathscr{F}}$$

where the second morphism is the unique frame map. Following Kock and Reyes, let

$$\mathrm{Clp}_{\mathscr{S}}(\psi_*\Omega_{\mathscr{F}}) = \{\, U \in \psi_*\Omega_{\mathscr{F}} \mid \top = \bigvee\{U \Leftrightarrow u \mid u \in e^*\Omega_{\mathscr{S}}\}\,\}\,.$$

$\mathrm{Clp}_{\mathscr{S}}(\psi_*\Omega_{\mathscr{F}})$ is the subobject of $\mathscr{S}$-complemented elements of $\psi_*\Omega_{\mathscr{F}}$.

Proposition 3.2.2 *For any localic geometric morphism* $\mathscr{F} \xrightarrow{\psi} \mathscr{E}$ *over* $\mathscr{S}$, *the morphism* $\psi_*(\tau_{\mathscr{F}})$ *factors through the subobject* $\mathrm{Clp}_{\mathscr{S}}(\psi_*\Omega_{\mathscr{F}})$.

Proof. By Proposition 7.3.12, for any flat function $e^*\Omega_{\mathscr{S}} \xrightarrow{p} \psi_*\Omega_{\mathscr{F}}$, we have

$$\top = \bigvee\{p(u) \mid u \in e^*\Omega_{\mathscr{S}}\} \leq \bigvee\{p(\top) \Leftrightarrow u \mid u \in e^*\Omega_{\mathscr{S}}\}\,.$$

The result now follows by Proposition 3.2.1. □

Theorem 3.2.3 *If* $\mathscr{F} \xrightarrow{\psi} \mathscr{E}$ *is a spread, then the frame* $\psi_*\Omega_{\mathscr{F}}$ *is sup-generated by its sublattice* $\mathrm{Clp}_{\mathscr{S}}(\psi_*\Omega_{\mathscr{F}})$.

Proof. Apply Propositions 3.1.11 and 3.2.2. □

We have the following partial converse to Theorem 3.2.3.

Theorem 3.2.4 *Suppose that* $\mathscr{F} \xrightarrow{\psi} \mathscr{E}$ *is open and localic, and that* $\mathscr{E} \xrightarrow{e} \mathscr{S}$ *is subopen. If* $\psi_*\Omega_{\mathscr{F}}$ *is sup-generated by its sublattice* $\mathrm{Clp}_{\mathscr{S}}(\psi_*\Omega_{\mathscr{F}})$, *then* ψ *is a spread.*

Proof. If ψ is open, and $\tau_{\mathscr{E}} : e^*\Omega_{\mathscr{S}} \longrightarrow \Omega_{\mathscr{E}}$ is a monomorphism (e is subopen), then the factorization of $\psi_*(\tau_{\mathscr{F}})$ through $\mathrm{Clp}_{\mathscr{S}}(\psi_*\Omega_{\mathscr{F}})$ is an epimorphism. Indeed, if $U \in \mathrm{Clp}_{\mathscr{S}}(\psi_*\Omega_{\mathscr{F}})$, then the function $u \mapsto (U \Leftrightarrow u)$ is flat, and its value at $\top$ is U. The first condition of flatness (Definition 7.3.6) is immediately satisfied. For the second condition, first observe that for any $u, u' \in e^*\Omega_{\mathscr{S}}$, we have

$$u \Leftrightarrow u' = (\tau_{\mathscr{E}}u = \tau_{\mathscr{E}}u') = (u = u')$$

in the frame $\psi_*\Omega_{\mathscr{F}}$. The first equality holds because ψ is open, and the second because e is subopen. Then

$$(U \Leftrightarrow u) \wedge (U \Leftrightarrow u') = (U \Leftrightarrow u) \wedge (u \Leftrightarrow u') \leq (u \Leftrightarrow u') = (u = u')\,,$$

which concludes the proof. □

Exercises 3.2.5

1. *Show directly that for any topological space* X, *Sh*$(X) \longrightarrow$ *Set is a spread iff the clopens generate the topology in* X.

3.3 Components-generated Spreads

We shall show that our formal notion of spread captures the idea that informally a spread $\mathscr{F} \xrightarrow{\psi} \mathscr{E}$ with locally connected domain is defined by the property: if $\{X_i\}$ is a generating family for $\mathscr{E}$ over $\mathscr{S}$, then the connected components of the $\psi^* X_i$ generate $\mathscr{F}$ over $\mathscr{S}$. Our objective requires the formal introduction of such "families of components."

Definition 3.3.1 *Let $\mathscr{F} \xrightarrow{\psi} \mathscr{E}$ be over $\mathscr{S}$.*

1. A definable family *is a pair consisting of a morphism $D \xrightarrow{x} e^* I$ in $\mathscr{E}$, and a definable subobject $X \xrightarrow{m} \psi^* D$ in $\mathscr{F}$. We often depict such an object in a single diagram:*
$$X \xrightarrow{m} \psi^* D \xrightarrow{\psi^* x} f^* I \ ,$$
and sometimes we denote it more concisely just as (x, m).
2. *When $\mathscr{F}$ is locally connected, a definable family (x, m) is said to be* a family of components *if the transpose $f_! X \longrightarrow I$ of the composite $\psi^* x \cdot m$ with respect to $f_! \dashv f^*$ is an isomorphism.*
3. *A morphism of definable families, or of families of components, is a 3-tuple of morphisms $p : Y \longrightarrow X$, $b : E \longrightarrow D$ and $a : H \longrightarrow I$ making the obvious squares commute.*

Denote by FD_ψ the fibered category (over $\mathscr{S}$) of definable families, and by FC_ψ the full subfibration on families of components.

Intuitively, a family of components is a family with the property that for every $i \in I$, the fiber $\psi^*(x)^{-1}(i)$ contains a single component of $\psi^* D$.

We may associate with any topos $\mathscr{F}$ over $\mathscr{S}$ an $\mathscr{S}$-fibration, the glueing construction, which we denote (§ 1.3),

$$\mathscr{F}/f^* \longrightarrow \mathscr{S} \ .$$

Objects of $\mathscr{F}/f^*$ are morphisms $X \longrightarrow f^* I$ in $\mathscr{F}$. Then we have an $\mathscr{S}$-Cartesian forgetful functor

$$FD_\psi \longrightarrow \mathscr{F}/f^*$$

that associates with a definable family (x, m) the composite $\psi^* x \cdot m$. There is also an $\mathscr{S}$-Cartesian functor

$$FD_\psi \longrightarrow \mathscr{E}/e^*$$

sending (x, m) to x.

Proposition 3.3.2 *If $\mathscr{F}$ is locally connected, then the $\mathscr{S}$-fibered inclusion*

$$FC_\psi \hookrightarrow FD_\psi$$

has a right adjoint, which we shall call the FC-coreflection.

Proof. The FC-coreflection of a definable family (x, m) is formed as follows. Let $\hat{t}$ denote the transpose of the composite $t = \psi^* m{\cdot}x$ under $f_! \dashv f^*$. Consider the pullback in $\mathscr{E}$ in the diagram below (left square), and then obtain from it a pullback diagram in $\mathscr{F}$ by applying ψ^* (bottom right square).

$$\begin{array}{ccccccc}
 & & & & X & = & X \\
 & & & & \downarrow & & \downarrow \\
F & \xrightarrow{p} & D & \quad & \psi^* F & \xrightarrow{\psi^* p} & \psi^* D \\
\downarrow & & \downarrow & & \downarrow & & \downarrow \\
e^* f_! X & \xrightarrow{e^* \hat{t}} & e^* I & & f^* f_! X & \xrightarrow{f^* \hat{t}} & f^* I
\end{array}$$

(with $t : X \to f^* I$ the curved arrow on the right)

The subobject $X \overset{m}{\rightarrowtail} \psi^* D$ and the unit $\eta_X : X \longrightarrow f^* f_! X$ make the outer diagram at right commutative, inducing a subobject $X \rightarrowtail \psi^* F$ such that the top square commutes. $X \rightarrowtail \psi^* F$ is definable because p, and hence $\psi^* p$, is. The left vertical $X \rightarrowtail \psi^* F \longrightarrow f^* f_! X$ is the FC-coreflection of t. □

Remark 3.3.3 *We emphasize that the FC-coreflection functor is* not *$\mathscr{S}$-Cartesian.*

We shall say that an object $X \longrightarrow f^* I$ of $\mathscr{F}/f^*$ *covers* another $Y \longrightarrow f^* K$ if there is a commutative diagram

$$\begin{array}{ccc}
X & \twoheadrightarrow & Y \\
\downarrow & & \downarrow \\
f^* I & \xrightarrow{f^* \alpha} & f^* K
\end{array}$$

in $\mathscr{F}$. This is a morphism in $\mathscr{F}/f^*$ that we sometimes call *a collective epimorphism.*

We shall upgrade the comprehensive factorization to our present fibrational language. This will prepare us for our refined characterization of spread with locally connected domain (Theorem 3.3.5), which fulfills our objective set at the beginning of this section.

We define an $\mathscr{S}$-fibered category $\mathrm{Display}_\psi$ and an $\mathscr{S}$-Cartesian functor

$$\mathrm{Display}_\psi \longrightarrow \mathscr{E}/e^* ,$$

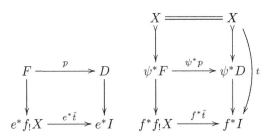

whose $\mathscr{S}$-fibers are discrete opfibrations. An object of $\mathrm{Display}_\psi$ is a pair (d, s), where $D \xrightarrow{d} e^* I$ is an object of $\mathscr{E}/e^*$, and $I \xrightarrow{s} f_! \psi^* D$ is a section of the transpose $\psi^* D \xrightarrow{\psi^* d} f^* I$. A morphism in this category is a morphism of (x, α) of $\mathscr{E}/e^*$ such that $f_! \psi^*(x)$ and α commute with sections.

Proposition 3.3.4 *There is an* $\mathscr{S}$*-Cartesian functor*

$$V : \mathrm{Display}_\psi \longrightarrow FC_\psi$$

over $\mathscr{E}/e^*$ *such that* $V(d, s)$ *is the following pullback.*

$$\begin{array}{ccccc} V(d,s) & \xrightarrow{m} & \psi^* D & & \\ \downarrow & & \downarrow \eta & \searrow^{\psi^* d} & \\ f^* I & \xrightarrow{f^* s} & f^* f_!(\psi^* D) & \xrightarrow{f^* t} & f^* I \end{array} \tag{3.1}$$

(Really $V(d, s)$ is the top row $V(d, s) \xrightarrow{m} \psi^* D \xrightarrow{\psi^* d} f^* I$.) *$V(d, s)$ is a family of components, and V is an equivalence. Therefore, and in particular, the* $\mathscr{S}$*-fibers of the* $\mathscr{S}$*-Cartesian functor* $FC_\psi \longrightarrow \mathscr{E}/e^*$ *are discrete opfibrations.*

Proof. In diagram (3.1) the composite morphism $\psi^* d \cdot m$ is equal to the left vertical since the bottom horizontal is the identity on $f^* I$. Thus the transpose of $V(d, s)$ under $f_! \dashv f^*$ is an isomorphism because it is equal to the left vertical of the transpose pullback.

$$\begin{array}{ccc} f_! V(d,s) & \xrightarrow{f_! m} & f_!(\psi^* D) \\ \downarrow & & \downarrow 1 \\ I & \xrightarrow{s} & f_!(\psi^* D) \end{array}$$

It follows easily that V is an equivalence. □

Theorem 3.3.5 *Suppose that* $\mathscr{F} \xrightarrow{\psi} \mathscr{E}$ *has locally connected domain and bounded codomain. Then the following are equivalent:*

1. *ψ is a spread.*
2. *Any object of* $\mathscr{F}/f^*$ *may be covered by a family of components* (after passing the family to $\mathscr{F}/f^*$ via composition).
3. *For any generating family* $G \xrightarrow{y} e^* K$ *of* $\mathscr{E}$ *over* $\mathscr{S}$*, any object* $X \longrightarrow f^* I$ *can be covered by a family of components* (x, m)*, where*

$$\begin{array}{ccc} D & \longrightarrow & G \\ \downarrow x & & \downarrow y \\ e^* I & \xrightarrow{e^* \alpha} & e^* K \end{array}$$

   is a pullback in $\mathscr{E}$*.*
4. *The pure factor of ψ is an inclusion.*

Proof. 1 $\Rightarrow$ 2. Let $X \longrightarrow f^*I$ be an arbitrary object of $\mathscr{F}/f^*$. There is a square

$$\begin{array}{ccc} P & \twoheadrightarrow & X \\ \downarrow & & \downarrow \\ \psi^*E & \longrightarrow & \psi^*e^*I \end{array}$$

in $\mathscr{F}$, where $P \rightarrowtail \psi^*E$ is a definable subobject. We have thus covered $X \longrightarrow f^*I$ by (the composite of) the definable family $P \rightarrowtail \psi^*E \longrightarrow f^*I$. The FC-coreflection of $P \rightarrowtail \psi^*E \longrightarrow f^*I$ is a family of components (but indexed by $f_!P$) that covers $X \longrightarrow f^*I$.

2 $\Rightarrow$ 3. This is clear from the definition of generating family, with the FC-coreflection as a last step.

3 $\Rightarrow$ 4. By Proposition 3.3.4, we may use FC_ψ instead of $\mathrm{Display}_\psi$ to describe the comprehensive factorization of ψ. The forgetful functor

$$FC_\psi \longrightarrow \mathscr{F}/f^*$$

is a flat functor. Thus, when we restrict it to the full subcategories $\mathbb{Y}$ of FC_ψ, and $\mathbb{C}$ of $\mathscr{E}/e^*$ defined by the given generating family for $\mathscr{E}$ over $\mathscr{S}$, we obtain a geometric morphism $\mathscr{F} \xrightarrow{\rho} P(\mathbb{Y})$. This geometric morphism is depicted in the following topos pullback diagram.

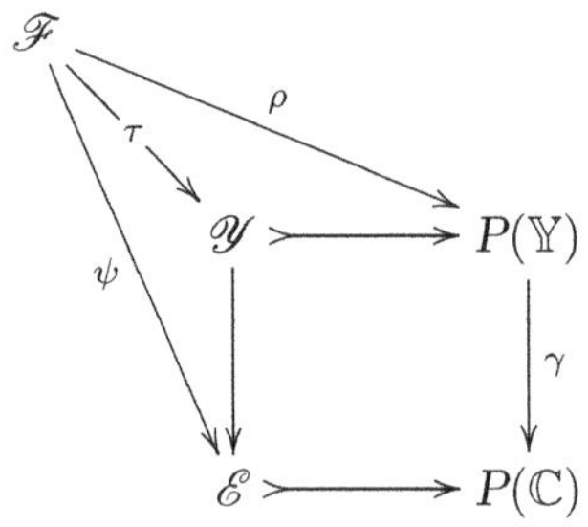

Our hypothesis implies that ρ is an inclusion. Hence the pure factor τ is also an inclusion.

4 $\Rightarrow$ 1. If τ is an inclusion, then so is ρ. By Lemma 3.1.13, the essential geometric morphism, called γ above, associated with a discrete opfibration is a spread. Therefore its composite with the inclusion ρ is a spread. Therefore, $\mathscr{F} \xrightarrow{\psi} \mathscr{E} \rightarrowtail P(\mathbb{C})$, and hence ψ, is a spread. □

Condition 3 of Theorem 3.3.5 formally expresses the informal idea stated at the start of this section.

Exercises 3.3.6

1. *Show that if for* $X \xrightarrow{n} \psi^* D \xrightarrow{\psi^* x} f^* I$, *$n$ is definable and the transpose is an isomorphism, then n is a monomorphism. Hence, (x, n) is a family of components.*
2. *Show that* FD_ψ, *and for $\mathscr{F}$ locally connected,* FC_ψ *are indeed $\mathscr{S}$-fibrations.*
3. *Show that the construction in the proof of Proposition 3.3.2 gives the FC-coreflection.*
4. *Show that the geometric morphism associated with a topological spread (Definition 2.1.2) is a spread.*

3.4 Amalgamation and Fibrations

In order to prepare for our explanation of completeness in § 3.5 we must revisit the construction of the comprehensive factorization from a fibrational point of view. This fibrational point of view is a valuable, even indispensable, tool. Instead of working with the display category, we wish to formulate the construction and the amalgamation topology in terms of the equivalent category FC_ψ, as in § 3.3.

We have already provided a fibrational description of the display category, denoted $\mathrm{Display}_\mu$ (§ 3.3). ($\mathrm{Display}_\mu$ is used to construct the complete spread associated with μ.) We remind the reader that an object of $\mathrm{Display}_\mu$ is a pair

$$(x, s) = (\, D \xrightarrow{x} e^* I, I \xrightarrow{s} \mu(D)\,)$$

where x is an object of $\mathscr{E}/e^*$, and where s is a section of

$$\mu^I(x) : \mu(D) \longrightarrow I.$$

Perhaps a word of explanation is needed here. Since μ preserves coproducts, $\mu(D)$ is indeed the domain object of $\mu^I(x)$, and we may identify $\mu^I(x)$ with

$$\mu(D) \xrightarrow{\mu(x)} \mu(e^* I) \longrightarrow I,$$

where $\mu(e^* I) \cong I \times \mu(1)$, so that we canonically identify the second morphism above with a projection. Morphisms in $\mathrm{Display}_\mu$ are as usual.

The distribution algebra $H = \mu_*(\Omega_{\mathscr{S}})$ is a partially ordered object in $\mathscr{E}$. We may consider H as an $\mathscr{S}$-fibered category, just as we do with $\mathscr{E}$ for instance. We denote this $\mathscr{S}$-fibered category

$$\mathbb{H}_\mu \longrightarrow \mathscr{S} .$$

Its objects are pairs

$$(x, S) = (\, D \xrightarrow{x} e^* I, S \rightarrowtail \mu(D)\,) ,$$

and a morphism $(x, S) \longrightarrow (y, T)$ is a morphism in $\mathscr{E}/e^*$ that under μ restricts to the subobjects S and T.

Proposition 3.4.1 *The full $\mathscr{S}$-Cartesian embedding*

$$\mathrm{Display}_\mu \longrightarrow \mathbb{H}_\mu$$

has a right adjoint (which is not $\mathscr{S}$-Cartesian).

Proof. Of course a section is a subobject, so there is such an embedding. The right adjoint assigns to a pair $(D \xrightarrow{x} e^*I, S \overset{r}{\rightarrowtail} \mu D)$ the pullback m

$$\begin{array}{ccc} F & \longrightarrow & D \\ {\scriptstyle m}\downarrow & & \downarrow{\scriptstyle x} \\ e^*S & \xrightarrow{e^*(tr)} & e^*I \end{array} \qquad \begin{array}{ccc} \mu F & \longrightarrow & \mu D \\ {\scriptstyle \mu^S(m)}\downarrow & & \downarrow{\scriptstyle t} \\ S & \xrightarrow{tr} & I \end{array}$$

in $\mathscr{E}$, where $t : \mu D \longrightarrow I$ is equal to $\mu^I(x)$. Then $\mu^S(m)$ is equipped with a section $s : S \longrightarrow \mu F$ induced by r and the pullback above right. □

We call the right adjoint in Proposition 3.4.1 *the D-coreflection.*

Proposition 3.4.2 *Let $\psi : \mathscr{F} \longrightarrow \mathscr{E}$ have locally connected domain, with $\mu = f_!\psi^*$. Then there is an equivalence $V' : \mathbb{H}_\mu \simeq FD_\psi$ over $\mathscr{S}$ that commutes with V* (Proposition 3.3.4). *Consequently, we may identify the D-coreflection and FC-coreflection.*

Proof. $V'(x, S)$ is the definable subobject of ψ^*D with characteristic morphism $\psi^*D \longrightarrow f^*\Omega_{\mathscr{S}}$ transposed from the characteristic morphism of

$$S \rightarrowtail f_!\psi^*D.$$

□

The $\mathscr{S}$-Cartesian forgetful functor

$$\mathbb{H}_\mu \longrightarrow \mathscr{E}/e^*$$

is a fibration, but

$$\mathrm{Display}_\mu \longrightarrow \mathscr{E}/e^*$$

is not (although the fibers of the latter are discrete opfibrations). We may therefore speak of the Cartesian lifting along $\mathbb{H}_\mu \longrightarrow \mathscr{E}/e^*$ of a morphism in $\mathscr{E}/e^*$: let

$$\begin{array}{ccc} D & \xrightarrow{m} & E \\ {\scriptstyle x}\downarrow & & \downarrow{\scriptstyle y} \\ e^*I & \xrightarrow{e^*(\alpha)} & e^*K \end{array}$$

be a morphism of $\mathscr{E}/e^*$ and (y, T) an object of $\mathbb{H}_\mu$. In order to lift (m, α) to $\mathbb{H}_\mu$ we must equip x with a subobject of $\mu(D)$. Of course this is done by pullback of the subobject T along $\mu(m)$.

We may now formulate the amalgamation covers (Definition 2.3.2) in our present fibrational language (Exercise 1).

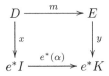

Definition 3.4.3 *A* μ-cover *of an object* (y,t) *of* $\mathrm{Display}_\mu$ *is the D-coreflection of the Cartesian lifting of a collective epimorphism* $(m,\alpha) : x \longrightarrow y$ *in* $\mathscr{E}/e^*$*. We sometimes refer to such a morphism as the* μ*-cover of* (y,t) *associated with the collective epimorphism* (m,α)*.*

Our easy final task in this section is to formulate μ-covers in the equivalent category FC_ψ, when we start with a geometric morphism ψ. We still keep the term μ-cover, but with the understanding that $\mu = f_!\psi^*$.

The domain object of the Cartesian lifting along $\mathbb{H}_\mu \longrightarrow \mathscr{E}/e^*$ of a morphism in $\mathscr{E}/e^*$ and a family of components must be a definable family, but it may not be a family of components. However, we may always consider the FC-coreflection of the domain object. Starting with a collective epimorphism (k,α) in $\mathscr{E}/e^*$ and a family of components $Y \rightarrowtail \psi^*E \longrightarrow f^*K$, the Cartesian lifting and FC-coreflection produce the following diagram.

$$
\begin{array}{ccccc}
X & = & X & \twoheadrightarrow & Y \\
\downarrow & & \downarrow & \lrcorner & \downarrow \\
\psi^*F & \xrightarrow{\psi^*p} & \psi^*D & \xrightarrow{\psi^*k} & \psi^*E \\
\downarrow & \lrcorner & \downarrow & & \downarrow \\
f^*f_!X & \xrightarrow{f^*\hat{t}} & f^*I & \xrightarrow{f^*\alpha} & f^*K
\end{array}
$$

The upper right and lower left squares are pullbacks. The perimeter

$$
\begin{array}{ccc}
X & \twoheadrightarrow & Y \\
\downarrow & & \downarrow \\
\psi^*F & \xrightarrow{\psi^*(kp)} & \psi^*E \\
\downarrow & & \downarrow \\
f^*f_!X & \xrightarrow{f^*(\alpha\hat{t})} & f^*B
\end{array}
$$

is a morphism in FC_ψ. The morphisms α and $\alpha\hat{t}$ are necessarily epimorphisms (Exercise 3.5.6, 6). In this last diagram, generally neither is the top square a pullback, nor is kp an epimorphism. The morphism of FC_ψ depicted in this last diagram is a μ-cover. We make this precise in the next proposition.

Proposition 3.4.4 *A morphism in* FC_ψ *corresponds under the equivalence* V *to a* μ*-cover iff it is the FC-coreflection of the Cartesian lifting along* $\mathbb{H}_\mu \longrightarrow \mathscr{E}/e^*$ *of a collective epimorphism* (k,α) *of* $\mathscr{E}/e^*$*.*

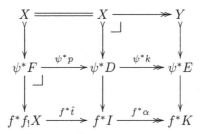

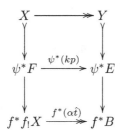

Exercises 3.4.5

1. *Show that the generating covers for the amalgamation topology (Definition 2.3.2) of a cosheaf corresponding to a distribution μ are precisely the μ-covers of Definition 3.4.3. Of course, we must restrict* Display$_\mu$ *to objects (x, s) such that x is from a small generating category of $\mathscr{E}$.*
2. *Prove Proposition 3.4.4.*

3.5 The Completeness Condition

In this section we introduce a notion of completeness for geometric morphisms with locally connected domain, and prove, when the codomain is bounded, that a geometric morphism is complete iff its pure factor is a surjection. We are able to do this within the same fibrational framework developed for Theorem 3.3.5.

Theorem 3.3.5 asserts that a geometric morphism with locally connected domain is a spread iff its pure factor in the comprehensive factorization of Theorem 2.4.8 is an inclusion. We seek to characterize geometric morphisms whose pure factor in the comprehensive factorization of Theorem 2.4.8 is a surjection. The completeness condition that we discover takes the form of a cover refinement property.

The proof of Theorem 3.3.5 begins to describe an 'FC_ψ-formulation' of the topos bipullback construction of the complete spread factor of a geometric morphism $\mathscr{F} \xrightarrow{\psi} \mathscr{E}$ with locally connected domain. In § 3.4 we had deepened our understanding of the FC_ψ-formulation by describing the μ-covers in FC_ψ. This prepares the way to define completeness and prove the main result of this section, Theorem 3.5.3.

Definition 3.5.1 A ψ-cover *of a family of components in FC_ψ is simply an $\mathscr{S}$-fibered collective epimorphism in $\mathscr{F}/\psi^*$. Explicitly, it is a diagram in $\mathscr{F}$ as follows.*

$$\begin{array}{ccc}
X & \twoheadrightarrow & Y \\
\downarrow & & \downarrow \\
\psi^* D & \xrightarrow{\psi^*(h)} & \psi^* E \\
\downarrow{\scriptstyle \psi^* x} & & \downarrow{\scriptstyle \psi^* y} \\
f^* I & \xrightarrow{f^*(\alpha)} & f^* K
\end{array}$$

(The morphism α in $\mathscr{S}$ is necessarily an epimorphism since the transpose $f_!(Y) \longrightarrow K$ is an isomorphism.)

Definition 3.5.2 *Let $\psi : \mathscr{F} \longrightarrow \mathscr{E}$ have locally connected domain. We say ψ is* complete *if in FC_ψ every ψ-cover can be refined by a μ-cover. The prism diagram*

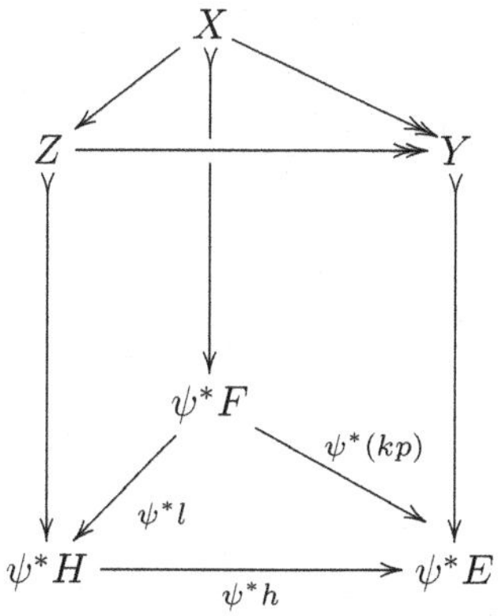

depicts such a refinement, where the front face of the prism is a typical ψ-cover, and the refining back right face is a μ-cover, associated with a collective epimorphism (k, α) in $\mathscr{E}/e^*$. We have not depicted the $\mathscr{S}$-fibering data in the above prism. We allow for reindexing over $\mathscr{S}$: ψ is complete if there is $\alpha : I' \twoheadrightarrow I$ in $\mathscr{S}$ such that α^* of the given ψ-cover, whose codomain is say $E \longrightarrow e^*I$, is refined by a μ-cover over I'.

Theorem 3.5.3 *Let $\psi : \mathscr{F} \longrightarrow \mathscr{E}$ have locally connected domain and bounded codomain. Then ψ is complete iff the pure factor of the comprehensive factorization of ψ is a surjection.*

Theorems 3.5.3 and 3.3.5 combine to give the following.

Corollary 3.5.4 *A geometric morphism with locally connected domain and bounded codomain is a complete spread* (Def. 2.4.6) *iff it is a spread and it is complete.*

Proof. A geometric morphism is a complete spread iff its pure factor is an equivalence iff the pure factor is an inclusion and a surjection iff the geometric morphism is a spread and complete. □

We occupy the rest of this section with a proof of Theorem 3.5.3. We begin with a recap of three kinds of covering morphisms we have defined:

1. collective epimorphisms in $\mathscr{E}/e^*$,
2. ψ-covers in FC_ψ,
3. μ-covers in FC_ψ - these are ψ-covers.

We repeat the construction of the complete spread factor of a geometric morphism $\psi : \mathscr{F} \longrightarrow \mathscr{E}$, but of course using FC_ψ instead of the equivalent category Display$_\psi$. We are assuming that $\mathscr{E}$ is bounded: let $\mathbb{C}$ denote the essentially small $\mathscr{S}$-fibered category whose objects are pullbacks

$$\begin{array}{ccc} X & \longrightarrow & G \\ \downarrow & & \downarrow \\ e^*I & \longrightarrow & e^*K \end{array}$$

of a given generating family $G \longrightarrow e^*K$. $\mathbb{C}$ is a full subcategory of $\mathscr{E}/e^*$. We consider just the objects (x, m) of FC_ψ for which x is an object of $\mathbb{C}$. This gives another essentially small $\mathscr{S}$-fibered category $\mathbb{Y}$, which is a full subcategory of FC_ψ. The functor $FC_\psi \longrightarrow \mathscr{E}/e^*$ restricts to an internal discrete opfibration $\mathbb{Y} \longrightarrow \mathbb{C}$ that corresponds to the cosheaf

$$\mathbb{C} \longrightarrow \mathscr{E} \xrightarrow{\mu} \mathscr{S} .$$

We have the following bipullback of toposes over $\mathscr{S}$, as well as the induced pure factor $\tau : \mathscr{F} \longrightarrow \mathscr{Y}$ over $\mathscr{E}$.

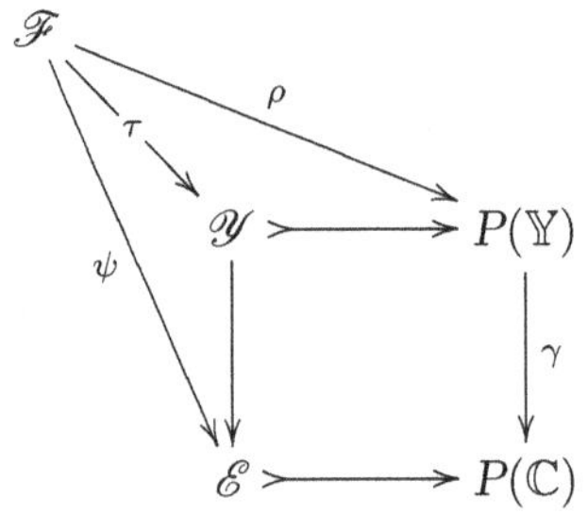

γ denotes the essential geometric morphism corresponding to the discrete opfibration $\mathbb{Y} \longrightarrow \mathbb{C}$. ρ corresponds to the flat functor that is the forgetful functor

$$FC_\psi \longrightarrow F/f^*$$

restricted to $\mathbb{Y}$.

Lemma 3.5.5 *We have the following.*

1. *If $D \longrightarrow e^*I$ is an object of $\mathbb{C}$, then the FC-coreflection of a definable family $X \rightarrowtail \psi^*D \longrightarrow f^*I$ is an object of $\mathbb{Y}$. In particular, a μ-cover associated with a collective epimorphism of $\mathbb{C}$ lies in $\mathbb{Y}$.*
2. *A μ-cover of a family of components in $\mathbb{Y}$ may be refined by a μ-cover associated with a collective epimorphism in $\mathbb{C}$.*

The three kinds of covering morphisms mentioned above give three kinds of sieves in $\mathbb{C}$ and $\mathbb{Y}$, for which we shall use the following terminology:

1. J-sieves - these form a topology in $\mathbb{C}$;
2. $\widetilde{J}$-sieves in $\mathbb{Y}$ - these form a topology in $\mathbb{Y}$, whose subtopos of $P(\mathbb{Y})$ is the image of ρ;
3. $\widehat{J}$-sieves - these generate the amalgamation topology in $\mathbb{Y}$.

Suppose that ψ is complete. Let S denote the $\widetilde{J}$-sieve on a family of components in $\mathbb{Y}$ engendered by a ψ-cover in $\mathbb{Y}$. We may refine this ψ-cover by a μ-cover, whose domain object is of course a family of components. The collective epimorphism associated with this μ-cover may not be in $\mathbb{C}$; however, we may apply Lemma 3.5.5, 2, and therefore refine the first μ-cover by a μ-cover whose associated collective epimorphism lies in $\mathbb{C}$. Hence, we may find a $\widehat{J}$-sieve that is contained in S, which shows that $\widetilde{J}$ is contained in the upclosure of $\widehat{J}$. Therefore $\widetilde{J}$ is contained in (hence equal to) the topology generated by $\widehat{J}$ in $\mathbb{Y}$ (this is the amalgamation topology). Thus, the image topos of ρ coincides with $\mathscr{Y}$, so that τ is a surjection.

Conversely, suppose that τ is a surjection. Suppose we are given a ψ-cover in FC_ψ. We may choose the generating $\mathbb{C}$ such that the collective epimorphism associated with the given ψ-cover lies in $\mathbb{C}$. Therefore, the ψ-cover lies in $\mathbb{Y}$. If τ is a surjection, then the $\widetilde{J}$-topology equals the amalgamation topology in $\mathbb{Y}$. By Corollary 2.3.4, for $\mathbb{Y} \longrightarrow \mathbb{C}$, the $\widetilde{J}$-sieve engendered by the given ψ-cover contains a $\widehat{J}$-sieve, which is engendered by a μ-cover. This μ-cover therefore refines the given ψ-cover, completing the proof of Theorem 3.5.3.

Exercises 3.5.6

1. *Suppose that* $G \xrightarrow{g} e^*B$ *is a generating family for* $\mathscr{E}$ *over* $\mathscr{S}$. *Let* $\mathbb{C}$ *denote the full subcategory of* $\mathscr{E}/e^*$ *on objects* $D \xrightarrow{d} e^*A$ *for which there is a pullback*

$$\begin{array}{ccc} D & \longrightarrow & G \\ {\scriptstyle d}\downarrow & & \downarrow{\scriptstyle g} \\ e^*A & \xrightarrow{e^*\alpha} & e^*B \end{array}$$

in $\mathscr{E}$. *Thus, every object of* $\mathscr{E}/e^*$ *can be covered by a collective epimorphism with domain in* $\mathbb{C}$. *Show that* $\mathbb{C}$ *is equivalent to a small (or internal or representable)* $\mathscr{S}$*-fibration:* $\mathbb{C}$ *is said to be* essentially small.

2. *Let* $\mathbb{Y}$ *and* $\mathbb{B}$ *denote the full subcategories of* Display_μ, *respectively* $\mathbb{H}_\mu$, *on the objects* (x, s), *respectively* (x, S), *such that* x *is an object of* $\mathbb{C}$, *as in the previous exercise. The usual functors form part of the category pullbacks.*

$$\begin{array}{ccc} \mathbb{Y} & \hookrightarrow & \text{Display}_\mu \\ \downarrow & & \downarrow \\ \mathbb{B} & \hookrightarrow & \mathbb{H}_\mu \\ \downarrow & & \downarrow \\ \mathbb{C} & \hookrightarrow & \mathscr{E}/e^* \end{array} \qquad U : \text{Display}_\mu \longrightarrow \mathscr{E}/e^*$$

Show that $\mathbb{Y}$ *and* $\mathbb{B}$ *are also essentially small, and that the functor* $\mathbb{Y} \longrightarrow \mathbb{C}$ *is internal to* $\mathscr{S}$. *Show that* U *is an* $\mathscr{S}$*-Cartesian functor whose fibers are discrete opfibrations, so that* $\mathbb{Y} \longrightarrow \mathbb{C}$ *is a discrete opfibration. Show that* $\mathbb{B} \longrightarrow \mathbb{C}$ *is an internal posetal fibration.*

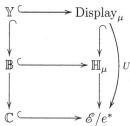

3. *As we know, the complete spread* ψ *associated with a distribution* μ *is the outer topos pullback*

$$\begin{array}{ccccc} & \mathcal{Y} & \rightarrowtail & P(\mathbb{Y}) & \\ & \downarrow & & \downarrow & \\ \psi \downarrow & \mathcal{E}^{H^{\mathrm{op}}} & \rightarrowtail & P(\mathbb{B}) & \downarrow \gamma \\ & \downarrow & & \downarrow & \\ & \mathcal{E} & \rightarrowtail & P(\mathbb{C}) & \end{array}$$

induced by the functor $\mathbb{Y} \longrightarrow \mathbb{B} \longrightarrow \mathbb{C}$. *Of course* $H = \mu_*(\Omega_{\mathcal{S}})$ *is the distribution algebra. Show that the bottom square is a pullback, and describe the pullback topology in* $\mathbb{H}_\mu$ *(or* $\mathbb{B}$*).*

4. *We know* ψ *(previous exercise) is localic, and* $\mathcal{Y}$ *may be considered as* $\mathrm{Sh}_{\mathcal{E}}(X_\mu)$, *where* X_μ *denotes the display locale in* $\mathcal{E}$. *Then the frame* $\mathcal{O}(X_\mu) = \psi_*(\Omega_{\mathcal{Y}})$. *Show that this frame may be described as a sheaf on* $\mathbb{C}$*: if* c *is any object of* $\mathbb{C}$, *then*

$$\mathcal{O}(X_\mu)(c) = \{T \rightarrowtail \gamma^*(h_c) \mid T \text{ is closed for } \mu\text{-covers}\}\ ,$$

where $\gamma^*(h_c)(c', s') = \mathbb{C}(c', c)$. *Show that the canonical order preserving map*

$$\psi_*(\tau) : H \longrightarrow \mathcal{O}(X_\mu)$$

in $\mathcal{E}$, *where* $\tau : y^*(\Omega_{\mathcal{S}}) \rightarrowtail \Omega_{\mathcal{Y}}$ *classifies* $y^*(\top_{\mathcal{S}})$ *in* $\mathcal{Y}$ *(*τ *is a subobject as* $\mathcal{Y}$ *is subopen) may be described explicitly as follows. For any object* c, *a subobject* $S \rightarrowtail \mu(h_c)$ *defines a subobject* $T \rightarrowtail \gamma^*(h_c)$ *that is closed for* μ*-covers:*

$$T(c', s') = \{c' \xrightarrow{m} c \mid \mu(m)(s') \in S\}\ .$$

5. *Continuing from Exercise 4, describe in* $\mathcal{E}$ *the covering families* $\{h_\alpha \le h\}$ *in* H *that give* $\mathcal{Y}$ *as* $\mathrm{Sh}_{\mathcal{E}}(H)$.
6. *If the codomain object of a* ψ*-cover is a family of components, then the morphism* $I \xrightarrow{\alpha} K$ *from* $\mathcal{S}$ *must be an epimorphism.*
7. *Prove Lemma 3.5.5.*
8. *Suppose* $\psi : \mathcal{F} \longrightarrow \mathcal{E}$ *over* $\mathcal{S}$ *has locally connected domain, and that* $\mathcal{E}$ *is bounded. Show* ψ *is a spread iff* $\mathcal{F}$ *is equivalent to the topos of sheaves over* $\mathcal{S}$ *on the topology determined by the* ψ*-covers on* FC_ψ.

Further reading: Bunge & Funk [BF96b, BF98], Bunge, Funk, Jibladze & Streicher [BFJS04a], Funk [Fun00], Kock [Koc90], Kock & Reyes [KR94], Streicher [Str03].

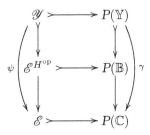

Part II

An Axiomatic Theory of Complete Spreads

4

Completion KZ-Monads

The symmetric topos of a given topos $\mathscr{E}$ over $\mathscr{S}$ classifies $\mathscr{S}$-valued Lawvere distributions on $\mathscr{E}$. It has been shown to exist by M. Bunge (1995)[Bun95] using forcing methods. Since distributions of $\mathscr{E}$ are generalized points - namely $\mathscr{S}$-cocontinuous functors $q : \mathscr{S} \longrightarrow \mathscr{E}$, not necessarily finite limits preserving, the symmetric topos of $\mathscr{E}$ can also be constructed, as done by M. Bunge and A. Carboni(1995) [BC95], by analogy with the symmetric algebra.

This chapter centers on the symmetric monad as a key model of a theory of completion Kock-Zöberlein monads, as we shall call them, which we abbreviate as (completion) KZ-monads.

Within the theory of a completion KZ-monad M we identify its Eilenberg-Moore algebras with cocomplete objects in a certain sense.

4.1 KZ-Adjointness

The notion of an adjointness in ordinary category theory may be extended to 2-functors between 2-categories by replacing identities involving units and counits by (coherent) 2-cells, required to be isomorphisms in the "pseudo" rather than "lax" case. Associated with a pair of lax adjoint functors there is a lax monad and corresponding lax algebras. In our context we shall need a specific sort of lax monad for which in the case of its algebras "structure is adjoint to units," so that in particular the so-called structure may instead be regarded as a property. The free addition of coproducts to a small category $\mathbb{C}$ is a typical case: we may express complete under coproducts in terms of a construction that freely adds coproducts. We simply regard the appropriate diagrams as the objects of a new category $\mathbb{C} \longrightarrow \mathbb{C}^+$. The left adjoint $\mathbb{C}^+ \longrightarrow \mathbb{C}$ assigns to a diagram its colimit in $\mathbb{C}$. The special situation just depicted involves also a special sort of lax adjointness and therefore also a special sort of lax monad, known in the literature as *a Kock-Zöberlein doctrine.*

Definition 4.1.1 *A* KZ-adjointness $F \dashv G$ *between 2-categories*

$$G : \mathscr{A} \longrightarrow \mathscr{B}\ ;\quad F : \mathscr{B} \longrightarrow \mathscr{A}$$

has units $\eta_B : B \longrightarrow GF(B)$ *and counits* $\epsilon_A : FG(A) \longrightarrow A$ *such that*

$$F(\eta_B) \dashv \epsilon_{F(B)}\ ;\quad G(\epsilon_A) \dashv \eta_{G(A)}\ . \tag{4.1}$$

We shall refer to $F(\eta_B) \dashv \epsilon_{F(B)}$ *as* the left adjointness of $F \dashv G$ *at* B, *and to* $G(\epsilon_A) \dashv \eta_{G(A)}$ *as* the right adjointness of $F \dashv G$ *at* A. *The unit of a left adjointness is a 2-cell in* $\mathscr{B}$, *and the counit of a right adjointness is a 2-cell in* $\mathscr{A}$. *As part of the definition, we assume that these 2-cells are isomorphisms.*

Example 4.1.2 *A colimit-completion construction is a typical example of a KZ-adjointness. For example, the free addition of finite coproducts to a small category is a KZ-adjointness. Let* B^+ *denote the finite coproduct completion of a small category* B. *It is well-known that the objects and morphisms of* B^+ *are finite families* $\{b_i\}$ *of objects and morphisms of* B. *The unit* η_B *carries an object* b *of* B *to* $\{b\}$ *in* B^+. *If* C *is a small category with finite coproducts, then the counit* ϵ_C *carries a finite family of objects in* C *to its coproduct. The right adjointness* $\epsilon_C \dashv \eta_C$ *is clearly satisfied: it expresses the universal of coproducts. The left adjointness* ${\eta_B}^+ \dashv \epsilon_{B^+}$ *is perhaps less clear. It follows from*

$${\eta_B}^+(\{b_1, \ldots, b_n\}) = \{\{b_1\}, \ldots, \{b_n\}\}\ ,$$

and

$$\epsilon_{B^+}(\{\{b_1^1, \ldots, b_{n_1}^1\}, \ldots, \{b_1^m, \ldots, b_{n_m}^m\}\}) = \{b_1^1, \ldots, b_{n_m}^m\}\ .$$

The unit of any left adjointness and counit of any right are isomorphisms.

The two conditions (4.1) are an appropriate strengthening of the notion of a lax adjointness. Consider three calculations:

1. Apply G to the inverse of any unit $1_{F(B)} \cong \epsilon_{F(B)}F(\eta_B)$. This produces an isomorphism
$$G(\epsilon_{F(B)})GF(\eta_B) \cong 1_{GF(B)}$$
that transposes to a 2-cell
$$GF(\eta_B) \Rightarrow \eta_{GF(B)}\ .$$
This 2-cell is not necessarily an isomorphism.
2. We have adjoints
$$FG(\epsilon_A) \dashv F(\eta_{G(A)}) \dashv \epsilon_{FG(A)}$$
in $\mathscr{A}$, and
3. adjoints
$$GF(\eta_B) \dashv G(\epsilon_{F(B)}) \dashv \eta_{GF(B)}$$
in $\mathscr{B}$.

A KZ-adjointness induces a special sort of 2-monad in $\mathscr{B}$ that we shall call *a KZ-monad* (traditionally, KZ-doctrine):

$$(M, \delta, \mu) = (GF, \eta, G\epsilon F) .$$

According to calculations 1 and 3 above we have a canonical 2-cell

$$M(\delta_B) \Rightarrow \delta_{M(B)} ,$$

and adjoints

$$M(\delta_B) \dashv \mu_B \dashv \delta_{M(B)} . \tag{4.2}$$

Proposition 4.1.3 *The following four conditions are equivalent for an adjointness $L \dashv R$:*

1. *L is faithful,*
2. *RL is faithful,*
3. *every unit of $L \dashv R$ is a monomorphism,*
4. *L reflects monomorphisms.*

For instance, if a geometric morphism $f^* \dashv f_*$ satisfies the conditions of Proposition 4.1.3, then it is called a *surjection.*

Proposition 4.1.3 has the following 2-dimensional generalization. A 1-cell $A \xrightarrow{f} B$ in a 2-category $\mathscr{K}$ is *full and faithful* if for every object K, the composition functor

$$f^{\#} = \mathscr{K}(K, f) : \mathscr{K}(K, A) \longrightarrow \mathscr{K}(K, B)$$

is a full and faithful functor. If a 2-functor is full and faithful on 2-cells, then we say that it is *locally full and faithful.*

Proposition 4.1.4 *Let $F \dashv G$ be a KZ-adjointness as in* (4.1.1)*. Then the following conditions are equivalent:*

1. *F is locally full and faithful,*
2. *GF is locally full and faithful,*
3. *for every object B in $\mathscr{B}$, the unit η_B is a full and faithful 1-cell,*
4. *F reflects full and faithful 1-cells.*

Proof. It is not hard to show that 1, 2 and 3 are equivalent, and that they imply 4. A KZ-adjointness has the property that for every object B of $\mathscr{B}$, $F(\eta_B)$ is a full and faithful 1-cell. Hence 4 implies 3. □

Definition 4.1.5 *If F is locally full and faithful, equivalently if $M = GF$ is, then we say that the associated KZ-monad (M, δ, μ) is* a locally full and faithful KZ-monad.

Remark 4.1.6 *By definition, in a KZ-adjointness any unit η_{GA} is a full and faithful 1-cell. Thus, if every 0-cell of $\mathscr{B}$ is a $G(A)$, then the associated KZ-monad is locally full and faithful. This is the case with the symmetric monad in $\boldsymbol{Top}_{\mathscr{S}}$.*

Exercises 4.1.7

1. *Prove Proposition 4.1.3.*

4.2 The Symmetric Monad

In order to construct the symmetric topos $\Sigma(\mathscr{E})$ of a topos $\mathscr{E}$ we shall employ the left exact, or lex completion of a small category, and moreover, the lex completion suffices in the case of a presheaf topos. We shall construct a left 2-adjoint Σ to the forgetful functor from Chapter 1, (1.1)

$$U : \mathbf{Frm}_{\mathscr{S}} \longrightarrow \mathbf{Coc}_{\mathscr{S}}$$

by lifting the lex completion of a small category in a suitable way. We may regard U as a functor into the 2-category $\mathbf{Coc}_{\mathscr{S}}$ of $\mathscr{S}$-toposes, cocontinuous functors, and natural transformations. The left adjoint Σ would give for each topos $\mathscr{E}$ a topos $\Sigma(\mathscr{E})$ classifying distributions on $\mathscr{E}$. Intuitively, Σ freely adds left exact structure to a distribution, so that the result is (the inverse image of) a geometric morphism. The lex completion of a small category is thus naturally involved in the construction of $\Sigma(\mathscr{E})$.

Remark 4.2.1 *We hope that the reader is not unnecessarily confused by the "symmetric topos (frame)" terminology for $\Sigma(\mathscr{E})$. It is suggested by algebra. If homomorphisms are taken as points of a commutative algebra A, then module maps may be taken as extensives, as an instance of the Riesz paradigm. In this case there is an algebra $\Sigma(A)$, called the "symmetric algebra" of A, whose points are the module maps, i.e., whose spectrum serves as a space of distributions. In algebra there are also non-commutative algebras and (in the graded case) anti-commutative ones, of which both categories also receive their left adjoints, so the term "symmetric" has been introduced in order to distinguish the commutative adjoint from those other two. However, cartesian product in toposes is automatically commutative, so the term "symmetric" may seem inappropriate in this context.*

Let **Lex** denote the 2-category of lex categories, lex functors, and all natural transformations. The lex completion of a small category $\mathbb{C}$ is a small lex category $\mathbb{C}^\star$ and a functor

$$\eta : \mathbb{C} \longrightarrow \mathbb{C}^\star \tag{4.3}$$

with the property that every functor from $\mathbb{C}$ to a lex category extends uniquely (up to unique isomorphism) to a lex functor from $\mathbb{C}^\star$. To put it another way, the lex completion is a left 2-adjoint to a forgetful 2-functor

$$U : \mathbf{Lex} \longrightarrow \mathbf{Cat}\ ;\ (\ \ )^\star : \mathbf{Cat} \longrightarrow \mathbf{Lex}\ , \tag{4.4}$$

where **Cat** denotes the 2-category of small categories. Using only the existence of lex completions of categories in $\mathscr{S}$, we shall show that the symmetric topos of any presheaf $\mathscr{S}$-topos exists and is again a presheaf $\mathscr{S}$-topos.

Proposition 4.2.2 *$\Sigma(P(\mathbb{C}))$ is equivalent to $P(\mathbb{C}^\star)$.*

Proof. Distributions $P(\mathbb{C}) \longrightarrow \mathscr{F}$ correspond to functors $\mathbb{C} \longrightarrow \mathscr{F}$, which in turn correspond to left exact functors $\mathbb{C}^\star \longrightarrow \mathscr{F}$. Such functors correspond to geometric morphisms $\mathscr{F} \longrightarrow P(\mathbb{C}^\star)$. This last assertion is the basic and well-known fact that the left Kan extension of a lex functor (between lex categories $\mathbb{C}$ and $\mathscr{F}$) is lex. Indeed, in this case the left Kan extension is a filtered colimit in $\mathscr{S}$, hence commutes with finite limits. (This is the version appropriate to the case where $\mathscr{S}$ is *Set*, but a suitable analogue can be given in the general case.) □

It is clear that $(\ \ )^\star \dashv U$ is a co-KZ-adjoint pair. We wish to lift this adjointness to one $\Sigma \dashv U$ between geometric morphisms and distributions. The following diagram presents a bird's eye view of the situation.

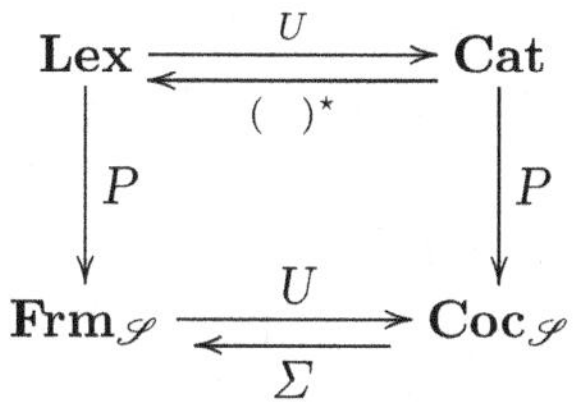

In this diagram P denotes the covariant presheaf 2-functor which sends a small category $\mathbb{C}$ to its topos of presheaves $P(\mathbb{C})$, and a functor F between two small categories to the left Kan extension $P(F) = f_! \dashv f^*$ between presheaf toposes. If F is left exact, then so is $P(F)$ so that it is the inverse image $P(F)^*$ of a geometric morphism $P(F)$. Thus, P commutes with the forgetful functors U. We already know by Proposition 4.2.2 that P also commutes with their left adjoints.

Proposition 4.2.3 *In the case of presheaf toposes, the 2-adjoint pair established in Proposition 4.2.2 is a co-KZ-adjointness.*

Proof. The 2-triangle conditions are preserved under a 2-functor, in this case P. Thus, they automatically hold for a presheaf topos: the unit $\delta_{P(\mathbb{C})}$ is an essential inclusion, and we have $\delta_*^{P(\mathbb{C})} = \mu^*_{P(\mathbb{C})}$. □

We have left out some details. For instance, Σ is a 2-functor on presheaf toposes and distributions: if $\lambda : P(\mathbb{C}) \longrightarrow P(\mathbb{D})$ is a distribution, then there is a geometric morphism $\Sigma(\lambda) : P(\mathbb{D}^\star) \longrightarrow P(\mathbb{C}^\star)$ corresponding to the distribution $\delta_!^{P(\mathbb{D})} \cdot \lambda$. The distribution $P(\eta)$ (4.3) is equal to $\delta_!^{P\mathbb{C}}$.

The way we extend Σ to an arbitrary $\mathscr{S}$-topos $\mathscr{E}$ depends on the fact that any such topos $\mathscr{E}$ is a coinverter of a diagram of presheaf toposes and distributions between them (Proposition 1.4.7), taking into account that as a left 2-adjoint Σ ought to preserve coinverters.

We can thus construct $\Sigma(\mathscr{E})$ in terms of a site-presentation $\langle \mathbb{C}, J\rangle$ of $\mathscr{E}$. Consider the following presheaf Z on the lex completion $\mathbb{C}^\star$:

$$Z(x) = \{(f, c, S) \mid x \xrightarrow{f} \eta(c), S \in J(c)\} ,$$

where $\mathbb{C} \xrightarrow{\eta} \mathbb{C}^\star$ is the unit functor. If $x' \xrightarrow{g} x$ is any morphism of $\mathbb{C}^\star$, then $Z(g)(f, c, S) = (fg, c, S)$. There is a natural transformation $Z \longrightarrow \Omega$ in $P(\mathbb{C}^\star)$ that at stage x sends (f, c, S) to the pullback along f of the sieve $\hat{S}$ on $\eta(c)$ generated by S. Technically, $\hat{S}$ is the image-subobject of the natural transformation

$$\delta_!^{P\mathbb{C}}(S) \longrightarrow \delta_!^{P\mathbb{C}}(c) \cong \eta(c) ,$$

where c and $\eta(c)$ are considered as representable presheaves. Let K denote the image-subobject of $Z \longrightarrow \Omega$. *Let $\Sigma(\mathscr{E})$ be the topos of sheaves on $\mathbb{C}^\star$ for the topology generated by K.* By Proposition 1.4.8, $\Sigma(\mathscr{E})$ is the following coinverter:

$$P(\mathbb{C}^\star)/K \underset{\Sigma_K}{\overset{d.\Sigma_K}{\rightrightarrows}} {\Downarrow} d.\Sigma_K \; P(\mathbb{C}^\star) \xrightarrow{i^*} \Sigma(\mathscr{E}) .$$

It remains to prove that this construction of $\Sigma(\mathscr{E})$ does indeed classify distributions on $\mathscr{E}$ with values in $\mathscr{S}$-toposes. At the same time we wish to establish that what we have is not just a 2-adjointness but a KZ-adjointness involving $\Sigma(\mathscr{E})$ and the forgetful 2-functor U.

It is now convenient to take explicitly the geometric point of view and regard Σ as a 2-functor

$$\Sigma : \mathbf{Coc}_{\mathscr{S}}{}^{\mathrm{op}} = \mathbf{Dist}_{\mathscr{S}} \longrightarrow \mathbf{Top}_{\mathscr{S}}$$

so that now we wish to establish $U \dashv \Sigma$. If this adjointness is a KZ-adjointness, then it induces a KZ-monad in $\mathbf{Top}_{\mathscr{S}}$:

$$(\mathrm{M}, \delta, \mu) = (\Sigma U, \eta, \Sigma\epsilon U) .$$

By convention, throughout we denote a KZ-monad by its functor part, in this case $\mathrm{M} = \Sigma U$. M is a locally full and faithful KZ-monad, since U is locally full and faithful.

It is not too difficult to show that indeed $U \dashv \Sigma$ is a KZ-adjointness. We begin with the following lemma.

Lemma 4.2.4 *Let $\langle \mathbb{C}, J\rangle$ be a site. Then the topology in $\mathbb{C}^\star$ generated by the object K defined above coincides with the least topology in $\mathbb{C}^\star$ that inverts $\delta_!^{P\mathbb{C}}(d)$, where $1 \xrightarrow{d} J$ is the top element.*

Proposition 4.2.5 *For any topos $\mathscr{X}$, there is an equivalence of categories*

$$\mathbf{\mathit{Dist}}_{\mathscr{S}}(\mathscr{X}, \mathscr{E}) \simeq \mathbf{\mathit{Top}}_{\mathscr{S}}(\mathscr{X}, \mathrm{M}(\mathscr{E})) , \qquad (4.5)$$

which is natural in the toposes $\mathscr{X}$ and $\mathscr{E}$.

Proof. A geometric morphism $\mathscr{X} \longrightarrow \Sigma(\mathscr{E})$ is equivalently given as a left exact $\mathscr{X}$-valued cosheaf on $\langle \mathbb{C}^{\star}, K \rangle$. We must show that such a cosheaf is equivalently given by a $\mathscr{X}$-valued cosheaf on $\langle \mathbb{C}, J \rangle$. We shall show that the equivalence between $\mathscr{X}$-valued functors on $\mathbb{C}$ and left exact ones on $\mathbb{C}^{\star}$, which is given by composition with the unit η, restricts to one between J-cosheaves on $\mathbb{C}$ and left exact K-cosheaves on $\mathbb{C}^{\star}$. There is a natural transformation $J \longrightarrow \delta^{*}_{P\mathbb{C}}(Z)$ that at stage c sends $S \in J(c)$ to $(1_c, c, S) \in Z(\eta(c))$. Hence there is a natural transformation $\zeta : J \longrightarrow \delta^{*}_{P\mathbb{C}}(K)$. In terms of total categories, we may regard this natural transformation as a functor $\zeta : \mathbb{J} \longrightarrow \mathbb{K}$ over η. Thus, we have a commutative coinverter diagram:

$$\begin{array}{ccccc} P(\mathbb{J}) & \overset{d.\Sigma_J}{\underset{\Sigma_J}{\underrightarrow{\overrightarrow{\;d.\Sigma_J \Downarrow\;}}}} & P(\mathbb{C}) & \xrightarrow{i^*} & \mathscr{E} \\ \downarrow \zeta_! & & \downarrow \delta_!^{P\mathbb{C}} & & \downarrow \delta_!^{\mathscr{E}} \\ P(\mathbb{K}) & \overset{d.\Sigma_K}{\underset{\Sigma_K}{\underrightarrow{\overrightarrow{\;d.\Sigma_K \Downarrow\;}}}} & P(\mathbb{C}^{\star}) & \xrightarrow{i^*} & \Sigma(\mathscr{E})\,. \end{array}$$

The commutativity of this diagram shows that a left exact K-cosheaf $\mathbb{C}^{\star} \longrightarrow \mathscr{X}$ restricts along η to a J-cosheaf $\mathbb{C} \longrightarrow \mathscr{X}$. It remains to show that if the restriction along η of a left exact functor F on $\mathbb{C}^{\star}$ is a J-cosheaf, then F is a K-cosheaf. (Equivalently, if a left exact distribution

$$P(\mathbb{C}^{\star}) \xrightarrow{p^*} \mathscr{X}$$

inverts the 2-cell $d.\Sigma_K \cdot \zeta_!$, then it inverts $d.\Sigma_K$.) But this follows by Lemma 4.2.4 since F is left exact. This establishes the equivalence. □

On account of Proposition 4.2.5 we can assert that $\mathrm{M}(\mathscr{E})$ is the topos classifier of distributions on $\mathscr{E}$.

Example 4.2.6 *Take $\mathscr{E}$ in (4.5) to be the base topos $\mathscr{S}$. Since*

$$\boldsymbol{Dist}_{\mathscr{S}}(\mathscr{X}, \mathscr{S}) = \mathbf{Coc}_{\mathscr{S}}(\mathscr{S}, \mathscr{X}) \simeq \mathscr{X}$$

canonically, we see that $\mathrm{M}(\mathscr{S})$ is the familiar object classifier topos. Now take $\mathscr{X}$ to be $\mathscr{S}$ in (4.5). Hence the points of $\mathrm{M}(\mathscr{E})$ correspond to distributions in $\mathscr{E}$.

We wish to show that $U \dashv \Sigma$ is a KZ-adjointness. Let us see what the left and right adjointnesses (4.1) amount to for $U \dashv \Sigma$. For a topos $\mathscr{E}$, the unit is a geometric morphism

$$\delta_{\mathscr{E}} : \mathscr{E} \longrightarrow \mathrm{M}(\mathscr{E})$$

that we call the Dirac delta. The counit is a distribution

$$\epsilon_{U(\mathscr{E})} : U(\mathscr{E}) \longrightarrow U(\mathrm{M}(\mathscr{E}))$$

$$\begin{array}{ccccc}
\delta_!^{\mathrm{M}(\mathcal{E})} & \dashv & \delta^*_{\mathrm{M}(\mathcal{E})} & \dashv & \delta_*^{\mathrm{M}(\mathcal{E})} \\
\mu_!^{\mathcal{E}} & \dashv & \mu^*_{\mathcal{E}} & \dashv & \mu_*^{\mathcal{E}} \\
\mathrm{M}(\delta_{\mathcal{E}})_! & \dashv & \mathrm{M}(\delta_{\mathcal{E}})^* & \dashv & \mathrm{M}(\delta_{\mathcal{E}})_*
\end{array}$$

Fig. 4.1. The adjoint line up

that according to the left adjointness of $U \dashv \Sigma$ for $\mathscr{E}$ is *left* adjoint to $U(\delta_{\mathscr{E}}) = \delta^*_{\mathscr{E}}$: we use the notation $\delta_!^{\mathscr{E}}$ for $\epsilon_{U(\mathscr{E})}$. The right adjointness of $U \dashv \Sigma$ for $\mathscr{E}$ says precisely that $\mu_{\mathscr{E}} \dashv \delta_{\mathrm{M}(\mathscr{E})}$ as geometric morphisms. Note: $\mu_{\mathscr{E}}$ denotes $\Sigma(\epsilon_{U(\mathscr{E})})$, which is the same as $\Sigma(\delta_!^{\mathscr{E}})$. The definition of a KZ-adjointness requires that for every topos $\mathscr{E}$, the unit for $U(\delta_{\mathscr{E}}) \dashv \epsilon_{U(\mathscr{E})}$, which is an adjointness in $\mathbf{Dist}_{\mathscr{S}}$, is an isomorphism. In the geometric notation this says that the unit for $\delta_!^{\mathscr{E}} \dashv \delta^*_{\mathscr{E}}$ is an isomorphism, which says that $\delta_{\mathscr{E}}$ is an inclusion geometric morphism. The requirement that the counit for $\mu_{\mathscr{E}} \dashv \delta_{\mathrm{M}(\mathscr{E})}$ is an isomorphism is a consequence of other requirements: it says only that $\delta_{\mathrm{M}(\mathscr{E})}$ is an inclusion.

Proposition 4.2.7 *For any $\mathscr{S}$-topos $\mathscr{E}$, the left adjointness of $U \dashv \Sigma$ at a topos $\mathscr{E}$ is equivalent to the condition that the Dirac delta $\delta_{\mathscr{E}}$ is an essential inclusion. The right adjointness of $U \dashv \Sigma$ at a topos frame $U(\mathscr{E})$ is equivalent to the condition that $\delta_*^{\mathrm{M}(\mathscr{E})} \cong \mu^*_{\mathscr{E}}$. We sum up the situation as follows. For any topos $\mathscr{E}$, we have adjoints as in Figure 4.1: functors in the same column are isomorphic.*

Each of the three geometric morphisms $\mathrm{M}(\delta_{\mathscr{E}}) \dashv \mu_{\mathscr{E}} \dashv \delta_{\mathrm{M}(\mathscr{E})}$ is essential. (Note: $\mathrm{M}(\delta_{\mathscr{E}}) = \Sigma(\delta^*_{\mathscr{E}})$.) *For any topos $\mathscr{E}$, there is a canonical 2-cell of geometric morphisms*

$$\mathrm{M}(\delta_{\mathscr{E}}) \Rightarrow \delta_{\mathrm{M}(\mathscr{E})} \,.$$

With Proposition 4.2.7 in mind, observe that the left exact distribution $i^*\delta^*_{P\mathbb{C}}$ inverts $\delta_!^{P\mathbb{C}}(d)$ since we have $i^*\delta^*_{P\mathbb{C}}(\delta_!^{P\mathbb{C}}(d)) \cong i^*(d)$, and since i^* inverts d. Therefore, $i^*\delta^*_{P\mathbb{C}}$ inverts the 2-cell $d.\Sigma_K$, which induces a left exact distribution $\delta^*_{\mathscr{E}} : \Sigma(\mathscr{E}) \longrightarrow \mathscr{E}$. The adjointness $\delta_!^{\mathscr{E}} \dashv \delta^*_{\mathscr{E}}$ holds because $\delta_!^{P\mathbb{C}} \dashv \delta^*_{P\mathbb{C}}$ holds. Moreover, $\delta_!^{\mathscr{E}}$ is full and faithful because $\delta_!^{P\mathbb{C}}$ is. Finally, it remains to show that $\delta^*_{\mathrm{M}(\mathscr{E})} \dashv \mu^*_{\mathscr{E}}$. By definition, $\mu_{\mathscr{E}}$ is the geometric morphism $\Sigma(\delta_!^{\mathscr{E}})$. Its inverse image functor $\mu^*_{\mathscr{E}}$ is the unique left exact distribution making the following diagram commute.

$$\begin{array}{ccc}
\mathscr{E} & \xrightarrow{\ \delta_!^{\mathscr{E}}\ } & \Sigma(\mathscr{E}) \\
\Big\downarrow{\scriptstyle \delta_!^{\mathscr{E}}} & & \Big\downarrow{\scriptstyle \mu^*_{\mathscr{E}}} \\
\Sigma(\mathscr{E}) & \xrightarrow{\ \delta_!^{\Sigma\mathscr{E}}\ } & \Sigma^2(\mathscr{E})
\end{array}$$

But again $\delta^*_{\mathrm{M}(\mathscr{E})} \dashv \mu^*_{\mathscr{E}}$ follows from the presheaf case $\delta^*_{\mathrm{M}(P\mathbb{C})} \dashv \mu^*_{P\mathbb{C}}$, where recall that $\mathrm{M}(P(\mathbb{C}))$ is constructed as the presheaf topos $P(\mathbb{C}^*)$. This completes our demonstration that $U \dashv \Sigma$ is a KZ-adjointness.

The KZ-adjointness $U \dashv \Sigma$ is not KZ-monadic in the sense that $\mathbf{Dist}_{\mathscr{S}}$ is equivalent to the 2-category of Eilenberg-MacLane algebras for M in $\mathbf{Top}_{\mathscr{S}}$. We will later characterize M-algebras in terms of a completeness condition, culminating in our 'Waelbroeck' theorem (Theorem 4.3.18). On the other hand, $\mathbf{Frm}_{\mathscr{S}} = \mathbf{Top}_{\mathscr{S}}{}^{\mathrm{op}}$ is KZ-monadic over $\mathbf{Coc}_{\mathscr{S}}$.

Exercises 4.2.8

1. *Show that the presheaf functor P reverses the direction of 2-cells.*
2. *(A. M. Pitts) The existence of the lex completion of a small category can be established in two steps. In the first step we freely add finite products by freely adding coproducts to the opposite category (Example 4.1.2), and then take opposites. The second step freely adds equalizers to a category with finite products.*
3. *Prove Lemma 4.2.4.*

4.3 Algebras and Cocompleteness

Our main goal in this section is to describe conditions on a KZ-monad M ("M is a completion KZ-monad") so that its algebras are characterized as cocomplete objects in an appropriate sense.

An algebra for a KZ-monad M in a 2-category is an object B with the property (not additional structure) that the unit $\delta_B : B \longrightarrow \mathrm{M}(B)$ has a reflection left adjoint $\sigma_B : \mathrm{M}(B) \longrightarrow B$. We shall refer to an Eilenberg-Moore algebra for M as an M-algebra.

We have defined a (locally full and faithful) KZ-monad $(\mathrm{M}, \delta, \mu)$ in $\mathbf{Top}_{\mathscr{S}}$ called the symmetric monad. An M-algebra in $\mathbf{Top}_{\mathscr{S}}$ is thus given by a topos $\mathscr{E}$ (over $\mathscr{S}$) for which the essential inclusion $\delta_{\mathscr{E}} : \mathscr{E} \longrightarrow \mathrm{M}(\mathscr{E})$ has a reflection left adjoint $\sigma_{\mathscr{E}} : \mathrm{M}(\mathscr{E}) \longrightarrow \mathscr{E}$ in $\mathbf{Top}_{\mathscr{S}}$, i.e., omitting obvious subscripts now, an M-algebra is a geometric morphism

$$\sigma : \mathrm{M}(\mathscr{E}) \longrightarrow \mathscr{E} \; ; \; \sigma \dashv \delta \,,$$

such that $\sigma \cdot \delta \cong id$. This says that $\delta^* \dashv \sigma^*$ since 2-cells in $\mathbf{Top}_{\mathscr{S}}$ are natural transformations between inverse image functors. Hence, $\mathscr{E}$ is an M-algebra iff δ_* has a right adjoint. This holds iff δ_* preserves $\mathscr{S}$-colimits. In particular, if $\mathscr{E}$ is an M-algebra, then $\delta_{\mathscr{E}}$ is what we have called a pure geometric morphism.

Usually algebras for a KZ-monad in a 2-category are objects that are in some sense *cocomplete.* In fact, adding cocomplete structure freely had motivated the very notion of a KZ-monad. For example, the free addition of filtered colimits to a category, its *Ind*-(co)completion, is a KZ-monad. Example 4.1.2 is another instance of this. However, there are other KZ-monads, such as the symmetric monad in $\mathbf{Top}_{\mathscr{S}}$, that are not obviously of the completion type in the ordinary sense of freely adding (co)limits.

Top$_{\mathscr{S}}$ has a certain kind of limit called a bicomma object. We record for future use an important result concerning bicomma objects in **Top**$_{\mathscr{S}}$, due to A. M. Pitts. Our explanation in § 6.5 of Theorem 4.3.1 involves complete spread geometric morphisms.

Theorem 4.3.1 ***Top***$_{\mathscr{S}}$ *has bicomma objects. If the lower morphism in a bicomma object of toposes over* $\mathscr{S}$ *is an* $\mathscr{S}$*-essential geometric morphism, then the geometric morphism opposite it is locally connected, and the Beck-Chevalley condition is satisfied.*

R. Street has studied cocompleteness as a generalized sheaf condition. By definition, an object in a 2-category is cocomplete (is a "sheaf") if it admits pointwise left extensions along 1-cells from a given class (the "dense" morphisms). If that class is small, then the forgetful functor from cocomplete objects is KZ-monadic, so that every object has a cocompletion.

Starting with a KZ-monad, we may try to identify a naturally associated class of 1-cells (said to have an M-adjoint) with respect to which we define cocompleteness. Then the algebras for a completion KZ-monad should coincide with the cocomplete objects.

Throughout, $\mathscr{K}$ shall denote an arbitrary 2-category, and (M, μ, δ) a KZ-monad in $\mathscr{K}$. We shall denote its 2-category of its Eilenberg-Moore algebras by $\mathscr{K}^{\mathrm{M}}$. The forgetful 2-functor $\mathscr{K}^{\mathrm{M}} \longrightarrow \mathscr{K}$ is locally full and faithful.

We start with the following definition. Our choice of terminology conforms with Grothendieck's "*Ind*-adjoint." Street has used the term "admissible" for this notion, meaning that the morphism satisfies a certain size condition.

Definition 4.3.2 *We shall say that a 1-cell* φ *in* $\mathscr{K}$ admits an M-adjoint *if* $M(\varphi)$ *has a right adjoint. If* φ *admits an* M*-adjoint, we denote the right adjoint of* $M(\varphi)$ *by* r_φ.

Remark 4.3.3 *The class of 1-cells that admit an* M*-adjoint is closed under composition. It includes all 1-cells with right adjoints. A unit* $E \xrightarrow{\delta_E} \mathrm{M}(E)$ *admits an* M*-adjoint, with* $r_{\delta_E} = \mu_E$*, because we have* $\mathrm{M}(\delta_E) \dashv \mu_E \dashv \delta_{\mathrm{M}E}$ *(4.2).*

If $E \xrightarrow{f} F$ is a 1-cell between M-algebras, then starting with the canonical isomorphism $\delta_F \cdot f \cong \mathrm{M}(f) \cdot \delta_E$ we may produce a canonical 2-cell ζ_f in the diagram

$$\begin{array}{ccc} \mathrm{M}(E) & \xrightarrow{\sigma_E} & E \\ {\scriptstyle \mathrm{M}f}\downarrow & \overset{\zeta_f}{\Rightarrow} & \downarrow{\scriptstyle f} \\ \mathrm{M}(F) & \xrightarrow{\sigma_F} & F \end{array}$$

using the adjointnesses $\sigma_E \dashv \delta_E$ and $\sigma_F \dashv \delta_F$. We say that f is *an* M-*homomorphism* if ζ_f is an isomorphism.

Proposition 4.3.4 *Suppose that φ admits an* M*-adjoint. Then r_φ is an* M*-homomorphism.*

Proof. Apply M to

$$\begin{array}{ccc} E & \xrightarrow{\delta_E} & \mathrm{M}(E) \\ {\scriptstyle f}\downarrow & & \downarrow{\scriptstyle \mathrm{M}f} \\ F & \xrightarrow{\delta_F} & \mathrm{M}(F) \end{array}$$

and then consider right adjoints. □

Occasionally we shall denote precomposition with a 1-cell $E \xrightarrow{\psi} F$ by $\psi^\sharp$. For a given object K of $\mathscr{K}$, we have

$$\psi^\sharp : \mathscr{K}(F, K) \longrightarrow \mathscr{K}(E, K) \; ; \; \psi^\sharp(q) = q \cdot \psi \, .$$

Definition 4.3.5 *We denote the left adjoint of $\psi^\sharp$, if it exists, by Σ_ψ. We refer to Σ_ψ as* left extension along *ψ (not to be confused with the symmetric topos construction Σ).*

We have the following characterization of 1-cells that admit M-adjoints in terms of algebras and homomorphisms.

Proposition 4.3.6 *A 1-cell $E \xrightarrow{\varphi} F$ admits an* M*-adjoint iff every* M*-algebra admits, and every* M*-homomorphism preserves, left extensions along φ. In this case, for a diagram*

$$\begin{array}{ccc} E & \xrightarrow{\varphi} & F \\ & {\scriptstyle p}\searrow & \\ & & A \end{array}$$

in which φ admits an M*-adjoint and (A, σ) is an* M*-algebra, the left extension $\Sigma_\varphi(p)$ is given as*

$$F \xrightarrow{\delta_F} \mathrm{M}(F) \xrightarrow{r_\varphi} \mathrm{M}(E) \xrightarrow{\mathrm{M}p} \mathrm{M}(A) \xrightarrow{\sigma} A \, .$$

We may retrieve the right adjoint r_φ as $\mu_E \cdot \mathrm{M}(\Sigma_\varphi \delta_E)$.

Proof. Assume that φ admits an M-adjoint. Let (A, σ) denote an arbitrary M-algebra. The structure σ is a reflection left adjoint to δ_A. We have the following natural bijections.

$$\frac{\dfrac{\dfrac{\dfrac{\dfrac{\dfrac{\dfrac{p \Rightarrow q \cdot \varphi}{\delta_A \cdot p \Rightarrow \delta_A \cdot q \cdot \varphi}}{\mathrm{M}p \cdot \delta_E \Rightarrow \mathrm{M}(q \cdot \varphi) \cdot \delta_E}}{\mathrm{M}p \Rightarrow \mathrm{M}(q \cdot \varphi)}}{\mathrm{M}p \cdot r_\varphi \Rightarrow \mathrm{M}q}}{\mathrm{M}p \cdot r_\varphi \cdot \delta_F \Rightarrow \mathrm{M}q \cdot \delta_F}}{\mathrm{M}p \cdot r_\varphi \cdot \delta_F \Rightarrow \delta_A \cdot q}}{\sigma \cdot \mathrm{M}p \cdot r_\varphi \cdot \delta_F \Rightarrow q}$$

Here we have used Proposition 4.3.4 and that $\mathrm{M}(E)$ and $\mathrm{M}(F)$ are free M-algebras, so that for any algebra B precomposition with δ_E is an equivalence of categories $\mathscr{K}(E,B) \simeq \mathscr{K}^{\mathrm{M}}(\mathrm{M}(E),B)$. The fact that homomorphisms preserve left extensions may be similarly established.

For the converse, by hypothesis we have the left extension $\Sigma_\varphi \delta_E$: $F \longrightarrow \mathrm{M}(E)$, since $\mathrm{M}(E)$ is an algebra. Let Z denote an arbitrary algebra. We have the following bijections, natural in the homomorphisms $\mathrm{M}(E) \xrightarrow{h} Z$ and $\mathrm{M}(F) \xrightarrow{z} Z$.

$$\frac{\dfrac{\dfrac{\dfrac{\dfrac{\dfrac{\dfrac{h \cdot \mu_E \cdot \mathrm{M}(\Sigma_\varphi \delta_E) \Rightarrow z}{h \cdot \mu_E \cdot \mathrm{M}(\Sigma_\varphi \delta_E) \cdot \delta_F \Rightarrow z \cdot \delta_F}}{h \cdot \mu_E \cdot \delta_{\mathrm{M}(E)} \cdot \Sigma_\varphi \delta_E \Rightarrow z \cdot \delta_F}}{h \cdot \Sigma_\varphi \delta_E \Rightarrow z \cdot \delta_F}}{\Sigma_\varphi (h \cdot \delta_E) \Rightarrow z \cdot \delta_F}}{h \cdot \delta_E \Rightarrow z \cdot \delta_F \cdot \varphi}}{h \cdot \delta_E \Rightarrow z \cdot \mathrm{M}\varphi \cdot \delta_E}}{h \Rightarrow z \cdot \mathrm{M}\varphi}$$

The forgetful 2-functor to $\mathscr{K}$ is locally full and faithful, so that $\mathrm{M}(\varphi) \dashv \mu_E \cdot \mathrm{M}(\Sigma_\varphi \delta_E)$ as homomorphisms. But then the adjointness must hold as 1-cells because any 2-functor preserves adjoint pairs (in this case the forgetful 2-functor to $\mathscr{K}$). □

Corollary 4.3.7 *Let E be an object of $\mathscr{K}$, and let (A,σ) denote an* M-*algebra. Then a homomorphism $\mathrm{M}(E) \xrightarrow{h} A$ is the left extension of its restriction to E. Explicitly, this means that precomposition with δ_E gives a bijection*

$$\frac{h \Rightarrow q}{h \cdot \delta_E \Rightarrow q \cdot \delta_E}$$

of 2-cells. (The 1-cell q is not required to be a homomorphism.)

Proof. By Proposition 4.3.6, the left extension of $h \cdot \delta_E$ along δ_E is given by

$$\Sigma_{\delta_E}(h \cdot \delta_E) = \sigma \cdot \mathrm{M}(h \cdot \delta_E) \cdot \mu_E \cdot \delta_{\mathrm{M}(E)} \cong \sigma \cdot \mathrm{M}(h \cdot \delta_E) \cong h \cdot \mu_E \cdot \mathrm{M}(\delta_E) \cong h\,.$$

□

Definition 4.3.8 *A span* $fE\psi : A \longrightarrow B$ *is a diagram*

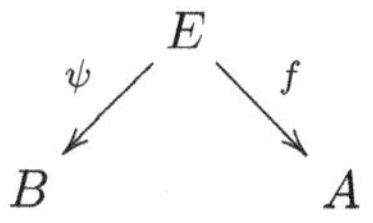

of 0-cells and 1-cells (usually in a 2-category). We think of f as the domain and ψ the codomain of the span.

A bicomma object *in a 2-category is a limit diagram*

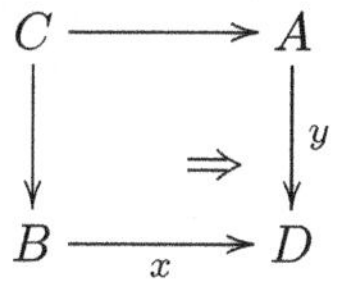

which is universal for spans $A \longrightarrow B$ and 2-cells at D. We sometimes denote such a bicomma object by $x \Downarrow y$, and refer to x as the lower leg.

Consider the following 'Yoneda strengthening' of left extension (Definition 4.3.5). The 1-cell h in a diagram

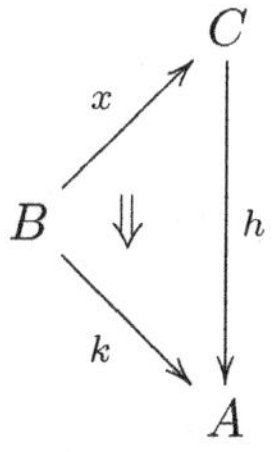

in a 2-category is *the pointwise left extension of k along x* if for every $D \xrightarrow{y} C$, $h \cdot y$ is the left extension of $k \cdot q$ along p, where

$$
\begin{array}{ccc}
E & \xrightarrow{\;p\;} & D \\
{\scriptstyle q}\downarrow & \Rightarrow & \downarrow{\scriptstyle y} \\
B & \xrightarrow{\;x\;} & C
\end{array}
$$

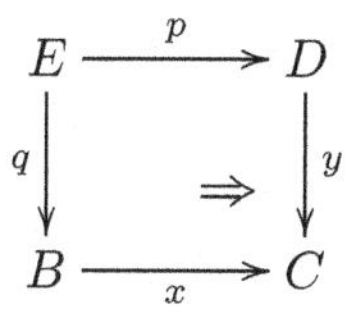

is a bicomma object.

Henceforth, we shall assume that a 2-category $\mathscr{K}$ has bicomma objects $x \Downarrow y$ for which the 1-cell x admits an M-adjoint.

Definition 4.3.9 *We shall say that an object of $\mathscr{K}$ is* cocomplete (for M) *if it admits pointwise left extensions along 1-cells that admit an M-adjoint. A 1-cell is said to be* cocontinuous *if it preserves pointwise left extension along 1-cells that admit an M-adjoint.*

Lemma 4.3.10 *In the following bicomma object, q has a left adjoint l such that $p \cdot l \cong f$.*

$$\begin{array}{ccc} E & \xrightarrow{p} & D \\ {\scriptstyle q}\downarrow & \Rightarrow & \downarrow{\scriptstyle 1} \\ C & \xrightarrow[f]{} & D \end{array}$$

Proposition 4.3.11 *A left extension into a cocomplete object along a 1-cell that admits an* M*-adjoint is a pointwise left extension. A 1-cell between cocomplete objects is cocontinuous iff it preserves left extensions along 1-cells that admit an* M*-adjoint.*

Proof. The first statement holds because by Exercise 4 a pointwise left extension and a left extension must be isomorphic. The second statement follows from the first. □

Proposition 4.3.12 *Assume that for bicomma objects*

$$\begin{array}{ccc} A & \xrightarrow{p} & B \\ {\scriptstyle q}\downarrow & \Rightarrow & \downarrow{\scriptstyle g} \\ C & \xrightarrow[f]{} & D \end{array}$$

in which f admits an M*-adjoint, p also admits an* M*-adjoint. Then we have the following.*

1. *An object X of $\mathscr{K}$ is cocomplete iff for any 1-cell f that admits an* M*-adjoint, the precomposition functor $f^\sharp$ for X has a left adjoint Σ_f, and for a bicomma object as above, the canonical natural transformation*

$$\Sigma_p \circ q^\sharp \Rightarrow g^\sharp \circ \Sigma_f \tag{4.6}$$

   is an isomorphism.
2. *If free* M*-algebras are cocomplete, then the canonical 2-cell*

$$\mathrm{M}(q) \cdot r_p \Rightarrow r_f \cdot \mathrm{M}(g)$$

   is an isomorphism.

Proof. 1. Let X denote an arbitrary object in $\mathscr{K}$. If the left adjoints Σ_f for X exist for any f admitting an M-adjoint, and (4.6) is an isomorphism, then for a 1-cell $C \xrightarrow{h} X$, the left extension $\Sigma_f h$ is pointwise, so that X is cocomplete. Conversely, assume that X is cocomplete. The Σ's exist because pointwise left extensions are left extensions (Exercise 4). It follows from the definition of pointwise left extension that (4.6) is an isomorphism.

2. Since we are assuming that free algebras are cocomplete, the left extension $\Sigma_f \delta_C$ for the free algebra MC must be pointwise (Proposition 4.3.11). By the cocompleteness of MC, and by 1, we have (Proposition 4.3.6)

$$\mathrm{M}q \cdot r_p \cdot \delta_B \cong \Sigma_p(\delta_C \cdot q) = \Sigma_p q^\sharp(\delta_C) \cong g^\sharp \Sigma_f(\delta_C) \cong r_f \cdot \delta_D \cdot g \cong r_f \cdot \mathrm{M}g \cdot \delta_B \, .$$

Since MB is free, and by Proposition 4.3.4, we have $\mathrm{M}q \cdot r_p \cong r_f \cdot \mathrm{M}g$. □

Our aim is to characterize M-algebras as cocomplete objects. Proposition 4.3.12 presents some necessary information. We introduce the following concept.

Definition 4.3.13 A completion KZ-monad (in $\mathscr{K}$) *is a KZ-monad for which the following bicomma object condition holds: the 2-category $\mathscr{K}$ has bicomma objects $f \Downarrow g$ of diagrams*

$$\begin{array}{ccc} & & B \\ & & \downarrow{\scriptstyle g} \\ C & \xrightarrow[f]{} & D \end{array}$$

in which f admits an M*-adjoint, for such a bicomma object*

$$\begin{array}{ccc} A & \xrightarrow{p} & B \\ {\scriptstyle q}\downarrow & \Rightarrow & \downarrow{\scriptstyle g} \\ C & \xrightarrow[f]{} & D \end{array}$$

the 1-cell p admits an M*-adjoint, and the canonical 2-cell*

$$\mathrm{M}(q) \cdot r_p \Rightarrow r_f \cdot \mathrm{M}(g) \tag{4.7}$$

is an isomorphism.

The main result of this section is the following.

Theorem 4.3.14 *Let* $(\mathrm{M}, \mu, \delta)$ *be a completion KZ-monad in a 2-category* $\mathscr{K}$*.*

1. *Then* M*-algebras are cocomplete, and* M*-homomorphisms are cocontinuous.*
2. *Conversely, if* M *is locally full and faithful, then cocomplete objects are* M*-algebras, and cocontinuous 1-cells between cocomplete objects are* M*-homomorphisms.*

Proof. By Proposition 4.3.6, the precomposition functors for algebras have left adjoints. We have the isomorphism (4.6) because of the isomorphism (4.7) of Definition 4.3.13 and by Proposition 4.3.6. Thus, by Proposition 4.3.12, algebras are cocomplete.

Let F denote an arbitrary cocomplete object of $\mathscr{K}$. Let $\mathrm{M}(F) \xrightarrow{\sigma} F$ denote the left extension $\Sigma_{\delta_F}(1_F)$. We will show that $\sigma \dashv \delta_F$, so that (F, σ) is an M-algebra. Let $H \xrightarrow{q} \mathrm{M}(F)$ and $H \xrightarrow{m} F$ be arbitrary 1-cells, and consider the following bicomma object.

$$\begin{array}{ccc} G & \xrightarrow{\varphi} & H \\ {\scriptstyle p}\downarrow & \Rightarrow & \downarrow{\scriptstyle q} \\ F & \xrightarrow[\delta_F]{} & \mathrm{M}(F) \end{array}$$

Then φ admits an M-adjoint, and we have the following natural bijections. The unit δ_F has been abbreviated to δ. Since we are assuming that M is locally full and faithful, every unit δ is a full and faithful 1-cell.

$$\begin{array}{c}
\sigma \cdot q = q^\sharp(\Sigma_\delta 1_F) \Rightarrow m \\ \hline
\Sigma_\varphi p^\sharp 1_F \Rightarrow m \\
\text{by (4.6) for } F \text{ and the} \\
\text{above bicomma object} \\ \hline \hline
p \Rightarrow \varphi^\sharp m \\ \hline
\delta \cdot p \Rightarrow \delta \cdot \varphi^\sharp m \\
(\delta \text{ is full and faithful}) \\ \hline \hline
p^\sharp \delta \Rightarrow \varphi^\sharp(\delta \cdot m) \\ \hline
\Sigma_\varphi p^\sharp \delta \Rightarrow \delta \cdot m \\ \hline
q^\sharp(\Sigma_\delta \delta) \Rightarrow \delta \cdot m \\
\text{by (4.6) for } \mathrm{M}(F) \text{ and the} \\
\text{above bicomma object} \\ \hline
q = q^\sharp(1_{\mathrm{M}(F)}) \Rightarrow \delta \cdot m \ (\text{Ex. } 2)
\end{array}$$

Here we have used that free algebras are cocomplete, which we have established in the first paragraph.

Let $F \xrightarrow{\psi} F'$ denote an arbitrary cocontinuous 1-cell between cocomplete objects (equivalently, between algebras). We must show that the canonical 2-cell $\zeta : \sigma' \cdot \mathrm{M}\psi \longrightarrow \psi \cdot \sigma$ is an isomorphism. There are canonical 2-isomorphisms

$$\sigma' \cdot \mathrm{M}\psi \cdot \delta_F \cong \sigma' \cdot \delta_{F'} \cdot \psi \cong \psi$$

to which we apply Σ_{δ_F} giving the top row of the following diagram. This diagram can be seen to commute by unraveling the adjoints.

$$\begin{array}{ccccc} \Sigma_{\delta_F}(\sigma' \cdot \mathrm{M}\psi \cdot \delta_F) & \longrightarrow & \Sigma_{\delta_F}(\sigma' \cdot \delta_{F'} \cdot \psi) & \longrightarrow & \Sigma_{\delta_F}\psi \\ \downarrow & & & & \downarrow \\ \sigma' \cdot \mathrm{M}\psi & \xrightarrow{\zeta} & \psi \cdot \sigma & \longrightarrow & \psi \cdot \Sigma_{\delta_F}(1_F) \end{array}$$

The left vertical arrow is the counit of $\Sigma_{\delta_F} \dashv {\delta_F}^{\#}$; it is an isomorphism because both σ' and $\mathrm{M}\psi$ are M-homomorphisms, whence cocontinuous, and since $\Sigma_{\delta_F}\delta_F \cong 1_{\mathrm{M}(F)}$. The right vertical arrow is an isomorphism because ψ is assumed to be cocontinuous. The bottom right morphism is the isomorphism obtained by applying ψ to $\sigma \cong \Sigma_{\delta_F} 1_F$. This shows that ζ is an isomorphism. □

Example 4.3.15 *(M. Escardo) An example of a locally full and faithful KZ-monad that is not a completion one appears implicitly in separate works of Erker, Hoffman, and Escardo. The Eilenberg-Moore algebras for the filter monad in the category of topological spaces are the same as the injective spaces. On the other hand, an object is cocomplete for this KZ-monad (in which case it is called a right Kan space) iff it is what B. Banaschewski has called an essentially complete T_0 space. But injective spaces do not coincide with essentially complete T_0 spaces, so that the algebras do not coincide with the cocomplete objects. By Theorem 4.3.14, the filter monad in topological spaces cannot be a completion KZ-monad.*

Our main example is the symmetric monad $(\mathrm{M}, \mu, \delta)$ in $\mathbf{Top}_{\mathscr{S}}$. We have the following characterization of geometric morphisms that admit an M-adjoint.

Proposition 4.3.16 *An arbitrary geometric morphism over $\mathscr{S}$ admits an* M*-adjoint iff it is $\mathscr{S}$-essential.*

Proof. Let $\varphi : \mathscr{E} \longrightarrow \mathscr{F}$ denote a geometric morphism over $\mathscr{S}$. Assume that φ is $\mathscr{S}$-essential, with left adjoint $\varphi_!$. In terms of distributions, $\mathrm{M}(\varphi)$ corresponds to composition with φ^*, so that the right adjoint r_φ corresponds to composition with $\varphi_!$.

Now assume that φ admits an M-adjoint, with ${r_\varphi}^* \dashv \mathrm{M}(\varphi)^*$. Then ${r_\varphi}^*\delta_{\mathscr{E}!}$ is left adjoint to ${\delta_{\mathscr{E}}}^*\mathrm{M}(\varphi)^*$ over $\mathscr{S}$. Therefore, $\varphi^*{\delta_{\mathscr{F}}}^*$ has a left adjoint. Since $\delta_{\mathscr{F}}$ is an inclusion, it follows that φ^* has a left adjoint (over $\mathscr{S}$). □

Theorem 4.3.17 *The symmetric monad is a locally full and faithful completion KZ-monad.*

Proof. This fact is in a new guise the theorem of Pitts (Theorem 4.3.1). □

A characterization of the Eilenberg-Moore algebras for the symmetric KZ-monad follows and agrees with the notion of cocompleteness given by Street. We sum it up as follows.

Theorem 4.3.18 (Waelbroeck Theorem) *A topos is an Eilenberg-Moore algebra for the symmetric KZ-monad iff it admits pointwise left extensions along essential geometric morphisms. A geometric morphism between two such algebras is a homomorphism iff it preserves left extensions along essential geometric morphisms.*

Example 4.3.19

1. *Consider the case of the topos $\mathscr{S}$. In this case, the unit is*
$$\delta_{\mathscr{S}} : \mathscr{S} \longrightarrow \mathrm{M}(\mathscr{S}) \simeq \mathscr{R} \,,$$
where $\mathscr{R}$ is the object classifier in $\boldsymbol{Top}_{\mathscr{S}}$. The object of $\mathscr{S}$ that corresponds to the geometric morphism $\delta_{\mathscr{S}}$ is the terminal object 1, *i.e., the unit is the Dirac delta supported on* 1. *It can be verified directly that $\delta_*^{\mathscr{S}}$ is cocontinuous, so that $\mathscr{S}$ is an* M*-algebra (Ex. 7).*
2. *The line $\mathscr{R}$ is also an* M*-algebra, in fact it is a free one, and $\delta_{\mathscr{S}}$ is a homomorphism. The evaluation of a distribution can be understood in terms of the structure map $\int : \mathrm{M}(\mathscr{R}) \longrightarrow \mathscr{R}$ of this* M*-algebra. Indeed, if $\mathscr{G} \xrightarrow{p} \mathrm{M}(\mathscr{E})$ is the geometric morphism corresponding to a distribution $\mu : \mathscr{E} \longrightarrow \mathscr{G}$, and if $\varphi : \mathscr{E} \longrightarrow \mathscr{R}$ is the geometric morphism corresponding to an object X of $\mathscr{E}$, then the evaluation $\mu(X)$ is the object of $\mathscr{G}$ whose correponding geometric morphism is the composite*
$$\mathscr{G} \xrightarrow{p} \mathrm{M}(\mathscr{E}) \xrightarrow{\mathrm{M}(\varphi)} \mathrm{M}(\mathscr{R}) \xrightarrow{\int} \mathscr{R} \,.$$

M-algebras also have the following 'category cocompleteness' property.

Proposition 4.3.20 *Let $\mathscr{E}$ be an* M*-algebra in $\boldsymbol{Top}_{\mathscr{S}}$. Then for each topos $\mathscr{G}$ in $\boldsymbol{Top}_{\mathscr{S}}$, the category $\boldsymbol{Top}_{\mathscr{S}}(\mathscr{G}, \mathscr{E})$ is $\mathscr{S}$-cocomplete. If $\varphi : \mathscr{E} \longrightarrow \mathscr{F}$ is a homomorphism of* M*-algebras, then the induced functor*
$$\boldsymbol{Top}_{\mathscr{S}}(\mathscr{G}, \mathscr{E}) \longrightarrow \boldsymbol{Top}_{\mathscr{S}}(\mathscr{G}, \mathscr{F}) \qquad (4.8)$$
is $\mathscr{S}$-cocontinuous.

Proof. If $\mathscr{E}$ is an M-algebra, then $\mathscr{E}$ is a retract of $\mathrm{M}(\mathscr{E})$, hence $\mathbf{Top}_{\mathscr{S}}(\mathscr{G}, \mathscr{E})$ is a retract of $\mathbf{Top}_{\mathscr{S}}(\mathscr{G}, \mathrm{M}(\mathscr{E})) \simeq \mathbf{Coc}_{\mathscr{S}}(\mathscr{E}, \mathscr{G})$, which is $\mathscr{S}$-cocomplete. Hence $\mathbf{Top}_{\mathscr{S}}(\mathscr{G}, \mathscr{E})$ is $\mathscr{S}$-cocomplete. If $\varphi : \mathscr{E} \longrightarrow \mathscr{F}$ is a homomorphism, then (4.8) is obtained by restriction from the corresponding functor
$$\mathbf{Top}_{\mathscr{S}}(\mathscr{G}, \mathrm{M}(\mathscr{E})) \longrightarrow \mathbf{Top}_{\mathscr{S}}(\mathscr{G}, \mathrm{M}(\mathscr{F})) \,,$$
which corresponds to composition with φ^*. It follows that (4.8) is $\mathscr{S}$-cocontinuous. □

Remark 4.3.21 *Proposition 4.3.20 is not reversible. Of course, the Waelbroeck Theorem gives a characterization of* M*-algebras and their homomorphisms, but we may also note directly that the $\mathscr{S}$-category $\boldsymbol{Top}_{\mathscr{S}}(\mathscr{G}, \mathrm{M}(\mathscr{E}))$ cannot be the free $\mathscr{S}$-cocompletion of the $\mathscr{S}$-category $\boldsymbol{Top}_{\mathscr{S}}(\mathscr{G}, \mathscr{E})$ since, for instance, there exist (non-trivial) locally connected Grothendieck toposes without points. If $\mathscr{E} \longrightarrow \mathscr{S} = \mathrm{Set}$ is any such topos with $\pi_0 \dashv \Delta$, then $\boldsymbol{Top}_{\mathscr{S}}(\mathscr{S}, \mathscr{E})$ is vacuous, but the distributions on $\mathscr{E}$ must contain a copy of* Set*, such that a set I corresponds to the distribution $F \mapsto I \times \pi_0(F)$.*

Exercises 4.3.22

1. *Show that if bipullbacks exist, then we can build arbitrary bicomma objects from bicomma objects*
$$\begin{array}{ccc} A & \longrightarrow & K \\ \big\downarrow & \Rightarrow & \big\downarrow 1_K \\ X & \underset{1_K}{\longrightarrow} & K \end{array} .$$
2. *Deduce from Corollary 4.3.7 that for any object* E*, the left extension of* δ_E *along itself is the identity 1-cell on* $\mathrm{M}(E)$*:* $1_{\mathrm{M}(E)} \cong \Sigma_{\delta_E}(\delta_E)$*. Also deduce from Corollary 4.3.7 that if* (A, σ) *is an* M*-algebra, then* σ *is the left extension of* 1_A *along* δ_A*:* $\sigma \cong \Sigma_{\delta_A}(1_A)$*. Show that* σ *is also the right extension of* 1_A *along* δ_A*.*
3. *For any algebra* (B, σ) *and 1-cell* $E \xrightarrow{k} B$*, the left extension* $\Sigma_{\delta_E} k \cong \sigma \cdot \mathrm{M}k$ *is a homomorphism. Thus, show that left extension is the pseudo inverse of the equivalence*
$$\mathscr{K}(E, B) \simeq \mathscr{K}^{\mathrm{M}}(\mathrm{M}(E), B)$$
given by composition with δ_E*.*
4. *Prove Lemma 4.3.10, and use it to show that a pointwise left extension is a left extension when bicomma objects exist.*
5. *Show that the free addition of finite coproducts to a small category is a completion KZ-monad (Example 4.1.2).*
6. *Complete the proof of 4.3.14: show that homomorphisms are cocontinuous.*
7. *Show that an* $\mathscr{S}$*-topos* $\mathscr{E}$ *is an* M*-algebra iff the right adjoint* $\delta_*^{\mathscr{E}} : \mathscr{E} \longrightarrow \mathrm{M}(\mathscr{E})$ *is* $\mathscr{S}$*-cocontinuous.*

Further reading: Bunge [Bun95], Bunge & Carboni [BC95], Bunge & Fiore [BF00], Bunge & Funk [BF98], Bunge & Niefield [BN00], Bunge, Funk, Jibladze & Streicher [BFJS04a], Carboni & Johnstone [CJ95], Fiore [Fio95], Funk [Fun95], Johnstone [Joh77, Joh02], Street & Walters [SW73], Waelbroeck [Wae67].

5

Complete Spreads as Discrete M-fibrations

We begin by introducing notions of discrete M-fibrations and discrete M-opfibrations for an arbitrary KZ-monad M on a 2-category $\mathscr{K}$. We investigate a pullback stability property of discrete M-fibrations.

For the symmetric monad M in $\mathbf{Top}_{\mathscr{S}}$, the discrete M-fibrations are the complete spreads, and the discrete M-opfibrations are the local homeomorphisms. The stability theorem, which we prove in a general setting, may be used to show that complete spreads are stable under pullback along an $\mathscr{S}$-essential geometric morphism (not just under a locally connected geometric morphism).

We give an application regarding a new pullback preservation property of the presheaf topos functor.

5.1 Discrete (op)fibrations Relative to M

Throughout we assume that $(\mathrm{M}, \delta, \mu)$ is a KZ-monad in a 2-category $\mathscr{K}$. We assume that $\mathscr{K}$ has a pseudo-terminal object T in the sense that for every object K, the category $\mathscr{K}(K, T)$ is equivalent to the trivial category with a single morphism.

Definition 5.1.1 *We shall say that a span $fE\psi$ (Def. 4.3.8) admits an* M*-adjoint if the domain 1-cell f admits an* M*-adjoint. Let* $\mathrm{Span}_{\mathrm{M}}(A, B)$ *denote the 2-category of spans admitting an* M*-adjoint.*

Definition 5.1.2 An M-bifibration *is a span $fY\psi$ that arises in a bicomma object*

$$\begin{array}{ccc} Y & \xrightarrow{f} & A \\ {\scriptstyle\psi}\downarrow & \Rightarrow & \downarrow{\scriptstyle p} \\ B & \xrightarrow[\delta_B]{} & \mathrm{M}(B) \end{array}$$

for some p.

If an M-bifibration admits an M-adjoint, as in the case when M is a completion KZ-monad, then we have a functor

$$\Delta : \mathscr{K}(A, \mathrm{M}(B)) \longrightarrow \mathit{Span}_{\mathrm{M}}(A, B)$$

such that $\Delta(p) = \delta_B \Downarrow p$. On the other hand, we may associate with any span $fE\psi$ admitting an M-adjoint the 1-cell

$$\Sigma_f(\delta_B \cdot \psi) = \mathrm{M}(\psi) \cdot r_f \cdot \delta_A \,.$$

This 1-cell is the left extension as the notation indicates. This defines a functor

$$\Lambda : \mathit{Span}_{\mathrm{M}}(A, B) \longrightarrow \mathscr{K}(A, \mathrm{M}(B)) \,.$$

Proposition 5.1.3 *Let* M *be a completion KZ-monad that is locally full and faithful. Then* $\Lambda \dashv \Delta$, *and* Δ *is full and faithful.*

Thus, we may equivalently regard M-bifibrations as 1-cells $A \longrightarrow \mathrm{M}(B)$, at least in the case of a completion KZ-monad. These are precisely 1-cells $A \longrightarrow B$ in the so-called *Kleisli category for* M, denoted $\mathscr{K}_{\mathrm{M}}$.

We are mostly interested in the following two special kinds of M-bifibration.

Definition 5.1.4 A discrete M-opfibration *is a 1-cell* $X \xrightarrow{f} B$ *that appears in bicomma object*

$$\begin{array}{ccc} X & \xrightarrow{f} & B \\ {\scriptstyle x}\downarrow & \Rightarrow & \downarrow{\scriptstyle q} \\ T & \xrightarrow[\delta_T]{} & \mathrm{M}(T) \end{array}$$

for some 1-cell q. *The 1-cell* x *to the terminal* T *is essentially unique.*

A discrete M-fibration *is a 1-cell* $Y \xrightarrow{\psi} B$ *that appears in bicomma object*

$$\begin{array}{ccc} Y & \xrightarrow{y} & T \\ {\scriptstyle \psi}\downarrow & \Rightarrow & \downarrow{\scriptstyle p} \\ B & \xrightarrow[\delta_B]{} & \mathrm{M}(B) \end{array}$$

for some 1-cell p. *Usually we omit reference to the KZ-monad* M.

Corollary 5.1.5 *Let* M *be a locally full and faithful completion KZ-monad in a 2-category* $\mathscr{K}$. *Then the category of discrete* M*-fibrations over an object* B *is equivalent to* $\mathscr{K}_{\mathrm{M}}(T, B)$. *The category of discrete* M*-opfibrations over* B *is equivalent to* $\mathscr{K}_{\mathrm{M}}(B, T)$.

Remark 5.1.6 *A discrete (op)fibration for any* M *is an (op)fibration in the sense of Ross Street. This is because in a bicomma object*

$$\begin{array}{ccc} A & \xrightarrow{x} & B \\ {\scriptstyle\varphi}\downarrow & \Rightarrow & \downarrow \\ C & \longrightarrow & D \end{array}$$

in a 2-category, the 1-cell φ *is a fibration and* x *is an opfibration.*

If M is a completion KZ-monad then we may recover $T \xrightarrow{p} \mathrm{M}(B)$ from the corresponding discrete fibration ψ by the left-extension formula

$$p \cong \Sigma_y(\delta_B \cdot \psi) \cong \mathrm{M}(\psi) \cdot r_y \cdot \delta_T \ .$$

Similarly, a 1-cell $B \xrightarrow{q} \mathrm{M}(T)$ may be recovered from its discrete opfibration:

$$q \cong \Sigma_f(\delta_T \cdot x) \cong \mathrm{M}(x) \cdot r_f \cdot \delta_B \ .$$

There is an action of discrete opfibrations on discrete fibrations (and vice versa, but that action is less interesting): we may 'multiply' a discrete opfibration $X \xrightarrow{f} B$ on a discrete fibration $Y \xrightarrow{\psi} B$ to form a discrete fibration $f.\psi$ as follows. In the diagram

$$\begin{array}{ccc} \mathrm{M}(X) & \xrightarrow{\mathrm{M}f} & \mathrm{M}(B) \\ {\scriptstyle \mathrm{M}x}\downarrow & \Rightarrow & \downarrow{\scriptstyle \mathrm{M}q} \\ \mathrm{M}(T) & \xrightarrow[\mathrm{M}(\delta_T)]{} & \mathrm{M}^2(T) \end{array}$$

we have $\mathrm{M}(f) \dashv r_f$. Now define $f.\psi$ to be the discrete fibration corresponding to the 1-cell

$$f.\psi : T \xrightarrow{p} \mathrm{M}(B) \xrightarrow{r_f} \mathrm{M}(X) \xrightarrow{\mathrm{M}(f)} \mathrm{M}(B) \ , \tag{5.1}$$

where p corresponds to ψ.

5.2 Complete Spreads as Discrete Fibrations

Topos bicomma objects are related to complete spread geometric morphisms. Concretely, we shall prove the following.

Proposition 5.2.1 *Let* $\mathscr{Y} \xrightarrow{\psi} \mathscr{E}$ *be the complete spread with locally connected domain, associated with a point* f *of the symmetric topos. Then there is a bicomma object*

$$\begin{array}{ccc} \mathscr{Y} & \xrightarrow{\;y\;} & \mathscr{S} \\ {\scriptstyle\psi}\big\downarrow & \Rightarrow & \big\downarrow{\scriptstyle f} \\ \mathscr{E} & \xrightarrow[\;\delta\;]{} & \mathrm{M}(\mathscr{E}) \end{array}$$

in $\boldsymbol{Top}_{\mathscr{S}}$*. Moreover, the BCC holds for this diagram:*

$$\mathrm{M}(\psi) \cdot r_y \cong \mu_{\mathscr{E}} \cdot \mathrm{M}(f) \; .$$

In particular, the distribution associated with f *is* $\nu = f^* \cdot \delta_! \cong y_! \cdot \psi^*$*. Conversely, in any such bicomma object,* ψ *is a complete spread.*

Proof. A 2-cell

$$\begin{array}{ccc} \mathscr{G} & \xrightarrow{\;g\;} & \mathscr{S} \\ {\scriptstyle\kappa}\big\downarrow & \Rightarrow & \big\downarrow{\scriptstyle f} \\ \mathscr{E} & \xrightarrow[\;\delta\;]{} & \mathrm{M}(\mathscr{E}) \end{array}$$

between geometric morphisms amounts to a natural transformation

$$\kappa^* \cdot \delta^* \longrightarrow g^* \cdot f^* \; ,$$

which may be transposed to one

$$t : \kappa^* \longrightarrow g^* \cdot f^* \cdot \delta_! = g^* \cdot \nu \; .$$

Our argument proceeds precisely as in Proposition 2.4.2. In order to exhibit the factoring geometric morphism $\mathscr{G} \longrightarrow \mathscr{Y}$, it suffices to define a flat functor $F : \mathbb{Y} \longrightarrow \mathscr{G}$ with left extension ρ^* such that

$$\begin{array}{ccc} \mathscr{G} & \xleftarrow{\;\rho^*\;} & \mathrm{P}(\mathbb{Y}) \\ {\scriptstyle\kappa^*}\big\uparrow & & \big\uparrow \\ \mathscr{E} & \longleftarrow & \mathbb{C} \end{array} \qquad (5.2)$$

commutes, where $\mathbb{Y}$ denotes the amalgamation site for $\mathscr{Y}$. Let (c, x) denote an arbitrary object of $\mathbb{Y}$, so that $x \in \nu(h_c)$. We define $F(c, x)$ by the pullback

$$\begin{array}{ccc} F(c,x) & \longrightarrow & 1_{\mathscr{G}} \\ \big\downarrow & & \big\downarrow{\scriptstyle g^*(x)} \\ \kappa^*(h_c) & \xrightarrow[\;t_c\;]{} & g^*\nu(h_c) \end{array}$$

in $\mathscr{G}$. Then F is flat and (5.2) commutes. The BCC holds because $f^* \cdot \delta_! \cong y_! \cdot \psi^*$. □

Proposition 5.2.1 asserts that a complete spread geometric morphism with locally connected domain is *a discrete* M-*fibration* in the sense of Definition 5.1.4. Conversely, a discrete M-fibration is a complete spread: if $\psi = \delta \Downarrow f$, then ψ is the complete spread corresponding to the distribution $f^*\delta_!$.

Proposition 5.2.2 *Let* M *be a KZ-monad in* $\mathscr{K}$. *Consider a bipullback*

$$\begin{array}{ccc} E' & \xrightarrow{\kappa} & E \\ {\scriptstyle \psi'}\downarrow & & \downarrow{\scriptstyle \psi} \\ B' & \xrightarrow{\rho} & B \end{array}$$

in $\mathscr{K}$, *where* ψ *is a discrete* M-*fibration, and* ρ *admits an* M-*adjoint. Then* ψ' *is a discrete* M-*fibration. Furthermore, if* $\psi : E \longrightarrow B$ *is witnessed by* $q : T \longrightarrow \mathrm{M}(B)$, *then* $\psi' : E' \longrightarrow B'$ *is witnessed by* $r_\rho \cdot q : T \longrightarrow \mathrm{M}(B')$, *where* $\mathrm{M}(\rho) \dashv r_\rho$.

Proof. It suffices to prove that the diagram

$$\begin{array}{ccc} E' & \xrightarrow{e'} & T \\ {\scriptstyle \psi'}\downarrow & \Rightarrow & \downarrow{\scriptstyle r_\rho \cdot q} \\ B' & \xrightarrow[\delta_{B'}]{} & \mathrm{M}(B') \end{array}$$

forms a bicomma object in $\mathscr{K}$. Consider the composite diagram

$$\begin{array}{ccccc} E' & \xrightarrow{\kappa} & E & \xrightarrow{e} & T \\ {\scriptstyle \psi'}\downarrow & & \downarrow{\scriptstyle \psi} & \Rightarrow & \downarrow{\scriptstyle q} \\ B' & \xrightarrow[\rho]{} & B & \xrightarrow[\delta_B]{} & \mathrm{M}(B) \end{array}$$

of a bicomma on the right and a bipullback on the left; the composite is then a bicomma object in $\mathscr{K}$. For a 1-cell $p : X \longrightarrow B'$ (and $X \xrightarrow{x} T$), we have the following natural bijections of 2-cells:

$$\frac{\dfrac{\delta_{B'} \cdot p \Rightarrow r_\rho \cdot q \cdot x}{\mathrm{M}(\rho) \cdot \delta_{B'} \cdot p \Rightarrow q \cdot x}}{\delta_B \cdot \rho \cdot p \Rightarrow q \cdot x.}$$

In turn, since the composite diagram is a bicomma object, the last 2-cells are in natural bijection with 1-cells $X \longrightarrow E'$ over B' (and T). □

5.3 A Pullback-preservation Theorem for Presheaves

The results of this section are from work in progress with M. Jibladze and T. Streicher [BFJS04b].

Consider the presheaf topos 2-functor

$$P : \mathbf{Cat} \longrightarrow \mathbf{Top}_{\mathscr{S}} \; .$$

It is known that P does not preserve bipullbacks in general, but for instance P does preserves finite products. Also, it is not difficult to show that P preserves bipullback squares in which one leg is a discrete fibration (Exercise 2).

The following asserts that dually P preserves bipullback squares in which one leg is a discrete opfibration.

Proposition 5.3.1 *The presheaf 2-functor P preserves bipullback squares in which one pair of opposite sides are discrete opfibrations.*

Proof. Suppose that

$$\begin{array}{ccc} \mathbb{A} & \xrightarrow{G} & \mathbb{B} \\ {\scriptstyle V}\big\downarrow & & \big\downarrow{\scriptstyle U} \\ \mathbb{D} & \xrightarrow[F]{} & \mathbb{C} \end{array}$$

is a category bipullback such that U and V are discrete opfibrations. (The bipullback of a discrete opfibration is only equivalent to a discrete opfibration - this fine point has no bearing on our argument.) The reader may easily check that if $K : \mathbb{C} \longrightarrow \mathscr{S}$ corresponds to U, then the functor

$$\mathbb{D} \xrightarrow{F} \mathbb{C} \xrightarrow{K} \mathscr{S}$$

corresponds to V. We wish to show that

$$\begin{array}{ccc} P(\mathbb{A}) & \xrightarrow{g} & P(\mathbb{B}) \\ {\scriptstyle v}\big\downarrow & & \big\downarrow{\scriptstyle u} \\ P(\mathbb{D}) & \xrightarrow[f]{} & P(\mathbb{C}) \end{array}$$

is a topos bipullback, where $g = P(G)$, and so on. Since U is a discrete opfibration, there is a bicomma object

$$\begin{array}{ccc} P(\mathbb{B}) & \longrightarrow & \mathscr{S} \\ {\scriptstyle u}\big\downarrow & \Rightarrow & \big\downarrow{\scriptstyle p} \\ P(\mathbb{C}) & \underset{\delta}{\rightarrowtail} & M(P(\mathbb{C})) \end{array}$$

for an essentially unique point p satisfying the formula $p^* \delta_! h \cong K$, where $h : \mathbb{C} \longrightarrow P(\mathbb{C})$ denotes Yoneda. Likewise, since V is a discrete opfibration, there is a bicomma object

$$\begin{array}{ccc} P(\mathbb{A}) & \longrightarrow & \mathscr{S} \\ {\scriptstyle v}\downarrow & \Rightarrow & \downarrow{\scriptstyle q} \\ P(\mathbb{D}) & \underset{\delta}{\rightarrowtail} & M(P(\mathbb{D})) \end{array} \qquad (5.3)$$

for a unique point q such that $q^*\delta_! h \cong KF$.

Since f is essential, $M(f)$ has a right adjoint r_f (Prop. 4.3.16), so $r_f^* \dashv M(f)^*$. Now observe that the formula $q \cong r_f \cdot p$ holds. Indeed, we have $r_f^* \cong M(f)_!$, so that

$$(r_f p)^*\delta_! h \cong p^* r_f^* \delta_! h \cong p^* M(f)_! \delta_! h \cong p^* \delta_! f_! h \cong p^* \delta_! hF \cong KF \ .$$

But q is the unique point of $M(P(\mathbb{D}))$ that satisfies this condition.

Observe next that the composite diagram

$$\begin{array}{ccccc} P(\mathbb{A}) & \xrightarrow{g} & P(\mathbb{B}) & \longrightarrow & \mathscr{S} \\ {\scriptstyle v}\downarrow & & \downarrow{\scriptstyle u} & \Rightarrow & \downarrow{\scriptstyle p} \\ P(\mathbb{D}) & \underset{f}{\longrightarrow} & P(\mathbb{C}) & \underset{\delta}{\rightarrowtail} & \mathrm{M}(P(\mathbb{C})) \end{array} \qquad (5.4)$$

is a bicomma object. Indeed, for any diagram

$$\begin{array}{ccc} \mathscr{X} & \xrightarrow{x} & \mathscr{S} \\ {\scriptstyle \gamma}\downarrow & & \\ P(\mathbb{D}) & & \end{array}$$

of geometric morphisms we have natural bijections of 2-cells

$$\frac{\dfrac{\dfrac{\delta f \gamma \Rightarrow px}{\mathrm{M}(f)\delta\gamma \Rightarrow px}}{\delta\gamma \Rightarrow r_f px}}{\delta\gamma \Rightarrow qx \ ,}$$

where the last is in bijection with factorizations of γ through v, since (5.3) is a bicomma object. This shows that (5.4) is a bicomma object. But the right-hand square in (5.4) is also a bicomma object, so the square must therefore be a bipullback. □

The discrete opfibration corresponding to the inclusion of a small full subcategory

$$\mathbb{S}_\kappa \rightarrowtail \mathscr{S}$$

is $1/\mathbb{S}_\kappa \longrightarrow \mathbb{S}_\kappa$. Here κ denotes a cardinal number such that every object in $\mathbb{S}_\kappa$ has cardinality less than κ. This small category is assumed here to have been constructively defined in $\mathscr{S}$.

We shall call the discrete M-fibration $P(1/\mathbb{S}_\kappa) \longrightarrow P(\mathbb{S}_\kappa)$ *the generic κ-entire geometric morphism.*

Remark 5.3.2 *We have the following comments concerning generic entire geometric morphisms.*

1. *When the base topos* $\mathcal{S}$ *is Set, it is well-known that the generic* $\aleph_0$*-entire geometric morphism is*

$$P(1/Set_f) \longrightarrow P(Set_f) \; ,$$

where Set_f *denotes the (essentially small) full subcategory of Set of finite sets, as it classifies the entire geometric morphisms in* $\mathbf{Top}_{\mathcal{S}}$. *Explicitly, for any given Boolean algebra* B *in a Grothendieck topos* $\mathcal{E}$, *the associated geometric morphism*

$$\mathcal{E} \longrightarrow P(Set_f)$$

is induced by the flat functor $Set_f \longrightarrow \mathcal{E}$ *which sends* n *to* $\mathrm{Part}_n(B)$, *the object of* n*-partitions of the top element in* B.

2. *For any* κ *and arbitrary base topos* $\mathcal{S}$, *the generic* κ*-entire geometric morphism is a complete spread (with locally connected domain).*
3. *The pullback of*

$$P(1/\mathbb{S}_\kappa) \longrightarrow P(\mathbb{S}_\kappa)$$

along a geometric morphism

$$\mathcal{E} \longrightarrow P(\mathbb{S}_\kappa)$$

need not be a complete spread (with locally connected domain), not even if $\mathcal{S}$ *is Set. For example, if* X *is any Stone locale* ($\mathcal{O}(X)$ = *frame of ideals of a Boolean algebra), then there is a topos pullback*

$$\begin{array}{ccc} Sh(X) & \longrightarrow & P(1/Set_f) \\ \downarrow & & \downarrow \\ Set & \longrightarrow & P(Set_f) \end{array}$$

where again Set_f *denotes finite sets. But a complete spread* $\mathcal{Y} \longrightarrow$ *Set is constant, i.e., of the form Set*$/I \longrightarrow$ *Set.*

4. *On the other hand, the pullback of the generic* κ*-entire geometric morphism along an essential geometric morphism*

$$\mathcal{E} \longrightarrow P(\mathbb{S}_\kappa)$$

must be a complete spread (with locally connected domain). Moreover, any complete spread appears in such a pullback, as the proposition below shows.

Proposition 5.3.3 *Any complete spread* $\mathcal{Y} \xrightarrow{\psi} \mathcal{E}$ *is a topos bipullback of a generic* κ*-entire geometric morphism in the sense that it appears as a pullback*

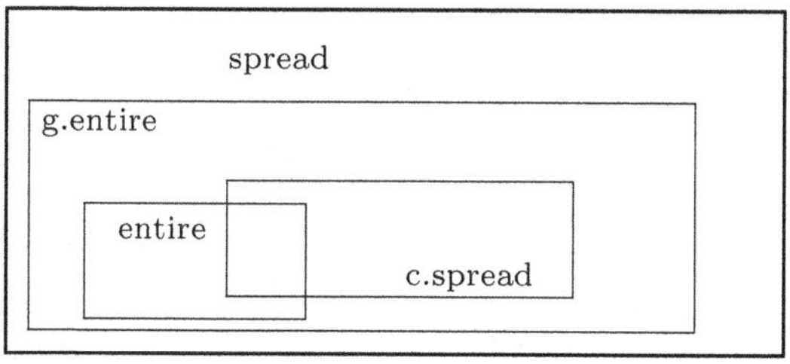

Fig. 5.1. A Venn diagram

$$\begin{array}{ccc} \mathscr{Y} & \longrightarrow & P(1/\mathbb{S}_\kappa) \\ {\scriptstyle \psi}\downarrow & & \downarrow \\ \mathscr{E} & \longrightarrow & P(\mathbb{S}_\kappa) \end{array}$$

for some $\mathbb{S}_\kappa$.

Proof. Let $\mathscr{Y} \xrightarrow{\psi} \mathscr{E}$ be a complete spread with distribution μ. Factor a restriction of μ to a site $\mathbb{C}$ through a full small subcategory $\mathbb{S}_\kappa$.

$$\begin{array}{ccc} \mathbb{C} & \longrightarrow & \mathbb{S}_\kappa \\ \downarrow & & \downarrow \\ \mathscr{E} & \xrightarrow[\mu]{} & \mathscr{S} \end{array}$$

Then the discrete opfibration $\mathbb{Y} \longrightarrow \mathbb{C}$ of μ appears in the following pullback of small categories and functors.

$$\begin{array}{ccc} \mathbb{Y} & \longrightarrow & 1/\mathbb{S}_\kappa \\ \downarrow & & \downarrow \\ \mathbb{C} & \longrightarrow & \mathbb{S}_\kappa \end{array}$$

By Proposition 5.3.1, the right square below is a topos pullback.

$$\begin{array}{ccccc} \mathscr{Y} & \rightarrowtail & P(\mathbb{Y}) & \longrightarrow & P(1/\mathbb{S}_\kappa) \\ {\scriptstyle \psi}\downarrow & & \downarrow & & \downarrow \\ \mathscr{E} & \rightarrowtail & P(\mathbb{C}) & \longrightarrow & P(\mathbb{S}_\kappa) \end{array}$$

By definition of complete spreads, the left square is a topos pullback, so the composite square is a pullback. □

The reader may find Figure 5.1 attractive: in this figure, by *generalized entire (g.entire, for short)* we mean the pullback in $\mathbf{Top}_{\mathscr{S}}$ of the generic κ-entire $P(1/\mathbb{S}_\kappa) \longrightarrow P(\mathbb{S}_\kappa)$, for some κ.

Exercises 5.3.4

1. *Prove that discrete* M*-opfibrations are stable under arbitrary bipullback.*
2. *Prove that the presheaf 2-functor* P *preserves bipullback squares in which one leg is a discrete fibration.*
3. *Explicitly describe the various elements of the bicomma object*

$$\begin{array}{ccc} P(\mathbb{B}) & \longrightarrow & \mathscr{S} \\ {\scriptstyle u}\downarrow & \Rightarrow & \downarrow{\scriptstyle p} \\ P(\mathbb{C}) & \underset{\delta}{\rightarrowtail} & \mathrm{M}(P(\mathbb{C})) \end{array}$$

associated with a discrete opfibration U, where $u = P(U)$,

$$\mathrm{M}(P(\mathbb{C})) \simeq P(\mathbb{C}^{\star})$$

and so on. What is the 2-cell? Thus, P *carries a discrete opfibration to a discrete* M*-fibration, a reflection of the fact that* P *reverses the direction of 2-cells.*

4. *(P. Johnstone) Discrete Conduché fibrations (unique factorization lifting) include both discrete fibrations and discrete opfibrations. Does* P *preserve bipullback squares in which two opposite sides are discrete Conduché fibrations?*
5. *Show that $1/Set$ is equivalent to the category of sets and partial maps.*
6. *Figure 5.1 implicitly conjectures that any generalized entire geometric morphism is a spread. Prove or refute this conjecture.*
7. *Prove the assertion made in 1 of Remark 5.3.2 and analyse its constructive status. In other words, examine the nature of the generic entire (or generic $\aleph_0$-entire) in the relative case where the base topos $\mathscr{S}$ is arbitrary.*
8. *Let N denote a monoid, and $\mathrm{P}(N)$ denote right N-sets. Let N also denote the single representable right N-set. Show that the free right N-set functor is given by $\Sigma_N \cdot \gamma^* \cong g^* \cdot \delta_!$, where $\Sigma_N \dashv N^*$, and where g denotes the geometric morphism corresponding to the object N of $\mathrm{P}(N)$ in the following bicomma object.*

$$\begin{array}{ccc} \mathrm{P}(N)/N & \longrightarrow & \mathrm{P}(N) \\ {\scriptstyle \gamma}\downarrow & \Rightarrow & \downarrow{\scriptstyle g} \\ Set & \underset{\delta}{\longrightarrow} & \mathrm{M}(Set) \end{array}$$

The symmetric topos M*(Set) coincides with the object classifier $\mathscr{R}$.*

Further reading: Bunge [Bun95, Bun74], Bunge and Funk [BF99], Johnstone [Joh99], Kock [Koc75], Street [Str, Str74], Zoeberlein [Zoe76].

6

Closed and Linear KZ-Monads

In this chapter we introduce additional axioms one may impose on a completion Kock-Zöberlein monad in order to further develop the theory of complete spreads in a more general context.

If a KZ-monad M is *closed* as we shall call it, we prove that any 1-cell whose domain admits an M-adjoint can be factored in an essentially unique way into a final 1-cell followed by a discrete M-fibration whose domain admits an M-adjoint. The final 1-cells for the symmetric monad M in $\mathbf{Top}_{\mathscr{S}}$ are precisely the pure geometric morphisms (relative to $\mathscr{S}$), and the discrete fibrations are precisely complete spread geometric morphisms with locally connected domain (§ 5.2).

In the context of a closed completion KZ-monad on $\mathscr{K}$ we discuss the existence of a *density functor* in connection with the validity of a *Gleason core* axiom. We consider a 'single universe' containing both the discrete M-fibrations and the discrete M-opfibrations, for any closed completion KZ-monad M in a 2-category $\mathscr{K}$. It has the required properties when the Gleason core axiom holds.

We also investigate what we call *additive and $\mathscr{K}$-equivariant KZ-monads*, and the nature of their M-algebras and M-homomorphisms. $\mathscr{K}$-equivariant KZ-monads may be used to establish Pitts' theorem on bicomma objects (§ 6.5).

6.1 Closed KZ-Monads and Comprehension

We shall prove two results about what we shall call a closed KZ-monad. The first result says that every closed, completion KZ-monad has an associated comprehensive factorization. The second shows that discrete fibrations compose for such a KZ-monad. Throughout, M is a KZ-monad in a 2-category with terminal object T.

Definition 6.1.1 *Let* $G \xrightarrow{p} D$ *be a 1-cell, and assume that both* $G \xrightarrow{g} T$ *and* $D \xrightarrow{d} T$ *admit an* M*-adjoint. We shall say that* p *is* a final 1-cell for M *if the canonical 2-cell* $\mathrm{M}(p) \cdot r_g \Rightarrow r_d$ *is an isomorphism.*

A 1-cell p is final iff $\mathrm{M}(p) \cdot r_g \cdot \delta_T \Rightarrow r_d \cdot \delta_T$ is an isomorphism because both 1-cells are homomorphisms.

Lemma 6.1.2 *Let* M *be a locally full and faithful KZ-monad. Let* $G \xrightarrow{p} D$ *be a 1-cell, and assume that* $G \xrightarrow{g} T$ *and* $D \xrightarrow{d} T$ *admit an* M*-adjoint. Then* p *is final iff for every 1-cell* $D \xrightarrow{b} X$ *and every 'constant'* $T \xrightarrow{a} X$*, composition with* p *gives a bijection*

$$\frac{b \Rightarrow a \cdot d}{b \cdot p \Rightarrow a \cdot d \cdot p}$$

of 2-cells, where $D \xrightarrow{d} T$ *is the unique 1-cell.*

Proof. If p is final, then because M is locally full and faithful we have the following natural bijections.

$$\begin{array}{c} b \Rightarrow a \cdot d \\ \hline \mathrm{M}b \Rightarrow \mathrm{M}a \cdot \mathrm{M}d \\ \hline \mathrm{M}b \cdot r_d \Rightarrow \mathrm{M}a \\ \hline \mathrm{M}b \cdot \mathrm{M}p \cdot r_g \Rightarrow \mathrm{M}a \\ \hline \mathrm{M}b \cdot \mathrm{M}p \Rightarrow \mathrm{M}a \cdot \mathrm{M}g \\ \hline b \cdot p \Rightarrow a \cdot g \cong a \cdot d \cdot p \end{array}$$

If the stated condition holds, then there are the following natural bijections. Here, h and k denote arbitrary M-homomorphisms with the appropriate domain and codomain.

$$\begin{array}{c} h \cdot \mathrm{M}p \cdot r_g \Rightarrow k \\ \hline h \cdot \mathrm{M}p \Rightarrow k \cdot \mathrm{M}g \cong k \cdot \mathrm{M}d \cdot \mathrm{M}p \\ \hline h \cdot \mathrm{M}p \cdot \delta_G \Rightarrow k \cdot \mathrm{M}d \cdot \mathrm{M}p \cdot \delta_G \\ \hline h \cdot \delta_D \cdot p \Rightarrow k \cdot \delta_T \cdot d \cdot p \\ \hline h \cdot \delta_D \Rightarrow k \cdot \delta_T \cdot d \\ \hline h \cdot \delta_D \Rightarrow k \cdot \mathrm{M}d \cdot \delta_D \\ \hline h \Rightarrow k \cdot \mathrm{M}d \end{array}$$

This shows that $\mathrm{M}(d) \dashv \mathrm{M}(p) \cdot r_g$ as M-homomorphisms, and hence as ordinary 1-cells. I.e., this shows that $\mathrm{M}(p) \cdot r_g \cong r_d$. Note: r_g is an M-homomorphism (4.3.4). □

Proposition 6.1.3 *A geometric morphism between locally connected toposes is final for the symmetric monad iff it is pure.*

Proof. A geometric morphism $\mathscr{F} \xrightarrow{\psi} \mathscr{E}$ is final for the symmetric monad iff

$$\mathrm{M}(\psi) \cdot r_f \cdot \delta_{\mathscr{S}} \Rightarrow r_e \cdot \delta_{\mathscr{S}}$$

is an isomorphism. This holds iff $f_! \cdot \psi^* \cong e_!$ iff ψ is pure. □

Proposition 6.1.4 *A geometric morphism* $\mathscr{F} \xrightarrow{\psi} \mathscr{E}$ *satisfies the condition in Lemma 6.1.2 iff* ψ_* *preserves* $\mathscr{S}$*-coproducts.*

Proof. Take for X in Lemma 6.1.2 the object classifier $\mathrm{M}(\mathscr{S})$. It follows that for every I in $\mathscr{S}$, the unit $e^*I \longrightarrow \psi_*\psi^*(e^*I)$ is an isomorphism. This is precisely the property that ψ_* preserves $\mathscr{S}$-coproducts. □

A 1-cell f in a 2-category is said to reflect isomorphisms if for every composable 2-cell t, ft invertible implies t invertible. Consider the following condition on a KZ-monad.

Definition 6.1.5 *We shall say that a KZ-monad* M *is* closed *if for every discrete* M*-fibration* ψ *the 1-cell* $\mathrm{M}(\psi)$ *reflects isomorphisms.*

As always, T denotes the terminal object in the 2-category.

Theorem 6.1.6 (Comprehensive factorization for KZ-monads)
Suppose that M *is a closed completion KZ-monad. Then every 1-cell whose domain admits an* M*-adjoint has an essentially unique factorization as a final 1-cell followed by a discrete* M*-fibration whose domain admits an* M*-adjoint.*

Proof. Let $A \xrightarrow{\varphi} B$ be an arbitrary 1-cell such that the essentially unique 1-cell $A \xrightarrow{a} T$ admits an M-adjoint. Consider the universal 1-cell $A \xrightarrow{p} D$, where D denotes

$$\delta_B \Downarrow \Sigma_a(\delta_B \cdot \varphi)$$

as in the following diagram.

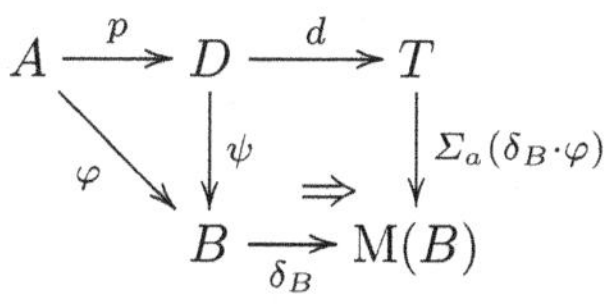

The 1-cell ψ is a discrete M-fibration, witnessed by $\Sigma_a(\delta_B \cdot \varphi)$. In order to show that p is final, let $i : \mathrm{M}(p) \cdot r_a \Rightarrow r_d$ denote the canonical 2-cell. We must show that i is an isomorphism. We know that

$$\Sigma_a(\delta_B \cdot \varphi) \cong \Sigma_d(\delta_B \cdot \psi) ,$$

and hence that

$$\mathrm{M}(\varphi) \cdot r_a \cdot \delta_T \cong \mathrm{M}(\psi) \cdot r_d \cdot \delta_T .$$

Therefore, the 2-cell

$$\mathrm{M}(\psi)\, i\, \delta_T : \mathrm{M}(\psi) \cdot \mathrm{M}(p) \cdot r_a \cdot \delta_T \Rightarrow \mathrm{M}(\psi) \cdot r_d \cdot \delta_T$$

is an isomorphism. Since $\mathrm{M}(T)$ is free, $\mathrm{M}(\psi)i$ is an isomorphism. Since M is closed, i is an isomorphism. This establishes the existence of the factorization.

Now suppose we have two final, discrete fibration factorizations of φ.

$$\begin{array}{ccc} A & \xrightarrow{p} & D \\ & \searrow^{\varphi} & \downarrow{\scriptstyle \psi} \\ & & B \end{array} \qquad\qquad \begin{array}{ccc} A & \xrightarrow{p'} & D' \\ & \searrow^{\varphi} & \downarrow{\scriptstyle \psi'} \\ & & B \end{array}$$

Let q and q' denote the 1-cells $T \longrightarrow \mathrm{M}(B)$ corresponding to ψ and ψ', respectively. Since p and p' are final, by reversing the steps of the previous paragraph we have

$$q \cong \Sigma_a(\delta_B \cdot \varphi) \cong q' \,.$$

Therefore, there is an equivalence of discrete fibrations over B as follows.

$$\begin{array}{ccc} D' & \xrightarrow{\kappa} & D \\ & {\scriptstyle \psi}\searrow \quad \swarrow{\scriptstyle \psi'} & \\ & B & \end{array}$$

We obtain $p' \cong \kappa \cdot p$ from the uniqueness of the universal property of the bicomma object defining ψ'. □

Remark 6.1.7 *The comprehensive factorization associated with a closed completion KZ-monad is 2-dimensional. Indeed, the construction in 6.1.6 shows that a 2-cell between 1-cells whose domains admit an* M*-adjoint has a unique decomposition of the following kind.*

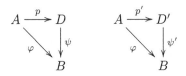

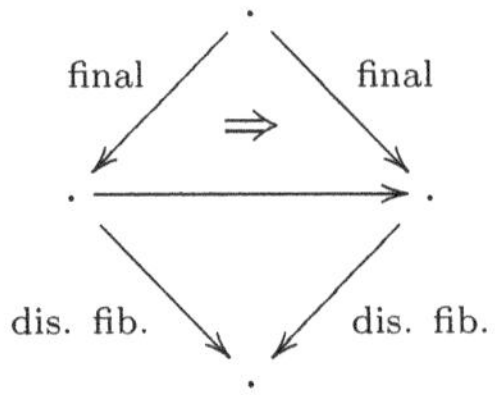

Theorem 6.1.8 *Let* M *be a locally full and faithful, closed, completion KZ-monad. Then the composite of two discrete* M*-fibrations is a discrete fibration. If a composite and its second factor are discrete* M*-fibrations, then its first factor is also.*

Proof. Throughout, we drop the prefix M from M-fibration. Let $G \xrightarrow{\psi} D$ and $D \xrightarrow{\varphi} E$ be two discrete fibrations. First, we write the bicomma object for φ as the following composite diagram.

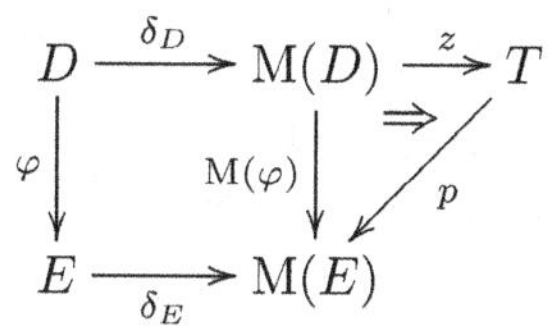

This can be done because there is a unique 2-cell $\mathrm{M}(\varphi) \Rightarrow p \cdot z$ corresponding to the 2-cell

$$\mathrm{M}(\varphi) \cdot \delta_D \cong \delta_E \cdot \varphi \Rightarrow p \cdot d \cong p \cdot z \cdot \delta_D \ .$$

This correspondence is by 4.3.7 applied to the M-homomorphism $\mathrm{M}(\varphi)$. As in Exercise 2.5.9, 1, if the composite $\varphi \cdot \psi$ is a discrete fibration, then its corresponding point $T \longrightarrow \mathrm{M}(E)$ must be $\mathrm{M}(\varphi) \cdot q$, where $T \xrightarrow{q} \mathrm{M}(D)$ is the point of ψ. Thus we should consider the bicomma object

$$K = \delta_E \Downarrow \mathrm{M}(\varphi) \cdot q \ ,$$

and the intervening final 1-cell $G \xrightarrow{h} K$.

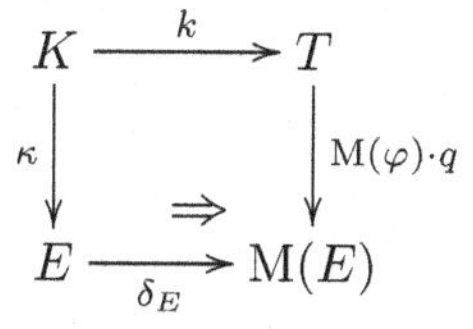

This is the comprehensive factorization of $\varphi \cdot \psi$. Our assumption that M is closed allows us to conclude that h is final. Now regard the bicomma object for φ. Since there is a 2-cell

$$\delta_E \cdot \kappa \Rightarrow \mathrm{M}\varphi \cdot q \cdot k \Rightarrow p \cdot z \cdot q \cdot k \cong p \cdot k \ ,$$

we can factor κ through φ by a 1-cell $K \xrightarrow{\gamma} D$. By the universal property of this bicomma object we also have $\gamma \cdot h \cong \psi$. Using this isomorphism we obtain a 2-cell

$$\delta_D \cdot \gamma \cdot h \cong \delta_D \cdot \psi \Rightarrow q \cdot g \cong q \cdot k \cdot h \ ,$$

which by Lemma 6.1.2 corresponds to a 2-cell

$$\delta_D \cdot \gamma \Rightarrow q \cdot k \ .$$

The bicomma object for ψ now gives a 1-cell $K \xrightarrow{j} G$ such that $\psi \cdot j \cong \gamma$, and it is a routine matter to show that h is an equivalence with pseudo-inverse j. This shows that $\varphi \cdot \psi$ is a discrete fibration, witnessed by $\mathrm{M}(\varphi) \cdot q$.

The second assertion of the theorem is a formal consequence of the first and the comprehensive factorization. □

Proposition 6.1.9 *The symmetric monad in* $\boldsymbol{Top}_{\mathscr{S}}$ *is closed.*

Proof. Let $\mathscr{F} \xrightarrow{\psi} \mathscr{E}$ be a spread with locally connected domain. We shall show that $\mathrm{M}(\psi)$ reflects isomorphisms. Let

$$t : p \Rightarrow q : \mathscr{X} \longrightarrow \mathrm{M}(\mathscr{F})$$

be a 2-cell such that $\mathrm{M}(\psi) \cdot t$ is an isomorphism. Equivalently, t is a natural transformation between ($\mathscr{X}$-valued) distributions

$$t : p^* \cdot \delta_! \Rightarrow q^* \cdot \delta_! \ ,$$

such that $t \cdot \psi^*$ is a natural isomorphism. Let $\{X_i\}$ be a generating family for $\mathscr{E}$ over $\mathscr{S}$. Then every component morphism $t_{\psi^* X_i}$ is an isomorphism. Since distributions preserve coproducts, it follows that for any component α of any $\psi^* X_i$, t_α is an isomorphism. We know from § 3.3 that the α generate $\mathscr{F}$ over $\mathscr{S}$, so that t must be an isomorphism. □

Remark 6.1.10 *Thus, the symmetric monad is closed (Proposition 6.1.9). An example of a different nature of a closed completion KZ-monad is the identity monad* Id *in a 2-category with bicomma objects. Consider the* lift monad $\mathbf{L}$ *associated with a domain structure* $(\mathbb{C}, \mathbb{D})$ *in the sense of M. Fiore. (The functor* $\mathbf{L}$ *is the right adjoint of the inclusion of the category* $\mathbb{C}$ *of "total maps" into the category* $\mathrm{p}(\mathbb{C}, \mathbb{D})$ *of "partial maps.") Algebras for* $\mathbf{L}$ *are "pointed objects," and their homomorphisms are "strict maps." Fiore's axiom states that in* $\mathbb{C}$ *every morphism with pointed domain factors as a strict map followed by an upper-closed monomorphism with pointed domain. We claim that this axiom can in fact be derived in the context of KZ-monads in view of the following remarks. An object of* $\mathbb{C}$ *has an* Id*-adjoint in* $\mathbb{C}^{\mathrm{op}}$ *iff its unique map to the terminal has a left adjoint iff it has a bottom, thus, it is pointed iff it is an algebra for the associated lift monad. The discrete* Id*-fibrations in* $\mathbb{C}^{\mathrm{op}}$ *are the principal upper-closed monomorphisms. A map between pointed objects is final for* Id *iff it preserves the bottom, in other words, iff it is strict. We can now apply the comprehension factorization in the context of closed KZ-monads to prove our claim.*

6.2 The Gleason Core and Density Axioms

The existence of a locally connected coreflection (or to use Lawvere's term 'Gleason core') was proved by Gleason (1963) for spaces, and by Funk (1999) for Grothendieck toposes. Although we know of no geometric morphism $\mathscr{E} \longrightarrow \mathscr{S}$ for which the Gleason core $\hat{\mathscr{E}} \longrightarrow \mathscr{E}$ does not exist, we do not know either whether the Gleason core always exists. Therefore, the assumption that every bounded topos over an arbitrary base topos $\mathscr{S}$ have a Gleason core may restrict the generality of our results. On the other hand, we do wish to retain the generality of $\mathscr{S}$ for other reasons. For instance, in Chapter 9 we consider $\mathscr{E}$-valued distributions over $\mathscr{E}$ in order to introduce a notion of index of a complete spread.

Throughout, M denotes a completion KZ-monad in a 2-category $\mathscr{K}$.

Definition 6.2.1 *An object X of $\mathscr{K}$ is said to admit a* Gleason core *relative to* M *if there exists a 1-cell $\hat{X} \xrightarrow{\varepsilon_X} X$ in $\mathscr{K}$, such that $\hat{X} \longrightarrow T$ has an* M*-adjoint, universal with this property. In other words, any 1-cell $Y \longrightarrow X$ in $\mathscr{K}$, such that $Y \longrightarrow T$ admits an* M*-adjoint, factors uniquely (up to 2-isomorphism) through $\hat{X} \xrightarrow{\varepsilon_X} X$.*

Proposition 6.2.2 *For an object X of $\mathscr{K}$, the Gleason core axiom relative to* M *holds for X iff the category $\mathscr{K}(T, \mathrm{M}(X))$ has a terminal object, denoted $\top$. The Gleason core is always a discrete* M*-fibration. Explicitly, the Gleason core of such an object X may be constructed by means of the bicomma object*

$$\begin{array}{ccc} \hat{X} & \xrightarrow{\hat{x}} & T \\ {\scriptstyle \varepsilon_X}\downarrow & \Rightarrow & \downarrow{\scriptstyle \top} \\ X & \xrightarrow[\delta_X]{} & \mathrm{M}(X) \end{array}$$

in $\mathscr{K}$.

Proof. The category $\mathscr{K}(T, \mathrm{M}(X)) = \mathscr{K}_M(T, X)$ is equivalent to the category of discrete M-fibrations over X, where the equivalence is given by the above bicomma object construction (Cor. 5.1.5). If $\mathscr{K}(T, \mathrm{M}(X))$ has a terminal, then $\hat{x}$ in the bicomma object admits an M-adjoint and ε_X is the Gleason core. On the other hand, if X has a Gleason core, then it is necessarily the bicomma of the left extension of $\delta_X \cdot \varepsilon_X$ along $\hat{x}$, denoted $\top$. In other words, the Gleason core must be a discrete M-fibration, and $\top$ is indeed the terminal object of $\mathscr{K}(T, \mathrm{M}(X))$. □

Proposition 6.2.3 *Let* M *be a locally full and faithful, closed, completion KZ-monad on a 2-category $\mathscr{K}$. Then for each object X of $\mathscr{K}$ for which the Gleason core axiom holds, composition with the discrete* M*-fibration $\hat{X} \xrightarrow{\varepsilon_X} X$ induces an equivalence between the categories of discrete* M*-fibrations over X and over $\hat{X}$.*

Proof. This result is a consequence of Theorem 6.1.8, the universal property of the bicomma objects defining discrete M-fibrations over X, and of the equivalence between $\mathscr{K}(T, \mathrm{M}(X))$ and discrete M-fibrations over X. □

Remark 6.2.4

1. *By Proposition 6.2.2 for the symmetric monad* M *in* ***Top***$_{\mathscr{S}}$*, the Gleason core axiom for an object $\mathscr{E}$ of* ***Top***$_{\mathscr{S}}$ *is equivalent to the existence of a terminal distribution on $\mathscr{E}$. We have $\mathrm{M}(\mathscr{E}) \simeq \mathrm{M}(\hat{\mathscr{E}})$.*

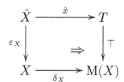

2. *We have* $\mathbf{E}(\mathscr{E}) \simeq \mathbf{E}(\hat{\mathscr{E}})$, *when* $\mathscr{E}$ *has a Gleason core. Thus, distributions on such an* $\mathscr{E}$ *are supported on its Gleason core.*
3. *There exist non-trivial Grothendieck toposes whose Gleason cores are trivial. For such a Grothendieck topos, no non-zero distributions exist - in particular, the terminal distribution agrees with the zero distribution.*

Proposition 6.2.5 *Assume that* M *is a completion KZ-monad. Then for any object* B, *the identity* 1_B *is a discrete* M*-fibration iff* $B \longrightarrow T$ *admits an* M*-adjoint. In this case, the 1-cell* $T \longrightarrow \mathrm{M}(B)$ *that corresponds to* 1_B *is* $r_b \cdot \delta_T$. *This 1-cell is the terminal object in* $\mathscr{K}(T, \mathrm{M}(B))$.

Proof. If 1_B is a discrete fibration, then by definition $B \longrightarrow T$ admits an M-adjoint. If $B \longrightarrow T$ admits an M-adjoint, then

$$\begin{array}{ccc} B & \xrightarrow{\ b\ } & T \\ {\scriptstyle 1}\downarrow & \Rightarrow & \downarrow{\scriptstyle r_b \cdot \delta_T} \\ B & \xrightarrow[\delta_B]{} & \mathrm{M}(B) \end{array}$$

is a bicomma object. □

Assume that B has the property described in Proposition 6.2.5. Let us denote the left extension $\Sigma_b(\delta_B) = r_b \cdot \delta_T$ by $\top$. If $X \xrightarrow{f} B$ is any discrete opfibration, corresponding to $\Sigma_f(\delta_T \cdot x)$, then taking $p = \top$ in (5.1) gives

$$f.\top = \mathrm{M}(f) \cdot r_f \cdot \top \cong \mathrm{M}(f) \cdot r_f \cdot r_b \cdot \delta_T \cong \mathrm{M}(f) \cdot r_x \cdot \delta_T = \Sigma_x(\delta_B \cdot f)\,.$$

Thus, this special case of the action provides a functor

$$\Phi_B : \mathscr{K}_{\mathrm{M}}(B, T) \longrightarrow \mathscr{K}_{\mathrm{M}}(T, B)\ ;\ f \mapsto f.\top \qquad (6.1)$$

given by 'inversion':

$$\Sigma_f(\delta_T \cdot x) \mapsto \Sigma_x(\delta_B \cdot f)\,.$$

Example 6.2.6 *For any topos* $\mathscr{E}$, $\mathrm{id}_{\mathscr{E}}$ *is a discrete fibration for the symmetric monad iff* $\mathscr{E}$ *is a locally connected topos. The inversion formula is another way of describing the Lawvere action* $F \mapsto F.\pi_0$, *such that* $F.\pi_0(X) = \pi_0(F \times X)$. *Distributions of the form* $F.\pi_0$ *are what Lawvere has termed* absolutely continuous (relative to π_0).

If B has a Gleason core, corresponding to a terminal 1-cell $\top : T \longrightarrow \mathrm{M}(B)$, then the functor $\Phi_B(f) = f.\top$ (6.1) still makes sense.

Definition 6.2.7 *Suppose that* B *has a Gleason core. We say that* B admits a density *(for* M*) if the inversion functor* Φ_B *has a right adjoint. We call this right adjoint the* density functor *for* B, *denoted* $\mathbf{d}_B$. *This defines a monad in* $\mathscr{K}_{\mathrm{M}}(B, T)$ *that we call* the density monad associated with M.

Proposition 6.2.8 *A locally connected topos admits a density for the symmetric monad.*

Proof. Let $\langle \mathbb{C}, J \rangle$ denote a locally connected site, with sheaves $Sh(\mathbb{C}, J)$. We have $\pi_0 \dashv \Delta$. Define

$$\mathbf{d}(\mu)(c) = \{\text{nat. trans. } h_c.\pi_0 \longrightarrow \mu\} .$$

Then $\mathbf{d}(\mu)$ is a sheaf, and $\mathbf{d}$ is right adjoint to inversion (Example 6.2.6). □

Remark 6.2.9

1. *Consider again the specific case of the symmetric monad in the 2-category* $\boldsymbol{Top}_{\mathscr{S}}$. *Then* $\boldsymbol{Top}_{\mathscr{S}}(\mathscr{E}, \mathrm{M}(\mathscr{S}))$ *is equivalent to the topos frame* $\mathscr{E}$, *and* $\boldsymbol{Top}_{\mathscr{S}}(\mathscr{S}, \mathrm{M}(\mathscr{E}))$ *is equivalent to the distribution category* $\mathbf{E}(\mathscr{E})$. *Thus, for a locally connected topos* $\mathscr{E}$, *the adjoint pair* $\Phi_{\mathscr{E}} \dashv \mathbf{d}_{\mathscr{E}} = \mathbf{d}$ *connects* $\mathscr{E}$ *with* $\mathbf{E}(\mathscr{E})$, *where*
$$\Phi_{\mathscr{E}} : \mathscr{E} \longrightarrow \mathbf{E}(\mathscr{E}) ,$$
associates with an object X *of* $\mathscr{E}$ *the distribution* $X.e_!$, *such that*
$$(X.e_!)(Y) = e_!(X \times Y) .$$
2. *The density of a distribution on a locally connected topos* $\mathscr{E}$ *coincides with the object of* $\mathscr{E}$*-points of the corresponding (localic) complete spread. Thus, we may generalize the density to any topos:*
$$\mathbf{d}(\mu) = \text{object of } \mathscr{E}\text{-points of the complete spread of } \mu .$$
If $\mathscr{E}$ *has a Gleason core, so that* $\mathbf{E}(\mathscr{E})$ *has a terminal object* $\mathbf{1}$, *then the above* $\mathbf{d}$ *is given by*
$$\mathbf{d}(\mu) = \mathrm{Hom}(\mathbf{1}, \mu) ,$$
and moreover $\mathbf{d}$ *has a left adjoint in this case:*
$$(\_).\mathbf{1} : \mathscr{E} \longrightarrow \mathbf{E}(\mathscr{E}) .$$
3. *Intuitively, the density monad is in some ways similar to the* regularization *monad in the lattice of open sets of a topological space, which associates with an open set the interior of its closure. But consider the example of the real numbers* R. *Regularization in* $\mathscr{O}(R)$ *is non-trivial, but the density monad in* $\mathrm{Sh}(R)$ *is the identity monad.*

Exercises 6.2.10

1. *Prove Proposition 5.1.3.*

2. *Suppose that $\mathscr{K}$ has bipullbacks, and that* M *is a locally full and faithful completion KZ-monad (as always). Show that the action $f.\psi$ of discrete opfibrations on discrete fibrations can be equivalently described:*

$$f.\psi = \Sigma_k(\delta_B \cdot f \cdot \pi_1) \ ,$$

where

$$\begin{array}{ccc} K & \xrightarrow{\pi_2} & Y \\ {\scriptstyle \pi_1}\downarrow & & \downarrow{\scriptstyle \psi} \\ X & \xrightarrow[f]{} & B \end{array}$$

is a bipullback.

3. *Investigate what are the algebras for the density monad in a locally connected topos.*
4. *Fill in the details of the proof of Proposition 6.2.3.*

6.3 The Twisted Single Universe

We define a category, *the twist category*, whose objects are twisted maps between M-bifibrations.

Definition 6.3.1 *For objects A and B in $\mathscr{K}$, denote by* $\mathrm{Tw}(A,B)$ *the following category. An* object *is a 3-tuple $(yE\psi, t, xD\varphi)$ where the spans $xD\varphi : A \longrightarrow B$ and $yE\psi : B \longrightarrow A$ are* M*-bifibrations, and $E \xrightarrow{t} D$ is a 1-cell for which the following diagram commutes (up to 2-isomorphism).*

$$\begin{array}{ccc} B & \xleftarrow{y} & E \\ {\scriptstyle \varphi}\uparrow & {\scriptstyle t}\swarrow & \downarrow{\scriptstyle \psi} \\ D & \xrightarrow[x]{} & \top \end{array}$$

We shall denote such an object more simply by (E, t, D). A morphism

$$(E, t, D) \longrightarrow (E', t', D')$$

is a pair (α, β) such that $\alpha : E \longrightarrow E'$ is a 1-cell of bifibrations $B \longrightarrow A$, $\beta : D \longrightarrow D'$ is a 1-cell of bifibrations $A \longrightarrow B$, such that there is a commutative diagram

$$\begin{array}{ccc} E & \xrightarrow{\alpha} & E' \\ {\scriptstyle t}\downarrow & & \downarrow{\scriptstyle t'} \\ D & \xrightarrow[\beta]{} & D' \end{array}$$

up to 2-isomorphism.

Under the assumption that $\mathscr{K}$ has a terminal object $\top$, we denote the category $\mathrm{Tw}(\top, B)$ *just by* $\mathrm{Tw}(B)$, *and call it* the twist category of B.

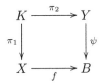

Assume now that M is a closed completion KZ-monad in a 2-category $\mathscr{K}$, and that B is an object of $\mathscr{K}$ that has a Gleason core relative to M. Using the existence of the density functor, we may now explicitate $\mathrm{Tw}(B)$ as a *single universe* for discrete M-fibrations and discrete M-opfibrations over B.

Theorem 6.3.2 *Let* M *be a closed completion KZ-monad in* $\mathscr{K}$ *and let* B *be any object such that* $B \xrightarrow{b} T$ *admits an* M*-adjoint. Assume also that* B *admits a density* $\mathbf{d}$*. Then* $\mathrm{Tw}(B)$ *is equivalent to the category* $\mathscr{K}_{\mathrm{M}}(B, \top) \downarrow \mathbf{d}$*, obtained by Artin glueing along* $\mathbf{d}$*. The glueing category* $\mathscr{K}_{\mathrm{M}}(B, \top) \downarrow \mathbf{d}$ *contains as full subcategories both* $\mathscr{K}_{\mathrm{M}}(B, \top)$ *and* $\mathscr{K}_{\mathrm{M}}(\top, B)$.

Proof. An object of $\mathrm{Tw}(B)$ is a 3-tuple (E, t, D) (Definition 6.3.1); such a t may be regarded as a morphism

$$t : \Sigma_{\psi}(\delta_B \cdot y) \longrightarrow \Sigma_x(\delta_B \cdot \varphi)\ .$$

Equivalently, for $q = \Sigma_y(\delta_\top \cdot \psi)$ and $r = \Sigma_x(\delta_B \cdot \varphi)$, we have $t : \Phi_B(q) \longrightarrow r$, which is given, suggestively, as

$$\Sigma_{x \cdot t}(\delta_B \cdot \varphi \cdot t) \longrightarrow \Sigma_x(\delta_B \cdot \varphi).$$

By the adjointness $\Phi_B \dashv \mathbf{d}_B$, this t corresponds uniquely to a morphism $\hat{t} : q \longrightarrow \mathbf{d}(r)$, i.e., to an object of the glueing category $\mathscr{K}_{\mathrm{M}}(B, \top) \downarrow \mathbf{d}$. This process is functorial and reversible, giving the desired equivalence. It follows from that the glueing category contains both full subcategories as claimed. □

Exercises 6.3.3

1. *Show that if* $X \xrightarrow{f} B$ *is a discrete opfibration, and* B *admits a Gleason core, then so does* X.
2. *Generalize Theorem 6.3.2, by assuming only that* B *has a Gleason core and a density.*
3. *The work of Carboni and Johnstone [CJ95] shows that the twist category* $\mathrm{Tw}(P(\mathbb{C}))$ *is again a presheaf topos, say* $\mathrm{P}(\mathbb{K})$*. Give an explicit description of* $\mathbb{K}$ *as a full subcategory of the category* $\mathrm{Tw}(P(\mathbb{C}))$*, using the definition of the latter. In particular, identify* $\mathbb{K}$ *with the* collage *of* $\mathbb{C}$.
4. *The category* ULF/$\mathbb{C}$ *of* unique lifting of factorizations, *also known as the category of discrete Giraud-Conduché fibrations over* $\mathbb{C}$*, is not a topos in general, but it is a topos for any* $\mathbb{C}$ *that is "paths linearizable" [BN00, BF00]. This too is a single universe for local homeomorphisms and complete spreads, as it can be generalized to toposes by the familiar amalgamation construction. Produce a comparison map between the two "single universes"* $\mathrm{Tw}(P(\mathbb{C}))$ *and* ULF/$\mathbb{C}$ *and study its properties. Notice that, unlike* ULF/$\mathbb{C}$*, the category* $\mathrm{Tw}(P(\mathbb{C}))$ *is a topos for any* $\mathbb{C}$.
5. *Show how the action of discrete fibrations on discrete opfibrations may be naturally described in* ULF/$\mathbb{C}$ *using the comprehensive factorization of functors.*

6.4 Linear KZ-Monads

Distributions have a basic additive property: two distributions

$$\mathscr{E} \xrightarrow{\mu} \mathscr{X} \ , \ \mathscr{F} \xrightarrow{\lambda} \mathscr{X}$$

may be paired into a single distribution $\langle \mu, \lambda \rangle$ on the coproduct $\mathscr{E} +_{\mathscr{S}} \mathscr{F}$, such that

$$\langle \mu, \lambda \rangle (E, F) = \mu(E) + \lambda(F) \ .$$

We remind the reader that the topos frame of $\mathscr{E} +_{\mathscr{S}} \mathscr{F}$ is the underlying category product $\mathscr{E} \times \mathscr{F}$ (of course not to be confused with the topos product $\mathscr{E} \times_{\mathscr{S}} \mathscr{F}$). The functor $\langle \mu, \lambda \rangle$ just defined is indeed a distribution because $\langle \mu, \lambda \rangle \dashv \langle \mu_*, \lambda_* \rangle$. It should be noted that a coproduct inclusion $\mathscr{E} \xrightarrow{\iota} \mathscr{E} +_{\mathscr{S}} \mathscr{F}$ is locally connected such that $\iota_!(E) = (E, 0)$. We recover μ by composing with $\iota_!$ (and similarly λ) thereby establishing an equivalence between such pairs of distributions and distributions on $\mathscr{E} +_{\mathscr{S}} \mathscr{F}$. Furthermore, a distribution on the coproduct is isomorphic to the pairing of its restriction to the summands, because the two distributions have the same right adjoint.

Since the symmetric monad classifies distributions we immediately conclude that there is a canonical equivalence

$$\langle r_{\mathscr{E}}, r_{\mathscr{F}} \rangle : \mathrm{M}(\mathscr{E} +_{\mathscr{S}} \mathscr{F}) \ \simeq \ \mathrm{M}(\mathscr{E}) \times_{\mathscr{S}} \mathrm{M}(\mathscr{F})$$

of toposes.

We are indicating here that the equivalence is given explicitly by pairing right adjoints $r_{\mathscr{E}}$ and $r_{\mathscr{F}}$, where $\mathrm{M}(\iota_{\mathscr{E}}) \dashv r_{\mathscr{E}}$ (geometric morphisms). Finally, we remark that since the coproduct inclusion ι is indeed an inclusion in the sense of geometric morphisms, we conclude that $r_{\mathscr{E}} \cdot \mathrm{M}(\iota) \cong id_{M(\mathscr{E})}$. We say that ι has *a coreflection* M*-adjoint.*

The additivity of the symmetric monad is reflected in the fact that its discrete fibrations, the complete spreads, may be summed. To be more precise, the '$\mathbf{Top}_{\mathscr{S}}$ is extensive' equivalence

$$\mathbf{Top}_{\mathscr{S}}/(\mathscr{E} +_{\mathscr{S}} \mathscr{F}) \ \simeq \ \mathbf{Top}_{\mathscr{S}}/\mathscr{E} \times \mathbf{Top}_{\mathscr{S}}/\mathscr{F}$$

restricts to complete spreads. This is intuitively plausible, but to prove it we must invoke the comprehensive factorization. This suggests and we prove that a similar result holds in the generic context (Theorem 6.4.6).

We may now posit what we mean by an additive KZ-monad in the generic context. It makes sense to study additivity in an extensive 2-category $\mathscr{K}$. Throughout, f^* denotes pullback along a 1-cell f in $\mathscr{K}$.

Definition 6.4.1 *A KZ-monad* M *in an extensive 2-category* $\mathscr{K}$ *is said to be* additive*, or to satisfy the* exponential law*, if*

1. *for any two objects* X *and* Y *of* $\mathscr{K}$*, the coproduct injections* $X \xrightarrow{\iota_X} X + Y$ *and* $Y \xrightarrow{\iota_Y} X + Y$ *have coreflection* M*-adjoints* r_X *and* r_Y*, and*

2. 'M$(X+Y) \simeq$ M$(X) \times$ M(Y)': *for any coproduct diagram in* $\mathscr{K}$, *below left,*

$$\begin{array}{ccc} & & X \\ & & \downarrow{\scriptstyle \iota_X} \\ Y & \xrightarrow[\iota_Y]{} & Z \end{array} \qquad \begin{array}{ccc} \mathrm{M}(Z) & \xrightarrow{r_X} & \mathrm{M}(X) \\ {\scriptstyle r_Y}\downarrow & & \\ \mathrm{M}(Y) & & \end{array}$$

the right-hand diagram is a product diagram.

Remark 6.4.2 *If* M *is additive, then for any two objects* X *and* Y *of* $\mathscr{K}$, *the functor*

$$\mathrm{M}(\iota_X)^* : \mathscr{K}/\mathrm{M}(X+Y) \longrightarrow \mathscr{K}/\mathrm{M}(X)$$

is just composition with r_X. *This is so because we are assuming that* r_X *is a coreflection:* $r_X \cdot \mathrm{M}(\iota_X) \cong id_{M(X)}$.

Lemma 6.4.3 *Let* $\mathscr{K}$ *be an extensive 2-category. Let* $G_0 \xrightarrow{p_0} D_0$ *and* $G_1 \xrightarrow{p_1} D_1$ *be two 1-cells in* $\mathscr{K}$. *Then the sum*

$$G_0 + G_1 \xrightarrow{p_0+p_1} D_0 + D_1$$

is M*-final iff each of the 1-cells* p_0, p_1 *is* M*-final.*

Proof. We may establish this using the characterization of M-final 1-cells given in Lemma 6.1.2. □

It is our aim in this section to introduce a notion of linear KZ-monad in an extensive 2-category $\mathscr{K}$ with pullbacks. Let K be an object of a 2-category $\mathscr{K}$. Let $K_! : \mathscr{K}/K \longrightarrow \mathscr{K}$ denote composition with $K \longrightarrow T$, where T denotes the pseudo-terminal object in $\mathscr{K}$. We have $K_! \dashv K^*$.

Definition 6.4.4 *A KZ-monad* M *in* $\mathscr{K}$ *is said to be* $\mathscr{K}$-equivariant, *or just* equivariant, *if it is given by the following data and conditions:*

1. *For every object* K, *a KZ-monad*

$$(\mathrm{M}^K, \delta^K, \mu^K)$$

in $\mathscr{K}/K$. *The case* $K = T$ *gives a KZ-monad* $(\mathrm{M}, \delta, \mu)$ *in* $\mathscr{K}$.

2. *We have* '$\mathrm{M}^K(K \times X) \simeq K \times \mathrm{M}(K)$'. *Precisely, we require a connecting pseudo-natural transformation*

$$\rho : K_! \circ \mathrm{M}^K \longrightarrow \mathrm{M} \circ K_!$$

such that for any object X *of* $\mathscr{K}$ *the right hand square below is a pullback and the left hand one commutes. The left hand square is therefore a pullback.*

$$\begin{array}{ccccc} K \times X & \xrightarrow{\delta^K_{K*X}} & \mathrm{M}^K(K \times X) & \xrightarrow{\pi_1} & K \\ {\scriptstyle \pi_2}\big\downarrow & & \big\downarrow{\scriptstyle \pi_X} & & \big\downarrow \\ X & \xrightarrow[\delta_X]{} & \mathrm{M}(X) & \longrightarrow & T \end{array}$$

We systematically abuse the notation slightly. For instance, $\mathrm{M}^K(K \times X) \xrightarrow{\pi_1} K$ *means*

$$\mathrm{M}^K \left(\begin{array}{c} K \times X \\ {\scriptstyle \pi_1}\big\downarrow \\ K \end{array} \right)$$

as an object of $\mathscr{K}/K$. π_X *denotes the composite 1-cell*

$$\mathrm{M}^K(K \times X) \xrightarrow{\rho_{K*X}} \mathrm{M}(K \times X) \xrightarrow{\mathrm{M}(\pi_2)} \mathrm{M}(X)\,.$$

3. *The* ρ*'s commute with the multiplications* μ^K *and* μ*: for any object* $X \longrightarrow K$ *of* $\mathscr{K}/K$*, the diagram*

$$\begin{array}{ccc} (\mathrm{M}^K)^2(X) & \xrightarrow{\mathrm{M}(\rho_X)\cdot\rho_{\mathrm{M}^K X}} & \mathrm{M}^2(X) \\ {\scriptstyle \mu^K_X}\big\downarrow & & \big\downarrow{\scriptstyle \mu_X} \\ \mathrm{M}^K(X) & \xrightarrow[\rho_X]{} & \mathrm{M}(X) \end{array}$$

commutes in $\mathscr{K}$.

4. *Finally, we require that if a 1-cell* $A \xrightarrow{q} Y$ *over* K *admits an* M^K*-adjoint* $\mathrm{M}^K(q) \dashv r^K_q$*, then it admits an* M*-adjoint* $\mathrm{M}(q) \dashv r_q$ *in* $\mathscr{K}$*, and the canonical 2-cell*

$$\rho_A \cdot r^K_q \Rightarrow r_q \cdot \rho_Y$$

is an isomorphism.

Definition 6.4.5 *A* linear *KZ-monad in an extensive 2-category* $\mathscr{K}$ *with pullbacks is one that is both additive (Definition 6.4.1) and* $\mathscr{K}$*-equivariant (Definition 6.4.4).*

Theorem 6.4.6 *Let* M *be a KZ-monad in an extensive 2-category with pullbacks, and assume that coproduct injections have coreflection* M*-adjoints. If* M *is closed, completion, and equivariant, then* M *is additive.*

Proof. Since $\mathscr{K}$ is extensive, the functor

$$\Phi : \mathscr{K}/X \times \mathscr{K}/Y \longrightarrow \mathscr{K}/(X+Y)\,,$$

given by coproduct, is an equivalence with pseudoinverse

$$\Psi : \mathscr{K}/(X+Y) \longrightarrow \mathscr{K}/X \times \mathscr{K}/Y$$

given by bipullback along the coproduct injections. For a completion KZ-monad M, there is for each object X a full and faithful functor

$$\mathscr{K}(T, \mathrm{M}(X)) \longrightarrow \mathscr{K}/X\ , \tag{6.2}$$

via a bicomma object constuction equating $\mathscr{K}(T, \mathrm{M}(X))$ with the full subcategory of $\mathscr{K}/X$ whose objects are the discrete M-fibrations (whose domain admits an M-adjoint). Furthermore, Ψ restricts to

$$\langle r_X, r_Y \rangle : \mathrm{M}(X+Y) \longrightarrow \mathrm{M}(X) \times \mathrm{M}(Y)\ ,$$

because discrete fibrations are pullback stable along 1-cells with M-adjoints.

We now claim that Φ restricts to a functor

$$\mathrm{M}(X) \times \mathrm{M}(Y) \longrightarrow \mathrm{M}(X+Y)\ .$$

This is the case iff the coproduct $A + B \xrightarrow{\varphi+\psi} X + Y$ of two discrete M-fibrations $A \xrightarrow{\varphi} X$ and $B \xrightarrow{\psi} Y$ is again a discrete M-fibration. We shall prove that this is so when M is closed.

Note that if $A \longrightarrow T$ and $B \longrightarrow T$ have M-adjoints, then $(A+B) \longrightarrow T$ has an M-adjoint. Consider the comprehensive factorization of $A + B \xrightarrow{\varphi+\psi} X + Y$. Observe that the M-final factor ρ is the sum of two 1-cells ρ_0 and ρ_1, each of which must be M-final, by Lemma 6.4.3. Our assumption that the given 1-cells φ and ψ are discrete M-fibrations implies that ρ_0 and ρ_1 are both isomorphisms, hence so is their sum ρ.

Finally, when M is equivariant (6.2) holds in any slice $\mathscr{K}/K$, so that we may essentially repeat the above argument for 'generalized points' $K \longrightarrow \mathrm{M}(X)$. □

Example 6.4.7

1. *We saw at the beginning of this section that the symmetric monad is additive in* $\boldsymbol{Top}_{\mathscr{S}}$*. It is also equivariant. For instance, the forward preservation of* M*-adjoints (requirement 4) amounts to the observation that for any geometric morphism* $\mathscr{T} \longrightarrow \mathscr{S}$*, a* $\mathscr{T}$*-essential geometric morphism over* $\mathscr{T}$ *is* $\mathscr{S}$*-essential as a geometric morphism over* $\mathscr{S}$*. The pullback equivalence* $\mathrm{M}^K(K \times X) \simeq K \times \mathrm{M}(X)$ *for toposes is in a new guise another important fact about distributions and change of base discovered by A. M. Pitts. This fact says that the topos pullback is universal for distributions, not just left exact distributions.*
2. *The lower bagdomain* B_L *and probability distributions classifier* T *are both closed, completion and equivariant KZ-monads in* $\boldsymbol{Top}_{\mathscr{S}}$*.* B_L *is additive since* $\langle \mu, \lambda \rangle$ *preserves pullbacks if both* μ *and* λ *do; however,* T *is not additive because* $\langle \mu, \lambda \rangle$ *will not preserve* 1 *if* μ *and* λ *do. This is consistent with Theorem 6.4.6, because coproduct inclusions in* $\boldsymbol{Top}_{\mathscr{S}}$ *have (coreflection)* B_L*-adjoints, but they do not have* T*-adjoints.*

Remark 6.4.8 *The present exposition of additive and equivariant KZ-monads is only the beginning of an interesting subject. For instance, a simpler characterization of completion KZ-monads is available in the $\mathscr{K}$-equivariant case. This leads to a slightly different characterization of the algebras, which holds for the symmetric monad, resulting in another Waelbroeck theorem with locally connected geometric morphisms in place of essential ones, and pullbacks in place of bicomma objects.*

Exercises 6.4.9

1. *Prove that the homomorphisms for the algebras in a linear KZ-monad* M *in an extensive 2-category $\mathscr{K}$ are linear in the usual sense, meaning that they preserve addition and scalar multiplication.*

6.5 Pitts' Theorem Revisited

We shall now explain Pitts' theorem (herein Theorem 4.3.1) in terms of a linear KZ-monad. Although this is not a direct explanation, we feel that it is worthwhile. For instance, we gain new information about the geometric morphism opposite the upper one in a topos bicomma object in which the lower geometric morphism is essential.

The following lemma is at the heart of the bicomma object construction in the presence of an equivariant KZ-monad: it shows how to construct a bicomma object by localizing.

Lemma 6.5.1 *Under the above notation, let* M *be a locally full and faithful, 'closed, equivariant KZ-monad in $\mathscr{K}$. Suppose that a 1-cell $X \xrightarrow{\varphi} K$ admits an* M*-adjoint, and let n denote the pairing $(1_K, r_\varphi \cdot \delta_K)$, where $\mathrm{M}(\varphi) \dashv r_\varphi$. Suppose that*

$$\begin{array}{ccc} A & \xrightarrow{q} & K \\ {\scriptstyle p}\downarrow & \Rightarrow & \downarrow{\scriptstyle n=(1_K, r_\varphi \cdot \delta_K)} \\ K \times X & \xrightarrow[\delta^K_{K*X}]{} & \mathrm{M}^K(K \times X) \end{array}$$

is a bicomma object in $\mathscr{K}/K$ such that q admits an M^K-adjoint and the BCC holds for M^K: $\mathrm{M}^K(p) \cdot r^K_q \cong \mu^K_{K*X} \cdot \mathrm{M}^K(n)$. Then

$$\begin{array}{ccc} A & \xrightarrow{q} & K \\ {\scriptstyle \pi_2 \cdot p}\downarrow & \Rightarrow & \downarrow{\scriptstyle 1_K} \\ X & \xrightarrow[\varphi]{} & K \end{array}$$

is a bicomma object in $\mathscr{K}$ and the BCC holds for M *in $\mathscr{K}$* (by equivariance, q admits an M-adjoint)*: $r_\varphi \cong \mathrm{M}(\pi_2 \cdot p) \cdot r_q$.*

Proof. Since M is locally full and faithful, 2-cells in a diagram

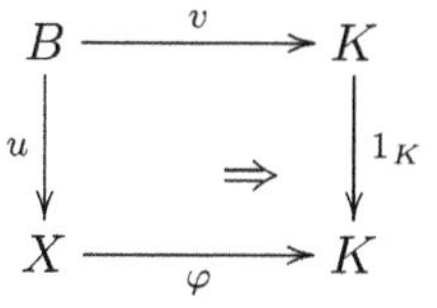

are in bijection with 2-cells

$$\frac{\mathrm{M}(\varphi) \cdot \delta_X \cdot u \cong \delta_K \cdot \varphi \cdot u \Rightarrow \delta_K \cdot v}{\delta_X \cdot u \Rightarrow r_\varphi \cdot \delta_K \cdot v\ .}$$

Now chase through the following diagram:

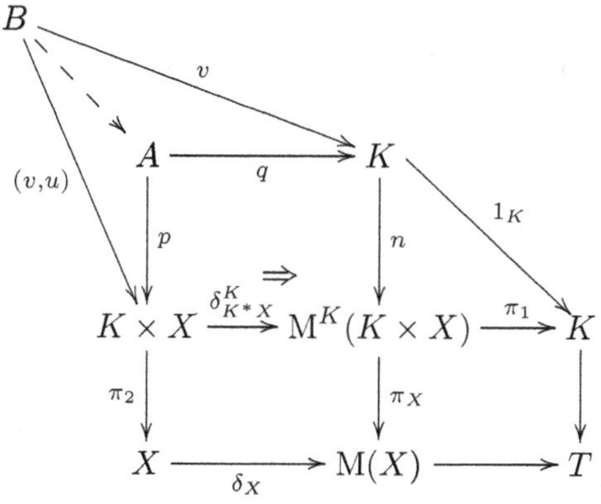

where $\pi_X \cdot n = r_\varphi \cdot \delta_K$.

As for the BCC for M in $\mathscr{K}$, it suffices to show

$$r_\varphi \cdot \delta_K \cong \mathrm{M}(\pi_2 \cdot p) \cdot r_q \cdot \delta_K\ ,$$

since both 1-cells in the BCC equation are M-homomorphisms. We have

$$\begin{aligned}
\mathrm{M}(\pi_2 \cdot p) \cdot r_q \cdot \delta_K &\cong \pi_X \cdot \mathrm{M}^K(p) \cdot r_q^K \cdot \delta_K^K \\
&\cong \pi_X \cdot \mu^K_{K*X} \cdot \mathrm{M}^K(n) \cdot \delta^K_K \\
&\cong \pi_X \cdot \mu^K_{K*X} \cdot \delta^K_{\mathrm{M}^K(K*X)} \cdot n \\
&\cong \pi_X \cdot n \\
&\cong r_\varphi \cdot \delta_K\ .
\end{aligned}$$

□

Remark 6.5.2 *The 1-cell p in the hypothesis of Lemma 6.5.1 is a discrete* M^K*-fibration.*

We turn to the existence of topos bicomma objects, and a proof of Pitts' bicomma object theorem.

By Lemma 6.5.1, any topos bicomma object

$$\begin{array}{ccc} \mathscr{A} & \xrightarrow{\;q\;} & \mathscr{T} \\ {\scriptstyle p}\big\downarrow & \Rightarrow & \big\downarrow{\scriptstyle \mathrm{id}_{\mathscr{T}}} \\ \mathscr{X} & \xrightarrow[\;\varphi\;]{} & \mathscr{T} \end{array} \tag{6.3}$$

in $\mathbf{Top}_{\mathscr{S}}$ in which φ is $\mathscr{S}$-essential can be obtained by localizing over $\mathscr{T}$. Indeed, the above bicomma object reduces to the bicomma object

$$\begin{array}{ccc} \mathscr{A} & \xrightarrow{\;q\;} & \mathscr{T} \\ \big\downarrow & \Rightarrow & \big\downarrow{\scriptstyle f} \\ \mathscr{X} \times_{\mathscr{S}} \mathscr{T} & \xrightarrow[\;\delta^{\mathscr{T}}\;]{} & \mathrm{M}^{\mathscr{T}}(\mathscr{X} \times_{\mathscr{S}} \mathscr{T}) \end{array} \tag{6.4}$$

in $\mathbf{Top}_{\mathscr{T}}$ for some point f. Thus, it suffices to construct bicomma objects $\delta \Downarrow f$ of diagrams

$$\begin{array}{ccc} & & \mathscr{S} \\ & & \big\downarrow{\scriptstyle f} \\ \mathscr{E} & \xrightarrow[\;\delta\;]{} & \mathrm{M}(\mathscr{E}) \end{array},$$

where now the usual $\mathscr{S}$ denotes the base topos. The geometric morphism opposite f in such a bicomma object is what we call a discrete M-fibration. We have seen in § 5.2 that such bicomma objects exist, and that by their very construction a discrete M-fibration is precisely a complete spread geometric morphism.

Remark 6.5.3

1. *The combination of bicomma objects (6.3) and the pullback stability of locally connected geometric morphisms (Exercise 2) gives Pitts' Theorem.*
2. *We gain the extra information that in (6.4)* $\mathscr{A} \xrightarrow{(p,q)} \mathscr{X} \times_{\mathscr{S}} \mathscr{T}$ *is a* $\mathscr{T}$*-complete spread.*
3. *For toposes* $\mathscr{E}$ *and* $\mathscr{G}$ *over* $\mathscr{S}$*, the functor which associates to a geometric morphism* $\mathscr{G} \xrightarrow{\rho} \mathrm{M}(\mathscr{E})$ *the complete spread* $(\gamma, \lambda) : \mathscr{Y} \longrightarrow \mathscr{G} \times_{\mathscr{S}} \mathscr{E}$ *for the bicomma object*

$$\begin{array}{ccc} \mathscr{Y} & \xrightarrow{\;\gamma\;} & \mathscr{G} \\ {\scriptstyle \lambda}\big\downarrow & \Rightarrow & \big\downarrow{\scriptstyle \rho} \\ \mathscr{E} & \xrightarrow[\;\delta\;]{} & \mathrm{M}(\mathscr{E}) \end{array}$$

is an equivalence of $\mathbf{Top}_{\mathscr{S}}(\mathscr{G}, \mathrm{M}(\mathscr{E}))$ *with the category of complete spreads over* $\mathscr{G} \times_{\mathscr{S}} \mathscr{E}$ *with locally connected* $\mathscr{G}$*-domain. It is reasonable to call the bifibration* (γ, λ) *a generalized complete spread over* $\mathscr{S}$*. Such a* λ *may not be localic, and* $\mathscr{Y} \longrightarrow \mathscr{S}$ *may not be locally connected, although* γ *is.*

Exercises 6.5.4

1. *Show that "admits an* M*-adjoint" is stable under bipullback.*
2. *Let* M *be* $\mathscr{K}$*-equivariant. Show that if* $X \longrightarrow T$ *admits an* M*-adjoint, then for any* K, $K \times X \longrightarrow K$ *admits an* M^K*-adjoint (and hence admits an* M*-adjoint) and the BCC holds for* M *and the bipullback*

$$\begin{array}{ccc} K \times X & \longrightarrow & K \\ \downarrow & & \downarrow \\ X & \longrightarrow & T \end{array}$$

   in $\mathscr{K}$.
3. *What is the point* f *in the bicomma object (6.3)?*

Further reading: Bunge [Bun95, Bun74, Bun04], Bunge & Funk [BF99], Bunge & Niefield [BN00], Bunge & Fiore [BF00], Carboni & Johnstone [CJ95], Johnstone [Joh99], Kock [Koc75], Street [Str, Str74], Waelbroeck [Wae67], Zöeberlein [Zoe76].

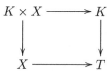

Part III

Aspects of Distributions and Complete Spreads

7

Lattice-Theoretic Aspects

In this chapter we discuss the notion of a distribution algebra, which is a lattice-theoretic concept associated with distributions. When the free distribution algebra on a sheaf is available, we prove that the opposite category of distributions on a topos is monadic over the topos. This generalizes Paré's monadicity theorem (1974), although we employ the latter in its proof. In §7.2 we characterize distribution algebras in $\mathscr{E}$ over $\mathscr{S}$ as the $\mathscr{S}$-(bi)complete $\mathscr{S}$-atomic Heyting algebras in $\mathscr{E}$.

A complete spread is a localic geometric morphism, so it may be regarded as a locale in the codomain topos, which we call the display locale. We examine a constructively finite approximation to the display locale of a complete spread via relative Stone locales, leading to the relative pure, entire factorization, with connections to work of P. T. Johnstone (1982), and in principle to the work of I. Moerdijk and J. J. C. Vermeulen (2000).

In the following investigation of a relative pure, entire factorization we work with geometric morphisms whose domain toposes are definable dominances. Recall that a geometric morphism is a definable dominance if the definable subobjects of an object form a dominance in the sense of G. Rosolini (1986). The notion of a definable dominance is weaker than local connectedness, and it holds automatically for any topos defined over a Boolean base topos $\mathscr{S}$.

The uniqueness part of the relative pure, entire factorization follows from an alternative construction of it via topos forcing in the sense of M. Tierney (1974).

This chapter is largely based on joint work with T. Streicher and M. Jibladze [BFJS02, BFJS04a].

7.1 Monadicity Theorem

A distribution on a bounded $\mathscr{S}$-topos must have a right adjoint. If $\mu : \mathscr{E} \longrightarrow \mathscr{S}$ is a distribution, then we denote its $\mathscr{S}$-indexed right adjoint by $\mu_* : \mathscr{S} \longrightarrow \mathscr{E}$.

Definition 7.1.1 *An object H in $\mathscr{E}$ is said to be a distribution algebra if for every object X of $\mathscr{E}$, $e_*(H^X)$ is a complete atomic Heyting algebra in $\mathscr{S}$. (In particular, every $e_*(H^X)$ is a frame in $\mathscr{S}$.) We require that every transition morphism $e_*(H^Y) \longrightarrow e_*(H^X)$ induced by a morphism $X \longrightarrow Y$ in $\mathscr{E}$ is a frame morphism. A morphism of distribution algebras is a morphism $m : H \longrightarrow H'$ in $\mathscr{E}$ such that every $e_*(m^X)$ is a frame morphism.*

The following result justifies our interest in distribution algebras.

Proposition 7.1.2 *An object H of $\mathscr{E}$ is a distribution algebra iff there is an order-preserving isomorphism $\mu_*(\Omega_{\mathscr{S}}) \cong H$ for some distribution μ on $\mathscr{E}$. $\mathbf{E}(\mathscr{E})^{\mathrm{op}}$ is equivalent to the category of distribution algebras.*

Proof. If we start with a distribution μ, then it is not difficult to see that

$$e_*(\mu_*(\Omega_{\mathscr{S}})^X) \cong {\Omega_{\mathscr{S}}}^{\mu X} \,,$$

which is a complete atomic Heyting algebra in $\mathscr{S}$. On the other hand, if for every X, $e_*(H^X)$ is a complete atomic Heyting algebra, then by Stone duality in an elementary topos, we have $e_*(H^X) \cong {\Omega_{\mathscr{S}}}^{\mu(X)}$, where $\mu(X) = \mathrm{Atoms}(e_*(H^X))$. From this it follows that the μ so defined extends to a distribution on $\mathscr{E}$, and that $H \cong \mu_*(\Omega_{\mathscr{S}})$. □

Remark 7.1.3 *We may think of $\mu \dashv \mu_*$ as a* generalized point *of $\mathscr{E}$, since μ may not preserve finite limits. In particular, a distribution algebra $H = \mu_*(\Omega_{\mathscr{S}})$ is a Heyting algebra, but may not be a frame.*

Remark 7.1.4 *If $\psi : \mathscr{F} \longrightarrow \mathscr{E}$ is a geometric morphism over $\mathscr{S}$ with locally connected domain, then $\Lambda(\psi) = f_!\psi^*$ is a distribution with right adjoint $\psi_* f^*$. Therefore, $\psi_* f^*(\Omega_{\mathscr{S}})$ is a distribution algebra in $\mathscr{E}$. Of course, we know that every distribution has the form $\Lambda(\psi)$. Therefore, every distribution algebra has the form $\psi_* f^*(\Omega_{\mathscr{S}})$.*

We have the following from the proof of Proposition 7.1.2.

Corollary 7.1.5 *If μ is a distribution on $\mathscr{E} = \mathrm{Sh}(\mathbb{C}, J)$ with corresponding cosheaf $M = \mu \cdot h$ on $\mathbb{C}$, then the object $\mu_*(\Omega_{\mathscr{S}})$ of $\mathscr{E}$ is given by the sheaf*

$${\Omega_{\mathscr{S}}}^{M(\_)} : \mathbb{C}^{\mathrm{op}} \longrightarrow \mathscr{S} \,.$$

We have seen that $\mathbf{E}(\mathscr{E})$ has the structure of an $\mathscr{E}$-indexed category.

Definition 7.1.6 *We shall refer to the $\mathscr{E}$-indexed functor*

$$\mathbf{U} : \mathbf{E}(\mathscr{E})^{\mathrm{op}} \longrightarrow \mathscr{E} \,,\; \mathbf{U}(\mu) = \mu_*(\Omega_{\mathscr{S}}) \,,$$

as the underlying sheaf functor. *We denote the $\mathscr{E}$-indexed left adjoint of $\mathbf{U}$, when it exists, by $\mathbf{F}$.*

Corollary 7.1.5 has the following consequence.

Lemma 7.1.7 *The underlying sheaf functor reflects isomorphisms.*

Proposition 7.1.8 *Assume that* $\mathscr{E} \xrightarrow{e} \mathscr{S}$ *is bounded. Then* $\mathbf{E}(\mathscr{E})$ *is a locally small* $\mathscr{E}$*-indexed category.*

Proof. We define the hom-object

$$\mathrm{Hom}(\mu, \lambda)(c) = \mathbf{E}(\mathscr{E})(h_c.\mu, \lambda) \ ,$$

which is a sheaf. There is a natural bijection

$$\frac{\text{morphisms } X \longrightarrow \mathrm{Hom}(\mu, \lambda) \text{ in } \mathscr{E}}{\text{morphisms } \mu \cdot \Sigma_X \longrightarrow \lambda \cdot \Sigma_X \text{ in } \mathbf{E}(\mathscr{E})^X.}$$

which expresses the representability of morphisms in $\mathbf{E}(\mathscr{E})$ by the objects $\mathrm{Hom}(\mu, \lambda)$ of $\mathscr{E}$. □

By Proposition 7.1.8, we see that the action of sheaves on a fixed distribution μ,

$$(\_).\mu : \mathscr{E} \longrightarrow \mathbf{E}(\mathscr{E}) \ ,$$

has a right adjoint:

$$\mathrm{Hom}(\_\,, \mu) : \mathbf{E}(\mathscr{E}) \longrightarrow \mathscr{E} \ .$$

Definition 7.1.9 *Suppose that* $\mathscr{E}$ *has a Gleason core. Equivalently, assume that* $\mathbf{E}(\mathscr{E})$ *has a terminal object* $\mathbf{1}$*. Let* $\mathbf{o}$ *denote the distribution* $e^*(\Omega_{\mathscr{S}}).\mathbf{1}$.

We have $\mathbf{o}(X) = \Omega_{\mathscr{S}} \times \mathbf{1}(X)$, for any object X of $\mathscr{E}$.

Lemma 7.1.10 *Suppose that* $\mathscr{E}$ *has a Gleason core. Then there is a natural isomorphism*

$$\mu_* \cong \mathrm{Hom}(\mu, e^*(\_).\mathbf{1}) \ .$$

In particular, the underlying sheaf functor is representable:

$$\mathbf{U}(\mu) = \mu_*(\Omega_{\mathscr{S}}) \cong \mathrm{Hom}(\mu, \mathbf{o}) \ .$$

Proof. Let I be an object in $\mathscr{S}$. Then we have the following bijections, natural in X.

$$\begin{array}{c} X \longrightarrow \mathrm{Hom}(\mu, e^*(I).\mathbf{1}) \\ \hline X.\mu \longrightarrow e^*(I).\mathbf{1} \\ \hline \mu(X \times \_) \longrightarrow \mathbf{1}(e^*(I) \times \_) \cong I \times \mathbf{1}(\_) \\ \hline \mu(X \times \_) \longrightarrow I \\ \hline \mu(X) \longrightarrow I \\ \hline X \longrightarrow \mu_*(I) \ . \end{array}$$

The third bijection holds because $\mathbf{1}$ is the terminal distribution. □

Remark 7.1.11 *The cotensor μ^X of a distribution μ by a sheaf X is defined by a natural bijection*

$$\frac{\lambda \longrightarrow \mu^X}{X.\lambda \longrightarrow \mu}$$

of natural transformations of distributions. Then the following natural bijections show that we must have $\mathbf{o}^X \cong \mathbf{F}(X)$.

$$\frac{\lambda \longrightarrow \mathbf{F}(X)}{\frac{X \longrightarrow \mathbf{U}(\lambda) \cong \mathrm{Hom}(\lambda, \mathbf{o})}{X.\lambda \longrightarrow \mathbf{o}}}$$

Thus, $\mathbf{F}(X)$ exists iff $\mathbf{o}^X$ does.

Lemma 7.1.12 *Let $\mathbb{C}$ be a small category. Then the underlying presheaf functor*

$$\mathbf{U}_{P(\mathbb{C})} : \mathbf{E}(P(\mathbb{C}))^{\mathrm{op}} \simeq (\mathscr{S}^{\mathbb{C}})^{\mathrm{op}} \longrightarrow P(\mathbb{C})$$

is monadic. Moreover, if $\langle \mathbb{C}, J \rangle$ is a site, then any precosheaf M is a cosheaf iff the presheaf $\mathbf{U}_{P(\mathbb{C})}(M)$ is a sheaf.

Proof. We start with Paré's Theorem [Par80], which asserts that the functor

$$\Omega_{\mathscr{S}}{}^{(-)} : \mathscr{S}^{\mathrm{op}} \longrightarrow \mathscr{S}$$

is monadic. Now exponentiate this functor by $\mathbb{C}^{\mathrm{op}}$ in order to deduce that $\mathbf{U}_{P(\mathbb{C})}$ is monadic. The second statement is straightforward to establish. □

For the Monadicity Theorem we shall assume that $\mathscr{E}$ has a site $\mathscr{E} \rightarrowtail P(\mathbb{C})$ for which the inclusion $\mathrm{Cosh}(\mathbb{C}, J) \rightarrowtail \mathscr{S}^{\mathbb{C}}$ has a right adjoint. We call this right adjoint *cosheafification*, and usually denote it $\mathbf{b}$.

Example 7.1.13 *There are two special cases when we know that cosheafification exists.*

1. *If $\mathscr{E}$ is bounded over Set, i.e., a Grothendieck topos, then the existence of an associated cosheaf is guaranteed because $\mathbf{E}(\mathscr{E})$ is locally presentable.*
2. *Cosheafification exists for any topos $\mathscr{E}$ defined over an arbitrary base topos $\mathscr{S}$ if $\mathscr{E}$ is an essential localization of a presheaf topos over $\mathscr{S}$, meaning that the inclusion $i : \mathscr{E} \rightarrowtail P(\mathbb{C})$ is $\mathscr{S}$-essential: i^* has an $\mathscr{S}$-indexed left adjoint $i_! \dashv i^*$. Then composition with these $\mathscr{S}$-cocontinuous functors induces an adjoint pair $\mathbf{E}(i^*) \dashv \mathbf{E}(i_!)$, and therefore $\mathbf{b} = \mathbf{E}(\mathbf{i}_!)$.*

Lemma 7.1.14 *Let $\mathscr{E}$ be a topos over $\mathscr{S}$ that has a site $\langle \mathbb{C}, J \rangle$ for which cosheafification exists. Then $\mathbf{E}(\mathscr{E})$ has all small limits* (including a terminal distribution $\mathbf{1}$), *and $\mathbf{U}$ has a left adjoint* (denoted $\mathbf{F}$).

Proof. Limits are constructed by cosheafifying the limit in $\mathscr{S}^{\mathbb{C}}$. $\mathbf{F}(X)$ is the associated cosheaf of the precosheaf $\Omega_{\mathscr{S}}{}^{X(-)}$. □

Theorem 7.1.15 *Let $\mathcal{E}$ be a topos bounded over $\mathcal{S}$ that has a small site for which cosheafification exists. Then the underlying sheaf functor*

$$\mathbf{U} : \mathbf{E}(\mathcal{E})^{\mathrm{op}} \longrightarrow \mathcal{E}$$

is monadic.

Proof. We resort to Beck's monadicity theorem. We have only to show that $\mathbf{E}(\mathcal{E})^{\mathrm{op}}$ has coequalizers of $\mathbf{U}$-contractible pairs, and that $\mathbf{U}$ preserves them. Let $g, h : M \longrightarrow N$ be two morphisms such that $\mathbf{U}(g)$ and $\mathbf{U}(h)$ are a contractible pair in $\mathcal{E}$. But contractible pairs are preserved by any functor, so g, h must be a contractible pair in $P(\mathbb{C})$. Let K denote the equalizer of g, h in $\mathcal{S}^{\mathbb{C}}$. Then the monadic $\mathbf{U}_{P(\mathbb{C})}$ must carry this equalizer diagram to the contractible coequalizer of $\mathbf{U}(g)$ and $\mathbf{U}(h)$ in $P(\mathbb{C})$, as in the following diagram:

$$\mathbf{U}(N) \underset{\mathbf{U}(h)}{\overset{\mathbf{U}(g)}{\rightrightarrows}} \mathbf{U}(M) \underset{d}{\overset{\mathbf{U}(f)}{\rightleftarrows}} \mathbf{U}(K)$$

where $\mathbf{U}(f)d = 1$. Thus, $\mathbf{U}(K)$ is a retract of $\mathbf{U}(M)$, from which it follows that $\mathbf{U}(K)$ is a sheaf. Therefore, the precosheaf equalizer K is in fact a cosheaf, and obviously $\mathbf{U}$ preserves it. □

Remark 7.1.16 *It may not seem surprising that the category of distribution algebras in $\mathcal{E}$ (Definition 7.1.1) is monadic over $\mathcal{E}$. However, we have shown that this is so under special assumptions on $\mathcal{E}$, which may not hold in general. It may be of interest to identify the class of $\mathcal{S}$-toposes for which cosheafification exists, and to determine the properties of cosheafification, by analogy with the associated sheaf functor.*

Exercises 7.1.17

1. *Let $\mu \dashv \mu_*$ be a distribution on $\mathcal{E}$ over $\mathcal{S}$. Show that the isomorphism $\mu(e^*(I)\times E) \cong I\times\mu(E)$ induces an isomorphism $\mu_*(\Omega_{\mathcal{S}}{}^{I}) \longrightarrow \mu_*(\Omega_{\mathcal{S}})^{e^*I}$.*

7.2 The Structure of Distribution Algebras

We investigate the intrinsic structure of distribution algebras.

Definition 7.2.1 *Let P be a poset in a topos $\mathcal{E}$ over $\mathcal{S}$.*

1. *We shall say that P is* $\mathcal{S}$-cocomplete *if for any definable morphism $\alpha : B \longrightarrow A$ in $\mathcal{E}$, the induced poset morphism*

$$\mathcal{E}(\alpha, P) : \mathcal{E}(A, P) \longrightarrow \mathcal{E}(B, P)$$

has a left adjoint $\bigvee_\alpha$ *satisfying the BBC. This requires that for every pullback square*

$$\begin{array}{ccc} C & \xrightarrow{\delta} & D \\ {\scriptstyle\beta}\downarrow & & \downarrow{\scriptstyle\gamma} \\ B & \xrightarrow[\alpha]{} & A \end{array}$$

in $\mathscr{E}$*, in which* α *(and hence* δ*) is definable, we have*

$$\bigvee_\delta \cdot \mathscr{E}(\beta, P) = \mathscr{E}(\gamma, P) \cdot \bigvee_\alpha .$$

Dually, we may define $\mathscr{S}$-complete *and we say that* P *is* $\mathscr{S}$-bicomplete *if it is* $\mathscr{S}$*-complete and* $\mathscr{S}$*-cocomplete.*

2. *We shall say that* P *is* $\Omega_{\mathscr{S}}$-cocomplete *if it has the property defined in 1 for definable monomorphisms* α*.*

If P is an $\mathscr{S}$-cocomplete poset, then for any $I \xrightarrow{\alpha} K$ in $\mathscr{S}$, there is a morphism $P^{e^*I} \longrightarrow P^{e^*K}$ in $\mathscr{E}$ that represents the externally defined functor $\bigvee_\alpha$, and is left adjoint in $\mathscr{E}$ to the transpose $P^{e^*K} \longrightarrow P^{e^*I}$ of

$$P^{e^*K} \times e^*I \longrightarrow P^{e^*K} \times e^*K \longrightarrow P .$$

We shall use the same symbol $\bigvee_\alpha$ to denote both the externally defined functor and its representing morphism in $\mathscr{E}$.

Proposition 7.2.2 *Let* P *be a poset in a topos* $\mathscr{E}$ *over* $\mathscr{S}$*. Then:*

1. P *is* $\mathscr{S}$*-cocomplete iff for every* $K \xrightarrow{\alpha} I$*, the formula*

$$\forall i \in e^*(I)\, \forall p \in P^{e^*\alpha^{-1}(i)}\, \exists h \in P\ (h = \sup(p))$$

holds in $\mathscr{E}$*, where "*$h = \sup(p)$*" is an abbreviation for a formula saying that* h *is the supremum of* p*;*

2. P *is* $\Omega_{\mathscr{S}}$*-cocomplete iff the formula*

$$\forall u \in e^*(\Omega_{\mathscr{S}})\, \forall p \in P^{\mathrm{ext}(u)}\, \exists h \in P\ (h = \sup(p))$$

is valid in $\mathscr{E}$*, where* $\mathrm{ext}(u) = e^*(\top)^{-1}(u)$*.*

Proposition 7.2.3 *An* $\Omega_{\mathscr{S}}$*-cocomplete poset* P *in* $\mathscr{E}$ *is* $\mathscr{S}$*-cocomplete iff for each object* I *of* $\mathscr{S}$*,*

$$\forall p \in P^{e^*(I)}\, \exists h \in P\ (h = \sup(p))$$

holds in $\mathscr{E}$*. The dual statement for completeness is analogously stated.*

Proof. In addition to topos semantics, the proof relies on the observation that for any morphism $I \longrightarrow K$ in $\mathscr{S}$ may be decomposed as $I \longrightarrow I \times K \longrightarrow K$. □

Definition 7.2.4 *An $\Omega_{\mathscr{S}}$-*Heyting algebra *is a Heyting algebra that is $\Omega_{\mathscr{S}}$-cocomplete.*

Any $\Omega_{\mathscr{S}}$-Heyting algebra H has both an action and a coaction by $\Omega_{\mathscr{S}}$ that we now define.

Definition 7.2.5 *The* action *of $\Omega_{\mathscr{S}}$ on an $\Omega_{\mathscr{S}}$-Heyting algebra H is defined by*

$$u \cdot p = \bigvee_{x \in \mathrm{ext}(u)} p .$$

Let $\|u\|$ denote $u \cdot 1_H$, which is the same as $\bigvee_{\top} 1_H$. Then $u \cdot p = \|u\| \wedge p$. The coaction *is given by $\|u\| \Rightarrow p$, denoted p^u.*

Definition 7.2.6 *For an $\Omega_{\mathscr{S}}$-Heyting algebra H, the morphism*

$$\mathrm{eq}_I : e^*(I \times I) \xrightarrow{e^*\chi} e^*(\Omega_{\mathscr{S}}) \xrightarrow{\|\cdot\|} H ,$$

where $\chi : I \times I \longrightarrow \Omega_{\mathscr{S}}$ classifies the diagonal $\nabla_I : I \longrightarrow I \times I$ is called its equality.

The equality of an $\Omega_{\mathscr{S}}$-Heyting algebra satisfies the axioms of Lawvere equality.

We shall need a name for the canonical isomorphism (Exercise 7.1.17, 1)

$$\iota_I : \mu_*({\Omega_{\mathscr{S}}}^I) \cong H^{e^*I} .$$

Proposition 7.2.7 *If $f : \mathscr{F} \longrightarrow \mathscr{S}$ is a definable dominance, then the canonical order-preserving map*

$$\tau : f^*(\Omega_{\mathscr{S}}) \longrightarrow \Omega_{\mathscr{F}}$$

preserves finite meets.

Proof. Since f is subopen, the pair $\langle f^*(\Omega_{\mathscr{S}}), f^*(\top)\rangle$ classifies definable subobjects in $\mathscr{F}$ (Exercise 1.5.7 4). We know that definable subobjects are pullback stable in $\mathscr{F}$; as f is a definable dominance, they also compose. The diagonal with codomain 1 in the following pullback square represents the meet of two given subobjects of 1:

$$\begin{array}{ccc} A \wedge B & \rightarrowtail & A \\ \downarrow & & \downarrow \\ B & \rightarrowtail & 1 \end{array}$$

and makes $f_*(f^*(\Omega_{\mathscr{S}})) \cong \mathrm{Sub}_{\mathrm{def}}(1)$ a sub meet-semilattice of $f_*(\Omega_{\mathscr{F}})$. □

Let $\psi : \mathscr{F} \longrightarrow \mathscr{E}$ be a geometric morphism over $\mathscr{S}$. Let H denote $\psi_*(f^*\Omega_{\mathscr{S}})$. H is a Heyting algebra without any conditions on f or ψ.

Corollary 7.2.8 *If f is a definable dominance, then the Heyting algebra $H = \psi_*(f^*\Omega_{\mathcal{S}})$ is a sub-Heyting algebra of the frame $\psi_*(\Omega_{\mathcal{F}})$.*

Proof. It follows from Proposition 7.2.7. □

Proposition 7.2.9 *Let $\psi : \mathcal{F} \longrightarrow \mathcal{E}$ be a geometric morphism over $\mathcal{S}$. Let H denote $\psi_*(f^*\Omega_{\mathcal{S}})$.*

1. *If f is a definable dominance, then H is an $\Omega_{\mathcal{S}}$-Heyting algebra.*
2. *If f is locally connected, then H is an $\mathcal{S}$-bicomplete Heyting algebra. In addition, $\mu_* = \psi_* f^*$ preserves existential and universal quantification modulo the isomorphisms ι_I.*

Proof. 1. Let $m : E \rightarrowtail F$ be definable in $\mathcal{E}$. $\bigvee_m : H^E \longrightarrow H^F$ arises as follows. A generalized element $X \longrightarrow H^E$ is the same as a definable subobject $S \rightarrowtail \psi^*(X \times E)$ in $\mathcal{F}$. We compose this with the definable subobject $\psi^*(X \times m)$ to produce a definable subobject of $\psi^*(X \times F)$, which is the same as a generalized element $X \longrightarrow H^F$.

2. $H \cong \mu_*(\Omega_{\mathcal{S}})$ is the distribution algebra of the distribution $\mu = f_!\psi^*$ (Proposition 7.1.2), and it follows that distribution algebras are $\mathcal{S}$-bicomplete Heyting algebras. The last assertion follows from the fact that for any $I \xrightarrow{\alpha} K$ in $\mathcal{S}$, the identities

$$\iota_K \cdot \mu_*(\exists_\alpha) = \bigvee_\alpha \cdot \iota_I \ ; \ \ \iota_K \cdot \mu_*(\forall_\alpha) = \bigwedge_\alpha \cdot \iota_I$$

hold, where $\bigvee_\alpha \dashv H^{e^*\alpha} \dashv \bigwedge_\alpha$. □

Corollary 7.2.10 *The right adjoint μ_* to a distribution μ on $\mathcal{E}$ preserves equality up to isomorphism.*

Proof. Equality in $\Omega_{\mathcal{S}}$ is an instance of existential quantification: it is the map

$$\exists_\nabla(\top) : I \times I \longrightarrow \Omega_{\mathcal{S}} .$$

If ${1_H}^I$ denotes the map $e^*I \longrightarrow 1 \xrightarrow{1_H} H$, then we have

$$\mathrm{eq}_I = \bigvee_{\nabla_I} {1_H}^I .$$

By Proposition 7.2.9, $\mu_*(\exists_\nabla(\top)) \cong \mathrm{eq}_I$, modulo the ι_I isomorphisms. □

Proposition 7.2.11 *For any distribution μ, and any map $I \xrightarrow{\alpha} K$ in $\mathcal{S}$, the suprema map $\bigvee_\alpha$ satisfies the formula*

$$\bigvee_\alpha q\,(k) = \bigvee_{i \in e^*I} q(i) \wedge \|e^*\alpha(i) = k\| ,$$

for all $q \in H^{e^*I}$ and $k \in e^*K$.

Definition 7.2.12 *Let I be an object of $\mathscr{S}$. The object* $\mathrm{Part}_I(H)$ *of I-*partitions *of an $\mathscr{S}$-bicomplete Heyting algebra H in $\mathscr{E}$ is defined to be the pullback*

$$\begin{array}{ccc} \mathrm{Part}_I(H) & \longrightarrow & 1 \\ \big\downarrow & & \big\downarrow{\scriptstyle \top} \\ H^{e^*I} & \xrightarrow[\theta_I]{} & H \end{array} ,$$

where

$$\theta_I(p) = \left(\bigvee_{i \,\in\, e^*I} p(i) \right) \wedge \left(\bigwedge_{i,j \,\in\, e^*I} (p(i) \wedge p(j)) \Rightarrow_H \|i = j\| \right)$$

for all $p \in H^{e^*I}$.

Intuitively, Definition 7.2.12 says that a partition on H is a map $p : e^*I \longrightarrow H$ whose values $p(i)$ cover H, and are pairwise disjoint, as least as far as H and $\mathscr{E}$ can tell. Then $\theta_I(p)$ is the extent in H to which p is a partition of H.

For any morphism $I \xrightarrow{\alpha} K$ in $\mathscr{S}$, the suprema morphism $\bigvee_\alpha$ restricts to partitions giving an $\mathscr{S}$-indexed functor

$$\mathrm{Part}(H) : \mathscr{S} \longrightarrow \mathscr{E} .$$

Lemma 7.2.13 *For any distribution μ, its right adjoint μ_* and* $\mathrm{Part}(H)$ *are isomorphic functors.*

Proof. Any object I of $\mathscr{S}$ fits in a pullback

$$\begin{array}{ccc} I & \longrightarrow & 1 \\ {\scriptstyle \{\cdot\}_I}\big\downarrow & & \big\downarrow{\scriptstyle \top} \\ {\Omega_{\mathscr{S}}}^I & \xrightarrow[\exists!_I]{} & \Omega_{\mathscr{S}} \end{array}$$

in $\mathscr{S}$. This pullback is sent by μ_* (modulo the ι isomorphisms) to the pullback

$$\begin{array}{ccc} \mu_*(I) & \longrightarrow & 1 \\ \big\downarrow & & \big\downarrow{\scriptstyle \top} \\ H^{e^*I} & \xrightarrow[\theta_I]{} & H \end{array}$$

in $\mathscr{E}$, where θ_I is defined in Definition 7.2.12. Thus, $\mu_*(I) \cong \mathrm{Part}_I(H)$. Of course, we must establish naturality also, but we leave this to the reader. □

Remark 7.2.14 *The morphisms $\theta_I \cong \mu_*(\exists!_I)$ in the definition of* $\mathrm{Part}(H)$ *are defined strictly in terms of the $\mathscr{S}$-bicomplete Heyting algebra structure that H carries. Thus, we see explicitly how to recover μ_* (and hence μ) from an $\mathscr{S}$-bicomplete Heyting algebra.*

Not every $\mathscr{S}$-bicomplete Heyting algebra arises as a distribution algebra. Our final goal in this section is to capture this distinction. If μ is a distribution, and $H = \mu_*(\Omega_{\mathscr{S}})$, then there are isomorphisms

$$\mathrm{Part}_{\Omega_{\mathscr{S}}{}^I}(H) \cong \mu_*(\Omega_{\mathscr{S}}{}^I) \cong H^{e^* I} .$$

The composite isomorphism, which we shall call σ_I, is equal to μ_* of the "big union" morphism (modulo the ι's):

$$\bigcup_I : \Omega_{\mathscr{S}}{}^{\Omega_{\mathscr{S}}{}^I} \longrightarrow \Omega_{\mathscr{S}}{}^I ,$$

but restricted to partitions in the domain. It turns out that σ_I has the following formula:

$$\sigma_I(p)(i) = \bigvee_{S \in e^*(\Omega_{\mathscr{S}}^I)} p(S) \wedge \|e^*(ev_I)(S, i)\| ,$$

where $ev_I : \Omega_{\mathscr{S}}{}^I \times I \longrightarrow \Omega_{\mathscr{S}}$ is evaluation.

A distribution algebra is $\mathscr{S}$-atomic in the following sense.

Definition 7.2.15 *An $\mathscr{S}$-bicomplete Heyting algebra H is said to be $\mathscr{S}$-*atomic *if for every I, the morphism*

$$\sigma_I : \mathrm{Part}_{\Omega_{\mathscr{S}}{}^I}(H) \longrightarrow H^{e^* I}$$

defined above is an isomorphism.

Informally, but perhaps more intuitively, we have:

$$\sigma_I(\{p_S | S \subseteq I\}) = \{ \bigvee_{i \in S} p_S | i \in I\} .$$

Let $\mathrm{caHA}_{\mathscr{S}}(\mathscr{E})$ denote the category of $\mathscr{S}$-bicomplete $\mathscr{S}$-atomic Heyting algebras in $\mathscr{E}$. We have a forgetful functor $\mathrm{caHA}_{\mathscr{S}}(\mathscr{E}) \longrightarrow \mathscr{E}$.

Theorem 7.2.16 *Let $\mathscr{E}$ be an $\mathscr{S}$-topos. Then there is an equivalence*

$$\Phi : \mathbf{E}(\mathscr{E})^{\mathrm{op}} \longrightarrow \mathrm{caHA}_{\mathscr{S}}(\mathscr{E})$$

that commutes with the underlying sheaf and forgetful functors over $\mathscr{E}$. The functor Φ sends a distribution μ to its corresponding distribution algebra $H = \mu_*(\Omega_{\mathscr{S}})$, and a natural transformation $\alpha : \mu \longrightarrow \nu$ to the evaluation at $\Omega_{\mathscr{S}}$ of the natural transformation $\nu_* \longrightarrow \mu_*$ induced from α by adjointness.

Proof. We have already shown that if μ is a distribution, then $H = \mu_*(\Omega_{\mathscr{S}})$ is an $\mathscr{S}$-bicomplete, $\mathscr{S}$-atomic Heyting algebra, and we recover μ_* as $\mathrm{Part}(H)$.

On the other hand, if we start with an H, then by Exercise 1, $\mathrm{Part}(H)$ preserves internal limits. By the fibred adjoint functor theorem, $\mathrm{Part}(H)$ has a left adjoint μ, which is a distribution on $\mathscr{E}$, whose $\mathscr{S}$-bicomplete, $\mathscr{S}$-atomic Heyting algebra is $\mathrm{Part}_{\Omega_{\mathscr{S}}}(H)$. But this latter is canonically isomorphic to H itself, via σ_1. □

Remark 7.2.17 *Summing up, we have seen that the morphisms $\exists!_I$, $\exists_\alpha$, and the 'big union' morphisms $\bigcup_I$ in $\mathscr{S}$ are important in the theory of distribution algebras. The application of μ_* to these give θ_I, $\bigvee_\alpha$, and σ_I respectively (modulo the ι isomorphisms).*

Exercises 7.2.18

1. *Prove that a Heyting algebra H is $\mathscr{S}$-atomic iff H is* weakly Boolean *in the sense that σ_1 is an isomorphism, and in addition the fibred functor* Part(H) *preserves internal limits.*

7.3 Relative Stone Locales

We develop a theory of Stone locales that is relative to a base topos. This theory is related to a relative pure, entire factorization whose existence and uniqueness is shown in the next two sections.

Definition 7.3.1 *An $\Omega_{\mathscr{S}}$-*ideal *of an $\Omega_{\mathscr{S}}$-cocomplete poset P in a topos $\mathscr{E}$ is a subobject of P such that:*

1. *its classifying map $\chi : P \longrightarrow \Omega_{\mathscr{E}}$ is order-reversing, in the sense that it satisfies $(p \leq q) \Rightarrow (\chi(q) \Rightarrow \chi(p))$, and*
2. *for any definable subobject $\alpha : X \rightarrowtail Y$ in $\mathscr{E}$, the diagram*

$$\begin{array}{ccc} P^X & \xrightarrow{\bigvee_\alpha} & P^Y \\ {\scriptstyle \chi^X}\downarrow & & \downarrow{\scriptstyle \chi^Y} \\ {\Omega_{\mathscr{E}}}^X & \xrightarrow[\bigwedge_\alpha]{} & {\Omega_{\mathscr{E}}}^Y \end{array}$$

commutes.

We denote the subobject of ${\Omega_{\mathscr{E}}}^P$ in $\mathscr{E}$ of all $\Omega_{\mathscr{S}}$-ideals of an $\Omega_{\mathscr{S}}$-cocomplete poset P by $Idl_{\Omega_{\mathscr{S}}}(P)$

Definition 7.3.2 An $\Omega_{\mathscr{S}}$-distributive lattice *is an $\Omega_{\mathscr{S}}$-cocomplete poset in which finite meets distribute over $\Omega_{\mathscr{S}}$-joins.*

Proposition 7.3.3 *Let P be an $\Omega_{\mathscr{S}}$-cocomplete poset. Then the following hold.*

1. *$\mathrm{Idl}_{\Omega_{\mathscr{S}}}(P)$ is an internally cocomplete poset in $\mathscr{E}$. Moreover, the map $\downarrow : P \longrightarrow \mathrm{Idl}_{\Omega_{\mathscr{S}}}(P)$ sending $x \in P$ to $\downarrow(x) = \{y \in P \mid y \leq x\}$ is universal among all $\Omega_{\mathscr{S}}$-cocontinuous poset morphisms to internally cocomplete posets.*

2. *If P is an $\Omega_{\mathcal{S}}$-distributive lattice, then the poset $Idl_{\Omega_{\mathcal{S}}}(P)$ is a frame. Moreover, in this case, if $P \xrightarrow{h} A$ is an $\Omega_{\mathcal{S}}$-cocontinuous lattice homomorphism into a frame A, then the induced internally cocontinuous morphism $Idl_{\Omega_{\mathcal{S}}}(P) \longrightarrow A$ is a frame homomorphism.*

Proof. 1. $\downarrow(x)$ is an $\Omega_{\mathcal{S}}$-ideal and $\downarrow(\_)$ preserves definable joins. In particular, $\downarrow(\_)$ preserves the action of $e^*(\Omega_{\mathcal{S}})$: for $u \in e^*(\Omega_{\mathcal{S}})$ and $x \in P$, $u \cdot \downarrow(x) = \downarrow(u \cdot x)$. Next observe that $Idl_{\Omega_{\mathcal{S}}}(P)$ is a full reflective subposet of $\downarrow(P)$, the frame of downclosed subobjects of P, since $Idl_{\Omega_{\mathcal{S}}}(P)$ is closed in $\downarrow(P)$ under (internal) intersections. Therefore, $Idl_{\Omega_{\mathcal{S}}}(P)$ is internally cocomplete. For any $\Omega_{\mathcal{S}}$-ideal A, we have

$$A = \bigvee_{p \in A} \downarrow(p)$$

so that any internally cocontinuous map from $Idl_{\Omega_{\mathcal{S}}}(P)$ is uniquely determined by its composition with $\downarrow(\_)$. In particular, the internally cocontinuous extension to $Idl_{\Omega_{\mathcal{S}}}(P)$ of an $\Omega_{\mathcal{S}}$-cocontinuous poset morphism $P \xrightarrow{h} L$ must be given by $A \mapsto \bigvee_{p \in A} h(p)$. Moreover, this map is indeed cocontinuous since it has a right adjoint given by $x \mapsto \{p \mid h(p) \leq x\}$, which is clearly an $\Omega_{\mathcal{S}}$-ideal.

2. In order to show that $Idl_{\Omega_{\mathcal{S}}}(P)$ is a frame it is enough to prove, since $Idl_{\Omega_{\mathcal{S}}}(P)$ is already known to be closed in $\downarrow(P)$ under internal infima, that if $B \in \downarrow(P)$ and $A \in Idl_{\Omega_{\mathcal{S}}}(P)$, then $[B \Rightarrow A] \in Idl_{\Omega_{\mathcal{S}}}(P)$. In turn, this reduces to the case $B = \downarrow(p)$ for $p \in P$, since for arbitrary B we have

$$[B \Rightarrow A] = \left[\left(\bigcup_{p \in B} \downarrow(p) \right) \Rightarrow A \right] = \bigcap_{p \in B} [\downarrow(p) \Rightarrow A] \,.$$

The claim now follows from the following considerations: $x \in [\downarrow(p) \Rightarrow A]$ iff $\downarrow(x) \subseteq [\downarrow(p) \Rightarrow A]$ iff $\downarrow(x) \cap \downarrow(p) \subseteq A$ iff $x \wedge p \in A$.

Finally, $\downarrow(\_)$ preserves binary meets, so the extension of $P \xrightarrow{h} A$ to $Idl_{\Omega_{\mathcal{S}}}(P)$ preserves them because h does. □

If we take $I = \Omega_{\mathcal{S}}$ in Definition 7.2.12, then this defines an $\Omega_{\mathcal{S}}$-partition of an $\mathcal{S}$-bicomplete Heyting algebra. We may also define an $\Omega_{\mathcal{S}}$-partition, for an $\Omega_{\mathcal{S}}$-Heyting algebra. As always, we have the unit $\|\cdot\|$ of $\Omega_{\mathcal{S}}$-Heyting algebra H. Also, $\mathrm{eq}_{\Omega_{\mathcal{S}}} : \Omega_{\mathcal{S}} \times \Omega_{\mathcal{S}} \longrightarrow \Omega_{\mathcal{S}}$ denotes the characteristic morphism of the diagonal $\Omega_{\mathcal{S}} \rightarrowtail \Omega_{\mathcal{S}} \times \Omega_{\mathcal{S}}$. Then $u = u'$ stands for $e^*(\mathrm{eq}_{\Omega_{\mathcal{S}}})(u, u')$.

Definition 7.3.4 An $\Omega_{\mathcal{S}}$-partition (henceforth, partition) *of an $\Omega_{\mathcal{S}}$-Heyting algebra H is a morphism $e^*(\Omega_{\mathcal{S}}) \xrightarrow{p} H$ such that:*

1.

$$\bigvee_{u \in e^*(\Omega_{\mathscr{S}})} p(u) = 1_H \ ,$$

and

2.

$$\forall u, u' \in e^*(\Omega_{\mathscr{S}}) : \ p(u) \wedge p(u') \ \le \|u = u'\| \ .$$

Let $\mathrm{Part}(H) \rightarrowtail H^{e^*\Omega_{\mathscr{S}}}$ *denote the subobject of partitions of* H*.*

Definition 7.3.5 *An* $\Omega_{\mathscr{S}}$*-Heyting algebra* H *is said to be an* $\Omega_{\mathscr{S}}$-Boolean algebra *if the morphism*

$$\sigma : \mathrm{Part}(H) \longrightarrow H$$

defined by

$$\sigma(p) = \bigvee_{u \in e^*(\Omega_{\mathscr{S}})} p(u) \wedge \|u\| \ ,$$

is an isomorphism.

We bring $e^*(\Omega_{\mathscr{S}})$ into a frame $\mathscr{O}(X)$ in $\mathscr{E}$ via the canonical morphisms

$$e^*(\Omega_{\mathscr{S}}) \xrightarrow{\tau} \Omega_{\mathscr{E}} \longrightarrow \mathscr{O}(X) \ ,$$

where the second is the unique frame morphism. Usually we ignore this morphism in the notation.

We prefer to use a special term for a partition of a frame. We repeat Definition 7.3.4 in this case.

Definition 7.3.6 *Let* $\mathscr{O}(X)$ *denote a frame in a topos* $\mathscr{E}$ *over* $\mathscr{S}$*.* A flat function in $\mathscr{O}(X)$ *is a morphism* $e^*(\Omega_{\mathscr{S}}) \xrightarrow{p} \mathscr{O}(X)$ *such that:*

1.

$$\bigvee_{u \in e^*(\Omega_{\mathscr{S}})} p(u) = 1_X \ ,$$

and

2.

$$\forall u, u' \in e^*(\Omega_{\mathscr{S}}) : \ p(u) \wedge p(u') \ \le u = u' \ .$$

Let $\mathrm{Flat}(\mathscr{O}(X))$ *denote the subobject of flat functions in* $\mathscr{O}(X)$*.*

Lemma 7.3.7 *Condition* (2) *of a flat function is equivalent to*

$$(2') \ \ p(u) \wedge p(u') \le \bigvee \{V \in \mathscr{O}(X) \mid (V = p(u)) \wedge (u = u')\} \ .$$

Proof. $2'$ implies 2 because

$$\bigvee\{V \in \mathscr{O}(X) \mid (V = p(u)) \wedge (u = u')\} \leq u = u'$$

If 2 holds, then

$$\begin{aligned} p(u) \wedge p(u') &\leq p(u) \wedge (u = u') \\ = p(u) \wedge \bigvee\{V \in \mathscr{O}(X) \mid u = u'\} &= \bigvee\{p(u) \wedge V \mid u = u'\} \\ &\leq \bigvee\{W \mid (W = p(u)) \wedge (u = u')\} . \end{aligned}$$

To see that the last inequality holds consider $W = p(u)$. □

Lemma 7.3.8 *A morphism $e^*\Omega_{\mathscr{S}} \xrightarrow{p} \mathscr{O}(X)$ is flat iff its unique suplattice extension p^* in*

$$\begin{array}{ccc} e^*\Omega_{\mathscr{S}} & \xrightarrow{\{\cdot\}} & {\Omega_{\mathscr{E}}}^{e^*\Omega_{\mathscr{S}}} \\ & \searrow^{p} & \downarrow{p^*} \\ & & \mathscr{O}(X) \end{array}$$

is a frame homomorphism, i.e., iff p^* preserves finite infima.

Proof. The reader may find Lemma 7.3.7 helpful for this purpose. □

Lemma 7.3.9 *The pointwise ordering of flat functions in $\mathscr{O}(X)$ is discrete: if $\forall u : p(u) \leq q(u)$, then $p = q$.*

Proof. A basic principle may be used to explain this. Local homeomorphisms classify global sections in the sense that

$$\mathbf{Top}_{\mathscr{S}}(\mathscr{X}, \mathscr{S}/A) \simeq \mathscr{X}(1, x^*A) ,$$

for any topos $\mathscr{X} \xrightarrow{x} \mathscr{S}$. This is a discrete category. In our situation we know that the poset of flat functions in $\mathscr{O}(X)$ is isomorphic to the category of locale morphisms $X \longrightarrow |e^*\Omega_{\mathscr{S}}|$ in $\mathscr{E}$, which in turn is equivalent to the discrete category $\mathbf{Top}_{\mathscr{E}}(Sh(X), \mathscr{E}/e^*\Omega_{\mathscr{S}})$. □

An $\mathscr{S}$-atomic $\mathscr{S}$-bicomplete Heyting algebra is an $\Omega_{\mathscr{S}}$-Boolean algebra. Our immediate goal is to show, more generally, that for any definable dominance $\mathscr{F}$, and any geometric morphism $\mathscr{F} \xrightarrow{\psi} \mathscr{E}$, the $\Omega_{\mathscr{S}}$-Heyting algebra $H = \psi_* f^* \Omega_{\mathscr{S}}$ is $\Omega_{\mathscr{S}}$-Boolean. In this situation, the diagram of canonical morphisms

$$\begin{array}{ccc} e^*(\Omega_{\mathscr{S}}) & \xrightarrow{\|\cdot\|} & H \\ {\scriptstyle\tau}\downarrow & & \downarrow \\ \Omega_{\mathscr{E}} & \longrightarrow & \psi_*(\Omega_{\mathscr{F}}) \end{array}$$

commutes, where the bottom horizontal is the unique frame morphism.

Lemma 7.3.10 *A function* $e^*(\Omega_{\mathscr{S}}) \xrightarrow{p} H$ *is an partition iff*

$$e^*(\Omega_{\mathscr{S}}) \xrightarrow{p} H \rightarrowtail \psi_*(\Omega_{\mathscr{F}})$$

is flat. We have an order preserving inclusion

$$\Sigma : \mathrm{Part}(H) \rightarrowtail \mathrm{Flat}(\psi_* \Omega_{\mathscr{F}}) .$$

given by composition with $H \rightarrowtail \psi_*(\Omega_{\mathscr{F}})$.

Proposition 7.3.11 *Let* $\psi : \mathscr{F} \longrightarrow \mathscr{E}$ *be any geometric morphism over* $\mathscr{S}$. *Then we have a canonical isomorphism*

$$H = \psi_* f^* \Omega_{\mathscr{S}} \cong \mathrm{Flat}(\psi_* \Omega_{\mathscr{F}}) .$$

Proof. For any $E \in \mathscr{E}$, we have natural bijections

$$\frac{\dfrac{\dfrac{\dfrac{\dfrac{\dfrac{\text{morphisms } E \longrightarrow \psi_* f^* \Omega_{\mathscr{S}}}{\text{morphisms } \psi^* E \longrightarrow f^* \Omega_{\mathscr{S}}}}{\text{geo. morphisms } \mathscr{F}/\psi^* E \longrightarrow \mathscr{F}/f^* \Omega_{\mathscr{S}} \text{ over } \mathscr{F}}}{\text{geo. morphisms } \mathscr{F}/\psi^* E \longrightarrow \mathscr{E}/e^* \Omega_{\mathscr{S}} \text{ over } \mathscr{E}}}{\text{frame morphisms } \Omega_{\mathscr{E}}^{e^*(\Omega_{\mathscr{S}})} \longrightarrow \psi_*({\Omega_{\mathscr{F}}}^{\psi^* E}) \cong \psi_*(\Omega_{\mathscr{F}})^E}}{\text{flat morphisms } e^* \Omega_{\mathscr{S}} \longrightarrow \psi_*(\Omega_{\mathscr{F}})^E}}{\text{morphisms } E \longrightarrow \mathrm{Flat}(\psi_* \Omega_{\mathscr{F}}) ,}$$

where Lemma 7.3.8 justifies the second last bijection. □

Starting with $\mathscr{F} \xrightarrow{\psi} \mathscr{E}$, we wish to show that partitions of $H = \psi_* f^* \Omega_{\mathscr{S}}$ correspond to flat functions in the frame $\psi_*(\Omega_{\mathscr{F}})$. We have some preliminary work to do for this.

Lemma 7.3.12 *Any flat function* $e^* \Omega_{\mathscr{S}} \xrightarrow{p} \mathscr{O}(X)$ *in a topos* $\mathscr{E}$ *satisfies:*

1.

$$p(\top) = \bigvee_{e^* \Omega_{\mathscr{S}}} p(u) \wedge u ,$$

and

2.

$$\forall u \in e^* \Omega_{\mathscr{S}} : p(u) \leq (p(\top) \Leftrightarrow u) .$$

Proof. 1. In the frame ${\Omega_{\mathscr{E}}}^{e^* \Omega_{\mathscr{S}}}$ we have

$$\{\top\} = \bigcup_{e^* \Omega_{\mathscr{S}}} \{u\} \cap u , \tag{7.1}$$

where $\{\cdot\} : e^*\Omega_{\mathcal{S}} \longrightarrow \Omega_{\mathcal{E}}{}^{e^*\Omega_{\mathcal{S}}}$. Indeed, referring to the pullback

$$\begin{array}{ccc} 1 & \longrightarrow & 1 \\ {\scriptstyle\top}\downarrow & & \downarrow{\scriptstyle\top} \\ e^*\Omega & \xrightarrow[\tau]{} & \Omega_{\mathcal{E}} \end{array} ,$$

for any $v \in e^*\Omega$, we have $v \in \{u\} \cap u$ for some $u \in e^*\Omega$ iff $\tau(u) = \top$ and $v = u$ for some u iff $u = \top$ and $v = u$ for some u iff $v = \top$ iff $v \in \{\top\}$. The conclusion now follows by applying the unique frame extension p^* to (7.1).

2. By 1, for any u we have $p(u) \leq (u \Rightarrow p(\top))$. On the other hand, $p(u) \wedge p(\top) \leq (u = \top) = u$, and therefore $p(u) \leq (p(\top) \Rightarrow u)$. Hence, $p(u) \leq (p(\top) \Leftrightarrow u)$. □

Lemma 7.3.13 *For any partition p, we have $\sigma(p) = p(\top)$, and Σ commutes with σ and the canonical isomorphism $H \cong \mathrm{Flat}(\psi_*\Omega_{\mathcal{F}})$.*

Lemma 7.3.14 *Let $f : \mathcal{F} \longrightarrow \mathcal{S}$ be subopen. Let Y be any object of $\mathcal{F}$. Then any flat function $\Omega_{\mathcal{S}} \longrightarrow f_*(\Omega_{\mathcal{F}}{}^Y)$ factors uniquely through $f_*(f^*(\Omega_{\mathcal{S}})^Y)$.*

Proof. Let $\Omega_{\mathcal{S}} \xrightarrow{p} f_*(\Omega_{\mathcal{F}}{}^Y)$ be flat. Then p corresponds to a definable subobject $D \rightarrowtail Y$. The correspondence can be given explicitly. Say D is witnessed by a pullback

$$\begin{array}{ccc} D & \rightarrowtail & Y \\ \downarrow & & \downarrow \\ f^*(K) & \rightarrowtail & f^*(I) \end{array} .$$

We regard the subobject $K \rightarrowtail I$ as a flat function $\Omega_{\mathcal{S}} \xrightarrow{q} \Omega_{\mathcal{S}}{}^I$. Then p is recovered as the following pullback.

$$\begin{array}{ccc} p(u) & \rightarrowtail & Y \\ \downarrow & & \downarrow \\ f^*(q(u)) & \rightarrowtail & f^*(I) \end{array}$$

But evidently every $p(u)$ is a definable subobject of Y, so p factors through $f_*(f^*(\Omega_{\mathcal{S}})^Y)$. □

Proposition 7.3.15 *Let $\psi : \mathcal{F} \longrightarrow \mathcal{E}$ have subopen domain. Then $H = \psi_* f^*\Omega_{\mathcal{S}}$ is an $\Omega_{\mathcal{S}}$-Boolean algebra.*

Proof. Let $e^*\Omega_{\mathcal{S}} \xrightarrow{p} (\psi_*\Omega_{\mathcal{F}})^E \cong \psi_*(\Omega_{\mathcal{F}}{}^{\psi^*E})$ be a flat function in $\psi_*(\Omega_{\mathcal{F}})$. We must show that p factors through H^E. The transpose $\Omega_{\mathcal{S}} \xrightarrow{\hat{p}} f_*(\Omega_{\mathcal{F}}{}^{\psi^*E})$ is a flat function, which corresponds to the same definable subobject of ψ^*E

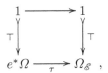

in $\mathscr{F}$. By Lemma 7.3.14, $\hat{p}$ factors through $f_*(f^*(\Omega_{\mathscr{S}})^{\psi^* E})$. Now transpose this to

$$e^*(\Omega_{\mathscr{S}}) \longrightarrow \psi_*(f^*(\Omega_{\mathscr{S}})^{\psi^* E}) \cong H^E$$

in $\mathscr{E}$, which is a factorization of the given p. □

Definition 7.3.16 *A locale in $\mathscr{E}$ is said to be* a Stone locale (relative to $\mathscr{S}$) *if its frame is isomorphic to $\mathrm{Idl}_{\Omega_{\mathscr{S}}}(H)$, for an $\Omega_{\mathscr{S}}$-Boolean algebra H. A geometric morphism $\psi : \mathscr{F} \longrightarrow \mathscr{E}$ over $\mathscr{S}$ is* entire (relative to $\mathscr{S}$) *if it is localic and defined by an Stone locale.*

If $\mathscr{S}$ is a Boolean topos, then this notion of Stone locale agrees with the usual. We shall usually omit the phrase 'relative to $\mathscr{S}$'.

We next discuss the existence of a relative pure, entire factorization.

Lemma 7.3.17 *Let $\{A_u\}$ be an $e^*\Omega_{\mathscr{S}}$-indexed family of $\Omega_{\mathscr{S}}$-ideals of an $\Omega_{\mathscr{S}}$-Heyting algebra H in $\mathscr{E}$. Let $q \in A_u$ for a given $u \in e^*\Omega_{\mathscr{S}}$. Then*

$$\forall v \in e^*\Omega_{\mathscr{S}} : \ (\|u = v\| \wedge q) \in A_v$$

holds.

Proof. Assume that $q \in A_u$ for a given $u \in e^*\Omega_{\mathscr{S}}$. Let χ_u denote the characteristic morphism of A_u. So $\chi_u(q)$ holds. Let χ_v denote the characteristic morphism of the $\Omega_{\mathscr{S}}$-ideal A_v, for $v \in e^*\Omega_{\mathscr{S}}$. Since A_v is an $\Omega_{\mathscr{S}}$-ideal, for all $w \in e^*\Omega_{\mathscr{S}}$ we have

$$\chi_v(w \cdot q) = (\tau(w) \Rightarrow \chi_v(q)) \,.$$

For w take $u = v$ in the above, and use $(u = v) \cdot q = \|u = v\| \wedge q$, and $\tau(u = v) \Leftrightarrow (u = v)$. Then

$$\chi_v(\|u = v\| \wedge q) \ = \ ((u = v) \Rightarrow \chi_v(q)) \,.$$

But as

$$((u = v) \Rightarrow \chi_v(q)) \Leftrightarrow ((u = v) \Rightarrow \chi_u(q))$$

and $\chi_u(q)$ holds by assumption, we conclude that $\chi_v(\|u = v\| \wedge q)$ holds, as claimed. □

Lemma 7.3.18 *An $\Omega_{\mathscr{S}}$-join of $\Omega_{\mathscr{S}}$-ideals A_u in an $\Omega_{\mathscr{S}}$-Heyting algebra H is given by*

$$\bigvee_{u \in e^*(\Omega_{\mathscr{S}})} A_u = \left\{ \bigvee_{u \in e^*(\Omega_{\mathscr{S}})} p_u \mid p \in \prod A_u \right\} .$$

Proof. The object on the right-hand side of the above claimed equation is already down-closed. Also, since each A_u is an $\Omega_{\mathscr{S}}$-ideal and since suprema in the product $\prod A_u$ are defined pointwise, it is clear that the right-hand side is an $\Omega_{\mathscr{S}}$-ideal of H. It remains to prove that each A_u is contained in the right-hand side:

$$A_u \subseteq \left\{ \bigvee_{u \in e^*(\Omega_{\mathscr{S}})} p_u \mid p \in \prod A_u \right\} .$$

This will establish the claim because any $\Omega_{\mathscr{S}}$-ideal containing every A_u must also contain the right-hand side; therefore it would indeed be the smallest. Let $q \in A_u$. We have

$$q = \bigvee_{v \in e^*(\Omega_{\mathscr{S}})} (\|u = v\| \wedge q) .$$

Define $p(q)$ by $p(q)_v = (\|u = v\| \wedge q)$. By Lemma 7.3.17, we have $p(q)_v \in A_v$, so $p(q) \in \prod A_u$. Hence q is a member of the right-hand side. □

Proposition 7.3.19 *Let H be an $\Omega_{\mathscr{S}}$-Heyting algebra in $\mathscr{E}$, with a top element 1_H. Then the morphism*

$$\vartheta : \mathrm{Part}(H) \longrightarrow \mathrm{Flat}(Idl_{\Omega_{\mathscr{S}}}(H))$$

induced by $\downarrow : H \longrightarrow Idl_{\Omega_{\mathscr{S}}}(H)$ is an isomorphism in $\mathscr{E}$.

Proof. ϑ is a monomorphism because $\downarrow$ is one. In order to show that θ is surjective, let $e^*\Omega_{\mathscr{S}} \xrightarrow{\beta} Idl_{\Omega_{\mathscr{S}}}(H)$ be a flat function. We have

$$\mathbf{1}_{Idl_{\Omega_{\mathscr{S}}}(H)} = \bigvee_{u \in e^*\Omega_{\mathscr{S}}} \beta(u) .$$

By Lemma 7.3.18, this implies that

$$H = \left\{ \bigvee_{u \in e^*\Omega_{\mathscr{S}}} p_u \mid p \in \prod \beta(u) \right\} .$$

In particular,

$$1_H = \bigvee_{u \in e^*\Omega_{\mathscr{S}}} \{r_u \mid r_u \in \beta(u)\}$$

for some family $\{r_u \mid r_u \in \beta(u)\}$. The function $\alpha(u) = \downarrow r_u$ is flat, and satisfies $\alpha \leq \beta$. By Lemma 7.3.9, we have $\alpha = \beta$, so that β factors uniquely through H. □

Theorem 7.3.20 *A geometric morphism over $\mathscr{S}$ whose domain is a definable dominance admits a pure, entire factorization. Moreover, the middle topos in the factorization is a definable dominance.*

Proof. Let $\mathscr{F} \xrightarrow{\psi} \mathscr{E}$ be a geometric morphism over $\mathscr{S}$, where $\mathscr{F}$ is a definable dominance. Let $H = \psi_* f^* \Omega_{\mathscr{S}}$, and let $\varphi : \mathscr{G} \longrightarrow \mathscr{E}$ denote the $\mathscr{E}$-topos of sheaves on the frame $Idl_{\Omega_{\mathscr{S}}}(H)$. By definition, φ is entire. By the universal property of the free frame on the $\Omega_{\mathscr{S}}$-distributive lattice H, the morphism $\psi_*\tau : H = \psi_* f^* \Omega_{\mathscr{S}} \longrightarrow \psi_* \Omega_{\mathscr{F}}$ induces a morphism

$$Idl_{\Omega_{\mathscr{S}}}(H) \longrightarrow \psi_* \Omega_{\mathscr{F}}$$

of frames. In turn, the latter induces a geometric morphism $\mathscr{F} \xrightarrow{\pi} \mathscr{G}$ such that $\psi \cong \varphi \cdot \pi$. To see that π is pure, we show that the unit of the adjunction $\pi^* \dashv \pi_*$ evaluated at $g^* \Omega_{\mathscr{S}}$ is an isomorphism for φ-global sections. This follows from the isomorphisms:

$$\varphi_* g^* \Omega_{\mathscr{S}} \cong \mathrm{Flat}(\varphi_* \Omega_{\mathscr{G}}) \cong \mathrm{Flat}(\mathrm{Idl}_{\Omega_{\mathscr{S}}}(H))$$

$$\cong \mathrm{Part}(H) \cong H = \psi_* f^* \Omega_{\mathscr{S}} \cong \varphi_* \pi_* \pi^* g^* \Omega_{\mathscr{S}},$$

where the first isomorphism holds by Proposition 7.3.11, the second by construction of φ, the third by Proposition 7.3.19, the fourth since H is $\Omega_{\mathscr{S}}$-Boolean, and the last by the natural isomorphisms $\psi \cong \varphi \cdot \pi$ and $f \cong g \cdot \pi$. For the general case we may localize and repeat this argument. Thus, π is pure. In particular, by Proposition 2.2.7, $\mathscr{G}$ is also a definable dominance. □

Remark 7.3.21 *A Stone locale is traditionally defined as a compact, zero-dimensional locale, whereas the notion that we have employed here is usually given instead as a characterization. We take zero-dimensional to mean the spread condition that we have defined in Chapter 3. Stone locales are spreads. The notion of weak compactness for locales, due to J. J. C. Vermeulen, together with the spread condition, provides the possibility of a characterization that parallels the traditional view. At the level of geometric morphisms, the 'compact, zero-dimensional' approach is explored in two different ways by I. Moerdijk and J. J. C. Vermeulen, also relative to an arbitrary base topos. We do not pursue this matter further here as it is beyond the scope of this book.*

Remark 7.3.22 *We have seen that a distribution on a topos $\mathscr{E}$ has associated with it a complete spread over $\mathscr{E}$, and a distribution algebra in $\mathscr{E}$. If μ is a distribution on $\mathscr{E}$, let X_μ denote the locale in $\mathscr{E}$ that defines the complete spread (as we know, spreads are localic). We call X_μ* the display locale *of μ. As usual, $\mathscr{O}(X_\mu)$ denotes the frame of this locale. We sometimes write the complete spread as $\mathrm{Sh}_{\mathscr{E}}(X_\mu) \longrightarrow \mathscr{E}$. There is a canonical inclusion $H \rightarrowtail \mathscr{O}(X_\mu)$ of the distribution algebra of μ into the display frame.*

Exercises 7.3.23

1. *Show that an $\Omega_{\mathscr{S}}$-Heyting algebra is an $\Omega_{\mathscr{S}}$-distributive lattice.*
2. *Prove that an $\Omega_{\mathscr{S}}$-ideal of an $\Omega_{\mathscr{S}}$-Heyting algebra H always contains 0_H.*

3. *Continuing in the notation of Remark 7.3.22, prove that the display locale X_μ is the largest sublocale X of the Stone locale of H (whose frame is $Idl_{\mathscr{S}}(H)$) such that $Sh_{\mathscr{E}}(X)$ is locally connected over $\mathscr{S}$, and the diagram*

$$\begin{array}{ccc} \mathrm{Flat}(\mathscr{O}(X)) & \xrightarrow{\ \sigma\ } & \mathscr{O}(X) \\ \downarrow & & \downarrow \\ H & \xrightarrow{\ \downarrow\ } & Idl_{\mathscr{S}}(H) \end{array}$$

commutes, where the left vertical is a canonical isomorphism.

7.4 On a version of the Heine-Borel Theorem

We have seen two alternative constructions of the complete spread associated with a distribution, one as a bicomma object (Theorem 6.1.6) and the other as a bipullback (Theorem 2.4.8), which we know are equivalent because the two constructions are both right adjoint to the same functor. This right adjoint provides the comprehensive factorization for geometric morphisms with a locally connected domain.

The bicomma construction of this factorization gives a transparent explanation of why the middle topos may be thought of as the 'space of cogerms' of the given map. Indeed, consider the universal property of the bicomma object

$$\begin{array}{ccc} \mathscr{Y} & \xrightarrow{\ y\ } & \mathscr{S} \\ {\scriptstyle\psi}\downarrow & \Rightarrow & \downarrow{\scriptstyle p} \\ \mathscr{E} & \underset{\delta}{\rightarrowtail} & \mathrm{M}(\mathscr{E}) \end{array}$$

in $\mathbf{Top}_{\mathscr{S}}$ associated with a point p of the symmetric topos $\mathrm{M}(\mathscr{E})$, or equivalently with a distribution μ on $\mathscr{E}$. A point of $\mathscr{Y}$ is given by a pair consisting of a point q of $\mathscr{E}$, and a natural transformation $q^* \cdot \delta^* \longrightarrow p^*$. The latter transposes to $t : q^* \longrightarrow p^* \cdot \delta_!$. The given μ is isomorphic to $p^* \cdot \delta_! \cong y_! \cdot \psi^*$, by Proposition 5.2.1 applied to the symmetric monad. When $\mathscr{E}$ is a localic topos $Sh(X)$, so is $\mathscr{Y} = Sh(Y)$. In particular, if X is a topological space, then the points of Y correspond (if X is sober) to pairs (x, t) consisting of an element $x \in X$ and a natural transformation $t : x^* \longrightarrow \mu$, where $x^* : Sh(X) \longrightarrow \mathscr{S}$ is the point-distribution corresponding to x. Of course, this Y is precisely what we have called the display space (of cogerms) of μ: a natural transformation $t : x^* \longrightarrow \mu$ is precisely a cogerm of ψ at x, since such a t amounts to a consistent choice $\{t_U \in \mu(U) \cong y_! \psi^*(U)\}$ of components of inverse images under ψ of neighborhoods U of x.

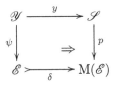

In topology we have seen (Definition 2.1.2 and Proposition 2.1.3) that a map is complete essentially when every cogerm of the map is the cogerm of a point, which is to say that "cogerms converge." This suggests that we introduce the following terminology. We shall say a geometric morphism's *cogerms*

converge if the pure factor of its comprehensive factorization is a surjection (assuming of course that the geometric morphism has locally connected domain). This is justified because, as we have just explained, the middle topos in the comprehensive factorizaton of a map is the space of cogerms of the map.

On the other hand, Definition 3.5.2 introduces a completeness notion that is a cover-refinement condition. Let us say that a geometric morphism (with locally connected domain and bounded codomain) is *cover-refinement complete* if it is complete in the sense of Definition 3.5.2.

Perhaps it is worthwhile at this point to see what cover-refinement complete comes down to in the simpler context of topological spaces. Consider a map $\psi : X \twoheadrightarrow B$, where X is locally connected. Let $U \subseteq B$ denote an open subset. Let $\alpha \subseteq \psi^{-1}(U)$ denote a connected component, which is an open subset as X is locally connected. As always, the notation (U, α) means such a pair. Then the completeness condition Definition 3.5.2 takes the following form for such a map ψ with locally connected domain. Associated with any pair (U, α) and a cover $\mathscr{U} = \{U_i \subseteq U \mid i \in I\}$ (so that $\bigcup U_i = U$) is the set

$$J(\mathscr{U}, \alpha) = \{(i, \beta) \mid i \in I,\ \beta \in \pi_0(\alpha \cap \psi^{-1}(U_i))\} ,$$

where as always π_0 denotes connected components. Then we say ψ *is complete* if for every (U, α), and every family $\{(V_k, \gamma_k) \mid k \in K\}$ with $V_k \subseteq U$, $\gamma_k \subseteq \alpha$, and such that $\bigcup \gamma_k = \alpha$, there is a cover

$$\mathscr{U} = \{U_i \subseteq U \mid i \in I\}$$

of U, and a function

$$f : J(\mathscr{U}, \alpha) \longrightarrow K ,$$

such that for every $(i, \beta) \in J(\mathscr{U}, \alpha)$, we have $U_i \subseteq V_{f(i,\beta)}$ and $\beta \subseteq \gamma_{f(i,\beta)}$. These two inclusions can be expressed with a single commutative diagram.

$$\begin{array}{ccc} \beta & \rightarrowtail & \gamma_{f(i,\beta)} \\ \downarrow & & \downarrow \\ \psi^{-1}(U_i) & \twoheadrightarrow & \psi^{-1}(V_{f(i,\beta)}) \end{array}$$

Note that cover-refinement complete does not refer to points.

We may now interpret Theorem 3.5.3 as a sort of Heine-Borel theorem in our context. (The Heine-Borel theorem in topology says for a subset of Euclidean space R^n that any open cover of the subset has a finite refinement iff the subset is closed and bounded.)

> Theorem (Heine-Borel): *A geometric morphism with locally connected domain and bounded codomain is cover-refinement complete iff its cogerms converge.*

The above theorem is a version of the Heine-Borel theorem that is not only 'infinitary' (in a sense made more precise in the context of distribution algebras in § 7.2), but also point-free. We should also emphasize that the proof of the above 'Heine-Borel' theorem relies on the equivalent pullback construction of the comprehensive factorization, as this construction explicitly reveals how to describe the 'space of cogerms' in terms of a Grothendieck topology.

We also comment in Exercises 7.4.1 on the usual (finitary) version of the Heine-Borel theorem in our context.

Exercises 7.4.1

1. *Let X be a locally connected space. Show that $X \longrightarrow 1$ and $\mathrm{id} : X \longrightarrow X$ are both cover-refinement complete maps.*
2. *Under possibly additional assumptions on a space B, show that a map $X \longrightarrow B$ of spaces is cover-refinement complete iff its cogerms converge.*
3. *We may say a geometric morphism with a definable dominance as domain and bounded codomain is 'entire-complete' if the pure factor of its pure, entire factorization is a surjection. What cover-refinement property characterizes entire-complete?*
4. *Since a locally connected geometric morphism is a definable dominance, we may factor a complete spread into its pure and entire parts.*

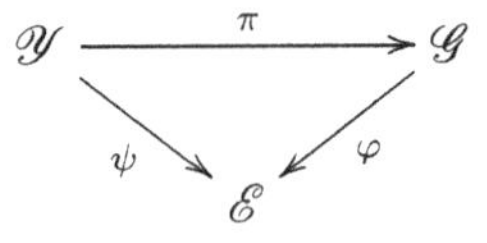

Show that the topos $\mathscr{G}$ is definable dominance, but not necessarily locally connected (Prop. 2.2.7). Argue that since ψ is a spread, so is π. Conclude that the pure spread π must be an inclusion (Lemma 3.1.12).

5. *Investigate the existence of a completion KZ-monad whose discrete fibrations are precisely the entire geometric morphisms with definable dominances as domains.*
6. *Using the above, deduce a (constructive, point-free) version of the usual Heine-Borel theorem in topology.*

7.5 The Entire Completion of a Spread via Forcing

We associate with a geometric morphism $\varphi : \mathscr{F} \longrightarrow \mathscr{E}$ over $\mathscr{S}$ the Heyting algebra $H = \varphi_* f^* \Omega_{\mathscr{S}}$. H comes equipped with a morphism

$$\eta_H = \eta : e^* \Omega_{\mathscr{S}} \longrightarrow H \ ,$$

called the unit for H. Indeed, this η is the unit of adjointness $\varphi^* \dashv \varphi_*$ evaluated at $e^* \Omega_{\mathscr{S}}$.

Consider the presheaf topos of H relative to $\mathscr{E}$:

$$\begin{array}{ccc} P(H) & \xrightarrow{\ \gamma\ } & \mathscr{E} \\ & \searrow \quad \swarrow_{e} & \\ & \mathscr{S} & \end{array}$$

As always, $h : H \longrightarrow P(H)$ denotes Yoneda, but over $\mathscr{E}$. Throughout, Ω_H denotes the subobject classifier of $P(H)$. We have a morphism $H \longrightarrow \gamma_*(\Omega_H)$ in $\mathscr{E}$ that sends an element of H to its down-closure.

In this section we shall produce a topos $\mathscr{E}[H] \xrightarrow{\psi} \mathscr{E}$ that best has the property that the $\mathscr{S}$-definable subobjects of an object $\psi^*(E)$ correspond to morphisms $E \longrightarrow H$ in $\mathscr{E}$. We construct $\mathscr{E}[H]$ as a subtopos of $P(H)$ in terms of a certain forcing condition. Ultimately, we shall show that if $H = \varphi_* f^* \Omega_{\mathscr{S}}$, where φ is a spread whose domain f is a definable dominance, then $\mathscr{E}[H] \longrightarrow \mathscr{E}$ is precisely the entire factor of φ.

Lemma 7.5.1 *For H as above, with unit η, the inequality*

$$\begin{array}{ccc} e^*\Omega_{\mathscr{S}} & \xrightarrow{\ \eta\ } & H \\ \downarrow & \leq & \downarrow \\ \Omega_{\mathscr{E}} & \longrightarrow & \gamma_*\Omega_H \end{array}$$

holds, where the bottom horizontal arrow in the above square denotes the unique frame morphism in $\mathscr{E}$.

Let us denote the transposes of the two morphisms $e^*\Omega_{\mathscr{S}} \longrightarrow \gamma_*\Omega_H$ in the above square as follows:

$$\vartheta : \gamma^*(e^*\Omega_{\mathscr{S}}) \xrightarrow{\gamma^*\eta} \gamma^* H \xrightarrow{\ \rho\ } \Omega_H$$

and

$$\tau : \gamma^*(e^*\Omega_{\mathscr{S}}) \longrightarrow \Omega_H \ ,$$

where τ is the classifying map of the subobject $\gamma^*\top : 1 \longrightarrow \gamma^*\Omega_{\mathscr{S}}$.

In order to introduce the forcing condition, consider the following pullbacks in $P(H)$.

$$\begin{array}{ccc} S & \longrightarrow & 1 \\ \downarrow & & \downarrow \\ Z & \longrightarrow & \gamma^*(e^*\Omega_{\mathscr{S}}) \\ \downarrow{\scriptstyle t} & & \downarrow{\scriptstyle \vartheta} \\ \gamma^* H & \xrightarrow{\ \rho\ } & \Omega_H \end{array}$$

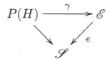

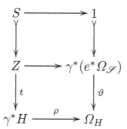

We wish to 'force' S to be a definable subobject of $\gamma^* H$ by forcing t to be an isomorphism. Thus, we introduce in $P(H)$ the smallest topology (over $\mathscr{E}$) for which t is bidense, denoting the corresponding subtopos of sheaves $j : \mathscr{E}[H] \rightarrowtail P(H)$. We shall call this *the definable topology* in H. This makes H a site in $\mathscr{E}$. Let

$$\psi = \psi_H = \gamma \cdot j : \mathscr{E}[H] \rightarrowtail P(H) \longrightarrow \mathscr{E}$$

denote the corresponding over $\mathscr{E}$. The geometric morphism ψ is localic since $\mathscr{E}[H]$ is a subtopos of $P(H)$.

We have defined a topos

$$\begin{array}{ccc} \mathscr{E}[H] & \xrightarrow{\psi} & \mathscr{E} \\ & {\scriptstyle k}\searrow \quad \swarrow{\scriptstyle e} & \\ & \mathscr{S} & \end{array}$$

in which we have forced the existence of a morphism

$$\zeta : \psi^*(H) \longrightarrow k^* \Omega_{\mathscr{S}} \, .$$

The morphism ζ classifies what we shall call the *generic definable subobject associated with* H: the subobject $S \rightarrowtail \psi^*(H)$ in the following diagram.

$$\begin{array}{ccccc} (E,x) & \longrightarrow & S & \longrightarrow & 1 \\ \downarrow & & \downarrow & & \downarrow{\scriptstyle \top} \\ \psi^*(E) & \xrightarrow[\psi^* x]{} & \psi^*(H) & \xrightarrow[\zeta]{} & k^* \Omega_{\mathscr{S}} \end{array}$$

Then we can associate with a generalized element $E \xrightarrow{x} H$ the definable subobject $(E,x) \rightarrowtail \psi^*(E)$ shown above. The transpose

$$\hat{\zeta} : H \longrightarrow \psi_*(k^* \Omega_{\mathscr{S}})$$

mediates the passage from an element $E \longrightarrow H$ to a definable subobject of $\psi^*(E)$. The following is immediate.

Lemma 7.5.2 *If a morphism* $Y \longrightarrow \Omega_H$ *in* $P(H)$ *factors through* $\gamma^*(H) \xrightarrow{\rho} \Omega_H$, *then the subobject it classifies is definable in* $\mathscr{E}[H]$ (after $\mathscr{E}[H]$-sheafification j^*).

Proposition 7.5.3 *The composite* $\zeta \cdot \psi^*(\eta)$ *is equal to the identity on* $k^* \Omega_{\mathscr{S}}$. *Hence* $\hat{\zeta}$ *and the unit* η *compose in* $\mathscr{E}$ *to give the unit of* $\psi^* \dashv \psi_*$ *at* $\Omega_{\mathscr{S}}$.

$$\begin{array}{ccc} e^* \Omega_{\mathscr{S}} & & \\ {\scriptstyle \eta}\downarrow & \searrow{\scriptstyle \text{unit}} & \\ H & \xrightarrow[\hat{\zeta}]{} & \psi_*(k^* \Omega_{\mathscr{S}}) \end{array}$$

Proof. Consider the pullback

$$\begin{array}{ccc} \psi^* Z & \xrightarrow{f} & k^* \Omega_{\mathscr{S}} \\ {\scriptstyle i}\downarrow & & \downarrow{\scriptstyle \psi^* \vartheta} \\ \psi^* H & \xrightarrow[j^* \rho]{} & j^* \Omega_H \end{array}$$

in $\mathscr{E}[H]$. The morphism i is an isomorphism. We have $\zeta = f \cdot i^{-1}$. A diagram chase using the induced pair

$$(\psi^* \eta, 1_{\Omega_{\mathscr{S}}}) : k^* \Omega_{\mathscr{S}} \longrightarrow \psi^* Z$$

gives the desired conclusion. □

SCHOLIUM: A geometric morphism $\varphi : \mathscr{F} \longrightarrow \mathscr{E}$ whose domain is a definable dominance always factors through $P(H)$, where $H = \varphi_*(f^* \Omega_{\mathscr{S}})$. If $\mathscr{F} \xrightarrow{q} P(H)$ denotes this factorization, then q is an inclusion iff φ is a spread.

An $\mathscr{S}$-site for $P(H)$ is the familiar category $\mathbb{H}$ (§ 3.4), whose objects are pairs (c, y), where c is an object of an $\mathscr{S}$-site $\mathbb{C}$ for $\mathscr{E}$, and $h_c \xrightarrow{y} H$ is a morphism in $\mathscr{E}$, which for brevity we denote $c \xrightarrow{y} H$. Let $\downarrow_c (y)$ denote

$$\{c \xrightarrow{z} H \mid z \leq y\} .$$

Lemma 7.5.4 *Suppose that the domain of a geometric morphism $\mathscr{F} \xrightarrow{\varphi} \mathscr{E}$ over $\mathscr{S}$ is a definable dominance. Let $H = \varphi_*(f^* \Omega_{\mathscr{S}})$ as usual. Then:*

1. *the object $q_* f^*(\Omega_{\mathscr{S}})$ may be described by the formula*

$$q_* f^*(\Omega_{\mathscr{S}})(c, y) \cong \downarrow_c (y) ,$$

where $(c, y) \in \mathbb{H}$;

2. *the counit*

$$\xi : \gamma^*(H) = \gamma^* \gamma_*(q_* f^*(\Omega_{\mathscr{S}})) \longrightarrow q_* f^*(\Omega_{\mathscr{S}})$$

may be described as follows: for $(c, y) \in \mathbb{H}$, we have

$$\gamma^*(H)(c, y) = \mathscr{E}(c, H) \longrightarrow \ \downarrow_c (y) \ ; \ z \mapsto y \wedge z \ ;$$

3. *the epimorphism, monomorphism factorization of ρ in $P(H)$ is the following.*

$$\begin{array}{ccc} \gamma^*(H) & \xrightarrow{\xi} & q_*(f^* \Omega_{\mathscr{S}}) \\ & {\scriptstyle \rho}\searrow \quad \swarrow{\scriptstyle m} & \\ & \Omega_H & \end{array}$$

Therefore, the transpose $H \longrightarrow \gamma_*(\Omega_H)$ of ρ is equal to $\gamma_*(m)$.

Proof. 1. Since (by assumption) f is subopen, the elements of

$$q_*(f^*(\Omega_{\mathscr{S}}))(c, y)$$

are in bijection with the definable subobjects of a definable subobject $Y = q^*(c, y) \rightarrowtail \varphi^*(c)$ in $\mathscr{F}$. Since in $\mathscr{F}$ definable subobjects compose, the definable subobjects of Y coincide with the definable subobjects of $\varphi^*(c)$ under Y.

2. This describes a morphism with the appropriate universal property.

3. The morphism m is given at (c, x) by sending a $c \xrightarrow{z} H$ for which $z \leq x$ to the subobject $(c, z) \rightarrowtail (c, x)$. The triangle clearly commutes. If z and z' produce in this way the same subobject of (c, x), then of course they define the same definable subobject of $q^*(c, x)$, so that $z = z'$. Thus, m is a monomorphism. Clearly ξ is an epimorphism. □

Proposition 7.5.5 *Assume that φ is a spread whose domain topos is a definable dominance. Then the canonical geometric morphism $\mathscr{F} \xrightarrow{q} P(H)$ (which is therefore an inclusion), factors through the inclusion $j : \mathscr{E}[H] \rightarrowtail P(H)$.*

Proof. We must show that $q^*(t)$ is an isomorphism. We have in $\mathscr{F}$ the following pullback.

$$\begin{array}{ccc} q^*(Z) & \xrightarrow{q^*r} & f^*\Omega_{\mathscr{S}} \\ {\scriptstyle q^*t}\downarrow & & \downarrow{\scriptstyle q^*\vartheta} \\ \varphi^*(H) & \xrightarrow[q^*\rho]{} & q^*(\Omega_H) \end{array}$$

The counit $\delta : \varphi^*(H) \longrightarrow f^*(\Omega_{\mathscr{S}})$ divides this square into two triangles, a top one and a bottom one. We claim that these two triangles commute, from which the desired conclusion follows immediately.

In order to see that the top triangle commutes, consider its transpose back in $P(H)$ depicted below.

$$\begin{array}{ccc} Z & \xrightarrow{r} & \gamma^*(e^*\Omega_{\mathscr{S}}) \\ {\scriptstyle t}\downarrow & & \downarrow \\ \gamma^*H & \xrightarrow[\xi]{} & q_*(f^*\Omega_{\mathscr{S}}) \end{array}$$

where ξ is the counit for γ, whose description is given in Lemma 7.5.4. Z is defined by the pullback below.

$$\begin{array}{ccc} Z & \xrightarrow{r} & \gamma^*(e^*\Omega_{\mathscr{S}}) \\ {\scriptstyle t}\downarrow & & \downarrow{\scriptstyle \vartheta} \\ \gamma^*H & \xrightarrow[\rho]{} & \Omega_H \end{array}$$

Observe that $\gamma^*(\eta)$ makes the following triangle commute. The morphism depicted horizontally is a unit

$$\begin{array}{ccc} \gamma^* e^*(\Omega_{\mathscr{S}}) & \longrightarrow & q_*(f^*\Omega_{\mathscr{S}}) \\ & {\scriptstyle \gamma^*(\eta)} \searrow \quad \swarrow {\scriptstyle \xi} & \\ & \gamma^* H & \end{array}$$

Thus, we have only to show that $\xi \cdot \gamma^*(\eta) \cdot r$ and $\xi \cdot t$ are equal. We have

$$m \cdot \xi \cdot \gamma^*(\eta) \cdot r = \rho \cdot \gamma^*(\eta) \cdot r = \vartheta \cdot r = \rho \cdot t = m \cdot \xi \cdot t \, .$$

But m is a monomorphism (Lemma 7.5.4).

To see that the triangle

$$\begin{array}{ccc} \varphi^*(H) & \xrightarrow{q^*\rho} & q_*(f^*\Omega_{\mathscr{S}}) \\ & {\scriptstyle \delta} \searrow \quad \swarrow {\scriptstyle q^*\vartheta} & \\ & f^*\Omega_{\mathscr{S}} & \end{array}$$

commutes, recall that ρ factors through the subobject $q_*(f^*\Omega_{\mathscr{S}})$ of Ω_H. ϑ denotes $\rho \cdot \gamma^*(\eta)$, so it must be shown that

$$q^*\xi \cdot \varphi^*\eta \cdot \delta = q^*\xi \, .$$

The transpose to $\mathscr{E}$ of this equation is the equation

$$\varphi_* q^*\xi \cdot \varphi_*\varphi^*\eta = \varphi_* q^*\xi \cdot \eta \, .$$

Since (by assumption) φ is a spread, it is easy to see that this equation holds. Indeed, $\mathscr{F} \xrightarrow{q} P(H)$ is an inclusion (above SCHOLIUM), so that the counit $q^*q_*(e^*\Omega_{\mathscr{S}}) \longrightarrow e^*(\Omega_{\mathscr{S}})$ is an isomorphism. The result of applying φ_* to this counit and then composing with $\varphi_* q^*\xi$ is equal to the counit $\varphi_*\varphi^* H \longrightarrow H$. This counit coequalizes η_H and $\varphi_*\varphi^*\eta$, so that $\varphi_* q^*\xi$ must do so as well. □

Under our standing hypothesis on φ and f, we have established a factorization

$$\mathscr{F} \xrightarrow{p} \mathscr{E}[H] \xrightarrow{\psi} \mathscr{E} \tag{7.2}$$

of φ, where $H = \varphi_*(f^*\Omega_{\mathscr{S}})$.

Proposition 7.5.6 *The geometric morphism p is a pure inclusion. Moreover, $\mathscr{E}[H]$ is the largest subtopos of $P(H)$ (whose domain is a definable dominance) containing $\mathscr{F}$ as a pure subtopos* ($\mathscr{E}[H]$ is 'pure-closed' in $P(H)$).

Proof. We know p is an inclusion just because $q : \mathscr{F} \rightarrowtail P(H)$ is one. We must show that the unit

$$k^*\Omega_{\mathscr{S}} \longrightarrow p_*(f^*\Omega_{\mathscr{S}})$$

is an isomorphism. Let us simplify the notation somewhat by not indicating k^*, f^* and so on. Let β denote the unit of $p^* \dashv p_*$. We claim that there is a morphism

$$\alpha : p_*(\Omega_{\mathscr{S}}) \longrightarrow \Omega_{\mathscr{S}}$$

in $\mathscr{E}[H]$ such that

$$p_*(\Omega_{\mathscr{S}}) \xrightarrow{\alpha} \Omega_{\mathscr{S}} \xrightarrow{\beta} p_*(\Omega_{\mathscr{S}})$$

is equal to the identity. In order to define α, we keep in mind that if j denotes the inclusion $\mathscr{E}[H] \rightarrowtail P(H)$, then by transposing under $j^* \dashv j_*$ elements $j^*(c,z) \longrightarrow p_*\Omega_{\mathscr{S}}$ may be equivalently described as elements $(c,z) \longrightarrow q_*\Omega_{\mathscr{S}}$. We then use the description of elements of $q_*\Omega_{\mathscr{S}}$ provided by Lemma 7.5.4. Given $c \xrightarrow{z} H$ such that $z \leq y$, we pass to the following composite morphism.

$$(c,y) \longrightarrow \psi^*(c) \xrightarrow{\psi^* z} \psi^*(H) \xrightarrow{\zeta} \Omega_{\mathscr{S}}$$

This morphism classifies the definable subobject $(c,z) \rightarrowtail (c,y)$ in $\mathscr{E}[H]$. It follows that $\beta \cdot \alpha$ is the identity.

To show that $\alpha \cdot \beta$ is also the identity, it will be enough to show that the square

$$\begin{array}{ccc} \Omega_{\mathscr{S}} & \xrightarrow{\psi^*\eta} & \psi^*(H) \\ {\scriptstyle\beta}\downarrow & & \downarrow{\scriptstyle\zeta} \\ p_*(\Omega_{\mathscr{S}}) & \xrightarrow[\alpha]{} & \Omega_{\mathscr{S}} \end{array}$$

commutes in $\mathscr{E}[H]$, since $\zeta \cdot \psi^*(\eta) = 1_{\Omega_{\mathscr{S}}}$, which we now show. Consider the counit of $\psi^* \dashv \psi_*$:

$$\epsilon : \psi^*(H) = \psi^*(\psi_*(p_*(\Omega_{\mathscr{S}}))) \longrightarrow p_*(\Omega_{\mathscr{S}}) .$$

We leave the equation $\epsilon \cdot \psi^*(\eta) = \beta$ as an exercise. To see that $\alpha \cdot \epsilon = \zeta$ holds we only have to remind ourselves how these three morphisms are described in terms of the $\mathscr{S}$-site $\mathbb{H}$. For example, the counit ϵ is derived (after $\mathscr{E}[H]$-sheafifying) from the counit ξ for $\gamma^* \dashv \gamma_*$, which is described in Lemma 7.5.4. We leave the remaining verifications to the reader.

Finally, suppose that $\mathscr{G} \rightarrowtail P(H)$ contains $\mathscr{F}$ as a pure subtopos, and that $\mathscr{G} \longrightarrow \mathscr{S}$ is a definable dominance. Then $\mathscr{G} \longrightarrow \mathscr{E}$ is a spread, and we may consider its factorization $G \rightarrowtail \mathscr{E}[H'] \longrightarrow \mathscr{E}$. But since $\mathscr{F} \rightarrowtail \mathscr{G}$ is pure, this H' is equal to H belonging to $\mathscr{F}$. Therefore, $\mathscr{E}[H'] \simeq \mathscr{E}[H]$, so $\mathscr{G}$ does indeed lie in $\mathscr{E}[H]$. □

Theorem 7.5.7 *Let $\varphi : \mathscr{F} \longrightarrow \mathscr{E}$ be a spread over $\mathscr{S}$ whose domain f is a definable dominance. Let $H = \varphi_*(f^*\Omega_{\mathscr{S}})$. Then the factorization* (7.2) *coincides* (up to equivalence) *with the pure, entire factorization* (Thm. 7.3.20).

Proof. We wish to show that there exists an equivalence

$$Sh_{\mathscr{E}}(Idl_{\Omega_{\mathscr{S}}}(H)) \simeq \mathscr{E}[H]$$

over $\mathscr{E}$ and under $\mathscr{F}$. Let $\Omega_{[H]}$ denote the subobject classifier of $\mathscr{E}[H]$. There is a $\Omega_{\mathscr{S}}$-distributive lattice morphism

$$H \longrightarrow \gamma_*(\Omega_H) \longrightarrow \psi_*(\Omega_{[H]}) ,$$

where the first map is down-closure, and the second is a frame morphism. By the universal property of the free frame on the $\Omega_{\mathscr{S}}$-distributive lattice H, we have a frame morphism

$$Idl_{\Omega_{\mathscr{S}}}(H) \longrightarrow \psi_*(\Omega_{[H]}) ,$$

corresponding to a geometric morphism (an inclusion)

$$\mathscr{E}[H] \rightarrowtail Sh_{\mathscr{E}}(Idl_{\Omega_{\mathscr{S}}}(H))$$

of subtoposes of $P(H)$. But by Proposition 7.5.6, $\mathscr{E}[H]$ is the largest subtopos of $P(H)$ that contains $\mathscr{F}$ as a pure subtopos. This concludes the proof. □

Exercises 7.5.8

1. *Show that the uniqueness of the pure, entire factorization follows from Theorem 7.5.7, in the spread case.*
2. *Show that a Stone locale is a spread.*

Further reading: Barr & Paré [BP80], Bunge [Bun95], Bunge & Funk [BF96b, BF98], Bunge, Funk, Jibladze & Streicher [BFJS00, BFJS02], Funk [Fun95], Gleason [Gle63], Johnstone [Joh82a, Joh82b], Kock & Reyes [KR94], Lawvere [Law68], Michael [Mic63], Mikkelsen [Mik76], Moerdijk & Vermeulen [MV00], Paré [Par74], Rosolini [Ros86], Tierney [Tie74] Vermeulen [Ver94], Willard [Wil68].

8

Localic and Algebraic Aspects

In this chapter we consider distributions on locales, and the lower power locale from a constructive point of view. We also consider factorizations other than the comprehensive one (or pure, complete spread), and compare them. The lower bagdomain B_L, and the probability distribution classifier T are two variants of the symmetric KZ-monad M; the equation $\mathrm{M} = \mathrm{B}_L \cdot \mathrm{T}$ offers a new perspective on distributions and complete spread geometric morphisms. Our notion of discrete complete spread structure provides yet another single universe for both local homeomorphisms and complete spreads. We illustrate some of the ideas discussed in this book with an example from algebraic geometry involving coschemes. We make a special analysis of distributions on the Jonsson-Tarski topos.

8.1 Distributions on Locales

Power domains (lower, upper, mixed) had been introduced in the 1970's in order to analyse the semantics of non-deterministic and parallel computation. Some computer scientists now believe that this is not an ideal solution to the problem, since infinite communicating processes are hardly ever determined by the finite (or partial) observations one can make about them. On the other hand, the lower bagdomains had emerged from efforts to make more accurate the model provided by the lower power domain, in which the 'partial information' about a database should not only be specified by individual partial records, but by an indexed family of such partial records (a 'bag'). Even if the domain from which one starts has only one point, the points of the bagdomain should correspond to the 'space' of all sets, and the refinement ordering on them, to arbitrary functions. The result is not a topological space, or even a locale. However, the space of all sets can easily be handled by passing to toposes by means of the object classifier $\mathscr{S}[U]$. The lower bagdomain has been constructed by S. Vickers, and put on a categorical foundation by P. T. Johnstone. We will return to bagdomains in § 8.2; however, power domains have other aspects that we shall address here.

We shall refer to a complete upper semilattice in an elementary topos $\mathcal{S}$ as *a suplattice*. Let **sl** denote the 2-category of of suplattices and sup-preserving maps, so $\mathbf{sl}(M, N)$ denotes the poset of sup-preserving maps from a suplattice M to another one N. We may consider distributions on a locale (or more generally, on a suplattice), in the following sense.

Definition 8.1.1 A distribution on a locale X *in* $\mathcal{S}$ *is a sup-preserving morphism* $\mathcal{O}(X) \longrightarrow \Omega_{\mathcal{S}}$, *where as always* $\mathcal{O}(X)$ *denotes the frame associated with the locale* X.

We denote by $\Sigma(M)$ *the symmetric frame of* M , defined to be the frame of the classifying locale for the theory of distributions on M , i.e., of sup-preserving maps $M \longrightarrow \Omega_{\mathcal{S}}$. Equivalently, the following universal property defines $\Sigma(M)$: for any frame $\mathcal{O}(X)$ in $\mathcal{S}$, there is an isomorphism

$$\mathbf{Fr}(\Sigma(M), \mathcal{O}(X)) \cong \mathbf{sl}(M, \mathcal{O}(X)) ,$$

of posets natural in $\mathcal{O}(X)$, where **Fr** denotes the 2-category of frames, and frame homomorphisms. In other words, Σ is left adjoint to the forgetful functor

$$U : \mathbf{Fr} \longrightarrow \mathbf{sl} \; ; \; \Sigma \dashv U \; .$$

We call $\Sigma(M)$ the *symmetric frame* of a suplattice M.

Just like our treatment of the symmetric topos, it is appropriate to take a 'geometric' point of view: we may regard $\Sigma \cdot U$ as an endofunctor of locales. If X is any locale, we define $\mathrm{P}_L(X)$ as the locale whose frame is

$$\mathcal{O}(\mathrm{P}_L(X)) = \Sigma(\mathcal{O}(X)) \, .$$

Of course we mean $\Sigma(U(\mathcal{O}(X))$ on the right, but we do not need to write U. Thus, $\Sigma(\mathcal{O}(X))$ is none other than the frame of opens of the *Hoare locale*, or *lower power locale* $\mathrm{P}_L(X)$ of X, as it is called in the literature.

Classically, the frame of the lower power locale $\mathrm{P}_L(X)$ is freely generated by symbols $\Diamond U$, $U \in \mathcal{O}(X)$, so that $U \mapsto \Diamond U$ preserves arbitrary joins. If the topos $\mathcal{S}$ is Boolean, it is known that a point of $\mathrm{P}_L(X)$, i.e., a sup-preserving map $\mathcal{O}(X) \longrightarrow \Omega_{\mathcal{S}}$, is completely determined by a closed sublocale of X. Before examining the extent of the validity of this assertion for an arbitrary topos $\mathcal{S}$, we give a construction of $\mathrm{P}_L(X)$ that parallels the construction of the symmetric topos.

The finite inf-completion $Q^{\bullet}$ of a poset Q can be given as the collection of equivalence classes $[S]$, where S is a (Kuratowski) finite subset of Q, and where $[S] = [S']$ iff S and S' generate the same upper set in Q. As a poset, $Q^{\bullet}$ has the partial order given by $[S] \leq [T]$ iff T is contained in the upper set generated by S.

Any frame $\mathcal{O}(X)$ is canonically presented as a coinverter

$$D(Q) \underset{d_1}{\overset{d_0}{\underset{\Downarrow}{\rightrightarrows}}} D(\mathcal{O}(X)) \longrightarrow\!\!\!\!\!\rightarrow \mathcal{O}(X)$$

in **sl**, where Q is the poset whose elements are pairs (R, U) such that $R \subseteq \downarrow U$, $U \in \mathscr{O}(X)$, and $\bigvee R = U$. The maps d_0 and d_1 are induced by the assignments $(R, U) \mapsto R$ and $(R, U) \mapsto \downarrow U$, respectively, where $\Rightarrow$ is the unique 2-cell from d_0 to d_1, i.e., $d_0 \leq d_1$. It is well-known that the free frame on an inf-semilattice Z is given by the frame $D(Z)$ of down-closed subsets of Z.

We have left the proof of the following as a exercise, since this proof proceeds by analogy with the construction of (topos-frame of) the symmetric topos. Note that the forgetful functor from frames to suplattices creates coinverters.

Proposition 8.1.2 *The symmetric frame $\Sigma(\mathscr{O}(X))$ is defined by the coinverter*

$$D(Q^\bullet) \underset{d_1^\bullet}{\overset{d_0^\bullet}{\rightrightarrows}} D(\mathscr{O}(X)^\bullet) \xrightarrow{i^*} \Sigma(\mathscr{O}(X))\,.$$

in **Fr** (created in **sl**), *where the parallel arrows $d_0^\bullet$, $d_1^\bullet$ are induced from the canonical suplattice presentation of $\mathscr{O}(X)$ via finite inf-completions at the level of the posets.*

We now turn to an identification of the points of $\mathrm{P}_L(X)$ where $\mathscr{S}$ is now an arbitrary topos. A locale morphism $f : Y \longrightarrow X$ in a topos $\mathscr{S}$ is said to be *strongly dense* if the canonical inequality $\omega \leq f_* f^* \omega$ is an equality, for every $\omega \in \Omega_{\mathscr{S}}$. (Notice the parallel with what we call a pure geometric morphism.) In particular, a strongly dense locale morphism is *dense* in the sense that $0 = f_* 0$. It turns out that f is strongly dense iff f is dense under pullback along every closed sublocale of the terminal locale (whose frame is $\Omega_{\mathscr{S}}$). Every locale inclusion may be factored uniquely into a strongly dense inclusion followed by *a weakly closed* sublocale. Tautologically speaking, we may say that an inclusion of locales $B \rightarrowtail X$ is weakly closed iff any strongly dense inclusion $B \rightarrowtail B'$ is an isomorphism, where $B' \rightarrowtail X$ is any sublocale.

Let $\mathrm{Sub}(X)$ denote the coframe of sublocales of a locale X. We denote by $\mathrm{W}(X)$ the poset of weakly closed sublocales of X. $\mathrm{W}(X)$ is a subcoframe of $\mathrm{Sub}(X)$ (Jibladze and Johnstone), and it contains $\mathrm{C}(X) = \mathscr{O}(X)^{\mathrm{op}}$ as a subcoframe. For an open $U \in \mathscr{O}(X)$, we use the same symbol U to denote the sublocale of X corresponding to the nucleus $U \Rightarrow (\_)$. This association $\mathscr{O}(X) \longrightarrow \mathrm{Sub}(X)$ (of a frame into a coframe) preserves arbitrary suprema and finite infima.

For any locale morphism $X \xrightarrow{f} Y$ and $B \in \mathrm{Sub}(X)$, we shall use $\|B\|_f$ to denote the image of B in Y under f. When no subscript is supplied, then the unique map to the terminal locale is intended. Consider the functor

$$\chi : \mathrm{Sub}(X) \longrightarrow \mathbf{sl}(\mathscr{O}(X), \mathrm{Sub}(1))$$

that carries a sublocale $B \rightarrowtail X$ to the suplattice map

$$\chi_B : U \mapsto \|B \wedge U\|\,.$$

Observe that χ has as right adjoint the functor that associates to a sup-preserving map $\mathscr{O}(X) \xrightarrow{f} \mathrm{Sub}(1)$ the sublocale

$$A_f = \bigwedge \{(X - U) \vee \gamma^\sharp(fU) \mid U \in \mathscr{O}(X)\} , \tag{8.1}$$

where $\gamma^\sharp$ denotes locale pullback along the unique locale morphism $\gamma : X \longrightarrow 1$. Moreover, observe that the sublocale A_f is weakly closed. This follows from the fact that every sublocale of 1 is weakly closed, so that $\gamma^\sharp(fU)$ is weakly closed by pullback stability, and from the fact that closed sublocales are weakly closed, using then the fact that $\mathrm{W}(X)$ is a subcoframe of $\mathrm{Sub}(X)$.

Let $\delta : 1' \longrightarrow 1$ denote the *splitting locale* of 1 , i.e., $\mathscr{O}(1')$ is the frame of nuclei on $\Omega_{\mathscr{S}}$, and δ^* is the frame morphism that associates to $\omega \in \Omega_{\mathscr{S}}$ the nucleus $\omega \vee (\_)$. Pullback along δ yields an isomorphism $\mathrm{Sub}(1) \longrightarrow \mathrm{C}(1')$. Our explanation of Theorem 8.1.4 below relies on the following result, for which we do not include a proof.

Proposition 8.1.3 *Let X be an arbitrary locale, and let*

$$\begin{array}{ccc} Z & \xrightarrow{\psi} & 1' \\ {\scriptstyle p}\downarrow & & \downarrow{\scriptstyle \delta} \\ X & \xrightarrow{\gamma} & 1 \end{array}$$

be a pullback. Then pullback along p gives an isomorphism $\mathrm{W}(X) \cong \mathrm{C}(Z)$.

Theorem 8.1.4 *For any locale X, the restriction of χ to $\mathrm{W}(X)$ yields an isomorphism*

$$\mathrm{W}(X) \cong \mathbf{sl}(\mathscr{O}(X), \mathrm{Sub}(1)) .$$

Proof. We employ the 'module' framework for frames and suplattices from the work of Joyal and Tierney. If M and N are $\mathscr{O}(Y)$-modules (suplattices that carry an $\mathscr{O}(Y)$-action in a suitable sense), then $\mathbf{sl}_{\mathscr{O}(Y)}(M, N)$ shall denote the poset of suplattice maps that preserve the $\mathscr{O}(Y)$-action. We start with the fact that for any locale morphism $X \xrightarrow{f} Y$, there are canonical isomorphisms

$$\mathrm{C}(X) \cong \mathbf{sl}_{\mathscr{O}(Y)}(\mathscr{O}(Y), \mathrm{C}(X)) \cong \mathbf{sl}_{\mathscr{O}(Y)}(\mathscr{O}(X), \mathrm{C}(Y)) . \tag{8.2}$$

This composite isomorphism sends a closed sublocale $B = X - W$ to the suplattice map

$$U \mapsto \bigvee \{V \in \mathscr{O}(Y) \mid U \leq f^*V \Rightarrow W\}$$

$$= \bigvee \{V \in \mathscr{O}(Y) \mid f^*V \leq U \Rightarrow W\} .$$

When written in terms of closed parts, this suplattice map is

$$U \mapsto \overline{\|B \wedge U\|_f} ,$$

on account of the identities

$$\bigwedge\{D \in C(Y) \mid B \wedge U \leq f^*D\} = \bigwedge\{D \in C(Y) \mid \|B \wedge U\|_f \leq D\}$$

and

$$\bigwedge\{D \in C(Y) \mid \|B \wedge U\|_f \leq D\} = \overline{\|B \wedge U\|_f} \,.$$

On combining Proposition 8.1.3 with (8.2) applied to the morphism $Z \xrightarrow{\psi} 1'$ of Proposition 8.1.3, we obtain

$$\mathrm{W}(X) \cong \mathrm{C}(Z) \cong \mathbf{sl}_{\mathcal{O}(1')}(\mathcal{O}(Z), C(1')) \,. \tag{8.3}$$

Since $\mathcal{O}(Z) = \mathcal{O}(1)' \otimes \mathcal{O}(X)$, by adjointness this is isomorphic to

$$\mathbf{sl}(\mathcal{O}(X), \mathrm{C}(1')) \cong \mathbf{sl}(\mathcal{O}(X), \mathrm{Sub}(1)) \,. \tag{8.4}$$

The isomorphism in (8.4) is composition with the isomorphism

$$\mathrm{C}(1') \longrightarrow \mathrm{Sub}(1) \,,$$

which carries a closed sublocale $E \rightarrowtail 1'$ to $\|E\|$, and furthermore, satisfies $\|\overline{I}\| = \|I\|$, for any sublocale $I \rightarrowtail 1'$, where $\overline{I}$ denotes the closure of I in $1'$. It remains to verify that the composite of (8.3) and (8.4) is indeed equal to χ restricted to $\mathrm{W}(X)$. By Proposition 8.1.3, and since B is weakly closed, we have

$$\|p^*(B \wedge U)\|_p = B \wedge U \,.$$

Then the composite of (8.3) and (8.4) sends $B \in \mathrm{W}(X)$ to the suplattice map

$$U \mapsto \| \, \overline{\|p^*B \wedge p^*U\|_\psi} \, \| = \| \, \|p^*(B \wedge U)\|_\psi \, \| \,,$$

which is equal to

$$U \mapsto \| \, \|p^*(B \wedge U)\|_p \, \| = \|B \wedge U\| = \chi_B(U) \,.$$

□

Remark 8.1.5 *In effect, χ is weak closure. Let χ^{-1} denote the right adjoint of χ (8.1). The notation is justified since by Theorem 8.1.4, this right adjoint is full and faithful. Thus, for any sublocale $S \rightarrowtail X$, $\chi^{-1}(\chi_S)$ is its weak closure, and S is weakly closed iff*

$$S = \bigwedge \{(X - U) \vee \gamma^\sharp \|S \wedge U\| \mid U \in \mathcal{O}(X)\} \,.$$

Let $\mathrm{Sub}_o(X)$, respectively $\mathrm{W}_o(X)$, denote the poset of sublocales, respectively weakly closed sublocales, of X *with open domain*, i.e., those $B \rightarrowtail X$ for which the unique locale morphism $B \longrightarrow 1$ to the terminal locale is open. By definition, $B \longrightarrow 1$ is open if the unique frame morphism $\Omega_{\mathcal{S}} \longrightarrow \mathcal{O}(B)$ has a left adjoint $\exists$. This is equivalent to the condition that for all $U \in \mathcal{O}(X)$,

$\|B \wedge U\|$ is an open sublocale of 1. It follows that the restriction of χ to sublocales with open domain yields the following commutative diagram.

$$\begin{array}{ccc} \mathrm{Sub}_o(X) & \xrightarrow{\chi_o} & \mathbf{sl}(\mathcal{O}(X), \Omega_{\mathcal{S}}) \\ \downarrow & & \downarrow \\ \mathrm{Sub}(X) & \xrightarrow{\chi} & \mathbf{sl}(\mathcal{O}(X), \mathrm{Sub}(1)) \end{array}$$

By definition, if a sublocale $B \rightarrowtail X$ has open domain, then $\chi_o(B)$ is the suplattice map

$$\mathcal{O}(X) \xrightarrow{b^*} \mathcal{O}(B) \xrightarrow{\exists} \Omega_{\mathcal{S}} .$$

The vertical arrow on the right is composition with the canonical lattice inclusion $\Omega_{\mathcal{S}} \longrightarrow \mathrm{Sub}(1)$. It is full and faithful. We now have the main result of this section.

Theorem 8.1.6 *For any locale X, the restriction of χ_o to $\mathrm{W}_o(X)$ yields an isomorphism*

$$\mathrm{W}_o(X) \cong \mathbf{sl}(\mathcal{O}(X), \Omega_{\mathcal{S}}) .$$

Proof. We have a commutative diagram

$$\begin{array}{ccc} \mathrm{W}_o(X) & \xrightarrow{\chi_o} & \mathbf{sl}(\mathcal{O}(X), \Omega_{\mathcal{S}}) \\ \downarrow & & \downarrow \\ \mathrm{W}(X) & \xrightarrow{\chi} & \mathbf{sl}(\mathcal{O}(X), \mathrm{Sub}(1)) \end{array}$$

where the vertical arrow on the right is full and faithful. But then we see that χ_o is an isomorphism. Indeed, χ_o is clearly full and faithful, and if $g \in \mathbf{sl}(\mathcal{O}(X), \Omega_{\mathcal{S}})$, then by Theorem 8.1.4, there is a weakly closed $B \rightarrowtail X$ such that

$$\forall U \in \mathcal{O}(X),\ g(U) = \|B \wedge U\| .$$

This says that $B \longrightarrow 1$ is open, i.e., that $B \in \mathrm{W}_o(X)$. □

Exercises 8.1.7

1. *Prove Proposition 8.1.2 by analogy with the proof of the corresponding theorem for the (frame of the) symmetric topos.*
2. *Prove that the following are equivalent, for any elementary topos $\mathcal{S}$.*
   (a) *$\mathcal{S}$ is Boolean.*
   (b) *$\mathrm{C}(X) = \mathrm{W}(X)$ for every locale X in $\mathcal{S}$.*
   (c) *$\mathrm{C}(I) = \mathrm{W}(I)$ for every object I of $\mathcal{S}$.*
   (d) *$\mathrm{W}_o(X) = \mathrm{W}(X)$ for every locale X in $\mathcal{S}$.*
   (e) *$\mathrm{W}_o(1) = \mathrm{W}(1)$.*

Hint: The equivalence of the first three conditions, and that they imply the last two is reasonably easy to establish. In order to prove that the last implies the first, observe that we always have $\mathrm{Sub}(1) = \mathrm{W}(1)$ *and* $\Omega_{\mathscr{S}} = \mathrm{Sub}_o(1) = \mathrm{W}_o(1)$ *. Thus, if* $\mathrm{W}_o(1) = \mathrm{W}(1)$ *, then* $\Omega_{\mathscr{S}} = \mathrm{Sub}(1)$*, i.e., then every sublocale of* 1 *is open. It is well-known that this implies* $\Omega_{\mathscr{S}}$ *is Boolean.*

3. *Show that for any locale* X *in* $\mathscr{S}$*, the topos* $Sh(\mathrm{P}_L(X))$ *of sheaves on the lower power locale of* X *is equivalent to the localic reflection of the symmetric topos* $\mathrm{M}(Sh(X))$*.*
4. *The hyperconnected geometric morphism*

$$h : \mathrm{M}(Sh(X)) \longrightarrow Sh(\mathrm{P}_L(X))$$

mediates support in a sense that can be described in terms of its action on geometric points. For any topos $\mathscr{F}$*, composite with* h *may be equivalently described as a functor*

$$\boldsymbol{Dist}_{\mathscr{S}}(\mathscr{F}, Sh(X)) \longrightarrow \mathbf{sl}(\mathscr{O}(X), f_*(\Omega_{\mathscr{F}}))$$

given by composition with the 'support' functor $\sigma : \mathscr{F} \longrightarrow \Omega_{\mathscr{F}}$*, which assigns to an* I*-indexed family* $F \longrightarrow f^*I$ *the characteristic map of its image, and with Yoneda* $\mathscr{O}(X) \longrightarrow Sh(X)$*.*
5. *Show that the lower power monad* P_L *falls within the theory of completion KZ-monads.*
6. *Directly establish the correspondence of Theorem 8.1.6 in terms of bicomma objects: if* $1 \xrightarrow{p} \mathrm{P}_L(X)$ *is a localic point corresponding to a weakly closed sublocale* $B \rightarrowtail X$ *with open domain, then there is a bicomma object*

$$\begin{array}{ccc} B & \longrightarrow & 1 \\ \downarrow & \leq & \downarrow{\scriptstyle p} \\ X & \xrightarrow[\delta]{} & \mathrm{P}_L(X) \end{array}$$

of locales.

8.2 Symmetric versus Lower Bagdomain

Given a topos $\mathscr{E}$ over $\mathscr{S}$, $\mathrm{B}_L(\mathscr{E})$ is a topos whose points are bags of points of $\mathscr{E}$. We remarked in the previous section that the lower bagdomain generalizes the lower power locale. The symmetric topos $\mathrm{M}(\mathscr{E})$ is also a sort of generalization of $\mathrm{P}_L(X)$ (Exercises 8.1.7). Also, we have

$$\mathrm{M}(\mathscr{S}) = \mathrm{B}_L(\mathscr{S}) = \mathscr{S}[U] .$$

We begin by examining $\mathrm{B}_L(\mathscr{E})$ from a symmetric viewpoint.

The essential inclusion $\delta : \mathscr{E} \rightarrowtail \mathrm{M}(\mathscr{E})$ factors through the lower bagdomain topos $\mathrm{B}_L(\mathscr{E})$ by essential inclusions as in the following diagram.

$$\begin{array}{ccc} & \mathrm{B}_L(\mathscr{E}) & \\ \nearrow & & \searrow \\ \mathscr{E} & \xrightarrow[\delta]{} & \mathrm{M}(\mathscr{E}) \end{array}$$

These morphisms are induced by corresponding site inclusions

$$\langle \mathbb{C}, J \rangle \longrightarrow \langle \mathbb{C}_{\mathrm{fp}}, J_{\mathrm{fp}} \rangle \longrightarrow \langle \mathbb{C}^{\star}, J^{\star} \rangle \, ,$$

where $\langle \mathbb{C}, J \rangle$ is a site definition of $\mathscr{E}$, $\mathbb{C}_{\mathrm{fp}}$ is the finite products completion of $\mathbb{C}$ is a site for $\mathrm{B}_L(\mathscr{E})$, and $\mathbb{C}^{\star}$ is the lex completion of $\mathbb{C}$, which we have seen is a site for $\mathrm{M}(\mathscr{E})$ in § 4.2. (We use the notation $J^{\star}$ for the topology in $\mathbb{C}^{\star}$, but it is not the lex completion of the total poset of J.)

In terms of the models of the theories that these toposes classify, the geometric morphism $\mathscr{E} \longrightarrow \mathrm{B}_L(\mathscr{E})$ corresponds to forgetting that a lex distribution $\mathscr{E} \longrightarrow \mathscr{X}$ preserves the terminal object, whereas $\mathrm{B}_L(\mathscr{E}) \longrightarrow \mathrm{M}(\mathscr{E})$ corresponds to forgetting that a pullback preserving distribution preserves pullbacks. $\mathrm{M}(\mathscr{E})$ classifies distributions, whereas $\mathrm{B}_L(\mathscr{E})$ classifies *bags of points* of $\mathscr{E}$, meaning a geometric morphism $\mathscr{S}/I \longrightarrow \mathscr{E}$. Equivalently, a bag of points is a pullback preserving distribution on $\mathscr{E}$.

Proposition 8.2.1 *The following conditions on a locally connected topos $\mathscr{E}$ are equivalent:*

1. *the connected components functor $e_!$ preserves pullbacks,*
2. *$\mathscr{E}$ has a pure 'bag' of points, in the sense that there is a diagram*

$$\begin{array}{ccc} \mathscr{S}/I & \xrightarrow{\;p\;} & \mathscr{E} \\ & \searrow \quad \swarrow_{e} & \\ & \mathscr{S} & \end{array}$$

in which p is pure.

The topos is also connected iff it has a pure point ($I = 1$ in this case).

Proof. 1 implies 2 because we may take $I = e_!(1)$. It follows that there is a geometric morphism p as above such that $e_! \cong \Sigma_I \cdot p^*$, which says that p is pure. 2 implies 1 because if p is pure, then $e_! \cong \Sigma_I \cdot p^*$. Thus, $e_!$ preserves pullbacks. □

The comparison between M and B_L can also be phrased in terms of complete spreads: M classifies complete spreads with locally connected domain, and B_L classifies complete spreads whose domain has totally connected components, in the following sense.

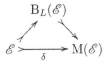

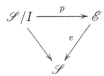

Definition 8.2.2 *We shall say that a locally connected topos* has totally connected components *if either of the conditions in 8.2.1 holds. If a topos is connected, locally connected, and has totally connected components, then we say it is* totally connected.

Proposition 8.2.3 $\mathbf{Top}_{\mathcal{S}}(\mathcal{X}, \mathrm{B}_L(\mathcal{E}))$ *is equivalent to the category of $\mathcal{X}$-complete spreads $\mathcal{Y} \longrightarrow \mathcal{X} \times_{\mathcal{S}} \mathcal{E}$, whose $\mathcal{X}$-domain has totally connected components. Moreover, a geometric morphism*

$$\mathcal{X} \xrightarrow{\rho} \mathrm{M}(\mathcal{E})$$

factors through $\mathrm{B}_L(\mathcal{E})$ *iff in the bicomma object*

$$\begin{array}{ccc} \mathcal{Y} & \xrightarrow{\gamma} & \mathcal{X} \\ {\scriptstyle \psi}\big\downarrow & \Rightarrow & \big\downarrow{\scriptstyle \rho} \\ \mathcal{E} & \underset{\delta}{\rightarrowtail} & \mathrm{M}(\mathcal{E}) \end{array}$$

the locally connected γ has totally connected components, in which case the resulting inside square with $\mathrm{B}_L(\mathcal{E})$ *is a bicomma object.*

Proof. The topos $\mathrm{B}_L(\mathcal{E})$ is the partial product of $\mathcal{S}[U]/U \longrightarrow \mathcal{S}[U]$ with $\mathcal{E}$, where U denotes the generic object of the object classifier $\mathcal{S}[U]$. It follows immediately that a geometric morphism $\mathcal{X} \longrightarrow \mathrm{B}_L(\mathcal{E})$ amounts to a pair consisting of an object F of $\mathcal{X}$ and a geometric morphism $\mathcal{X}/F \longrightarrow \mathcal{E}$ over $\mathcal{S}$, equivalently, to a pair consisting of an object F of $\mathcal{X}$ and a geometric morphism

$$\mathcal{X}/F \longrightarrow \mathcal{X} \times_{\mathcal{S}} \mathcal{E}$$

over $\mathcal{X}$. Now form the $\mathcal{X}$-comprehensive factorization of this geometric morphism.

We leave the second assertion of the proposition as an exercise. □

Remark 8.2.4 *Proposition 8.2.3 gives a characterization of those toposes $\mathcal{E}$ for which all distributions on it are* Riemman sums *by which we mean, in this context, bags of points. They are precisely those toposes $\mathcal{E}$ such that* $\mathrm{M}(\mathcal{E})$ *and* $\mathrm{B}_L(\mathcal{E})$ *coincide. Equivalently, they are the toposes $\mathcal{E}$ for which the domain topos of every complete spread over $\mathcal{E}$ is locally connected and has totally connected components (equivalently, locally connected and the connected components functor preserves pullbacks).*

We now turn to a third KZ-monad in $\mathbf{Top}_{\mathcal{S}}$: the classifier of probability distributions.

Definition 8.2.5 A probability distribution *on a topos is a distribution that preserves the terminal object. We denote the topos classifier of probability distributions on a topos $\mathcal{E}$ by* $\mathrm{T}(\mathcal{E})$*; the category of geometric morphisms $\mathcal{X} \longrightarrow \mathrm{T}(\mathcal{E})$ is naturally equivalent to the category of probability distributions $\mathcal{E} \longrightarrow \mathcal{X}$.*

Proposition 8.2.6 *For any topos $\mathscr{E}$ over $\mathscr{S}$, there is a subtopos $\mathrm{T}(\mathscr{E})$ of $\mathrm{M}(\mathscr{E})$ that classifies probability distributions. Furthermore, there is a factorization*

$$\begin{array}{ccc} \mathscr{E} & \xrightarrow{\ \delta\ } & \mathrm{M}(\mathscr{E}) \\ & {\scriptstyle\overline{\delta}}\searrow \quad \nearrow{\scriptstyle i} & \\ & \mathrm{T}(\mathscr{E}) & \end{array}$$

where $\overline{\delta}$ is essential and satisfies $\overline{\delta}_!(1) \cong 1$. A geometric morphism $\mathscr{X} \xrightarrow{\rho} \mathrm{M}(\mathscr{E})$ factors through $\mathrm{T}(\mathscr{E})$ iff in the bicomma object

$$\begin{array}{ccc} \mathscr{Y} & \xrightarrow{\ \gamma\ } & \mathscr{X} \\ {\scriptstyle\psi}\downarrow & \Rightarrow & \downarrow{\scriptstyle\rho} \\ \mathscr{E} & \xrightarrow[\ \delta\ ]{} & \mathrm{M}(\mathscr{E}) \end{array}$$

the locally connected γ is connected, in which case the resulting inside square with $\mathrm{T}(\mathscr{E})$ is a bicomma object.

Proof. Let $\mathrm{T}(\mathscr{E})$ denote the subtopos of $\mathrm{M}(\mathscr{E})$ given by the least topology forcing the morphism $\delta_!1 \longrightarrow 1$ to be an isomorphism. Then a geometric morphism $\mathscr{X} \xrightarrow{\rho} \mathrm{M}(\mathscr{E})$ factors (uniquely) through $\mathrm{T}(\mathscr{E})$ iff $\rho^*\delta_!1 \cong 1$ iff $\gamma_!1 \cong \gamma_!\psi^*1 \cong 1$ iff the locally connected γ is connected. Note that δ factors through $\mathrm{T}(\mathscr{E})$ since $\delta^*\delta_!1 \cong 1$. We have $\overline{\delta}^* = \delta^* i_*$, and since $i^*\delta_! \dashv \delta^* i_*$, $\overline{\delta}$ is essential with $\overline{\delta}_!1 = i^*\delta_!1 \cong 1$. □

In terms of models of the theories classified by these toposes, $\overline{\delta}$ corresponds to forgetting that a lex distribution $\mathscr{E} \longrightarrow \mathscr{X}$, i.e., a geometric morphism $\mathscr{X} \longrightarrow \mathscr{E}$, preserves pullbacks, and the second factor corresponds to forgetting that a probability distribution $\mathscr{E} \longrightarrow \mathscr{X}$, which corresponds to a geometric morphism $\mathscr{X} \longrightarrow \mathrm{T}(\mathscr{E})$, preserves 1.

Theorem 8.2.7 *Let $\mathscr{E}$ denote an arbitrary topos over $\mathscr{S}$. Then $\mathrm{M}(\mathscr{E}) \simeq \mathrm{B}_L(\mathrm{T}(\mathscr{E}))$, naturally in $\mathscr{E}$.*

Proof. This follows easily from universal properties. The category

$$\mathbf{Top}_{\mathscr{S}}(\mathscr{X}, \mathrm{B}_L(\mathrm{T}(\mathscr{E})))$$

is equivalent to the category of pairs $F \in \mathscr{X}$ and $\mathscr{X}/F \longrightarrow \mathrm{T}(\mathscr{E})$ over $\mathscr{S}$, as we had mentioned in Proposition 8.2.3. This data is equivalently given by an object $F \in \mathscr{X}$ and a probability distribution $\mathscr{E} \longrightarrow \mathscr{X}/F$. The category of such pairs is clearly equivalent to $\mathbf{Dist}_{\mathscr{S}}(\mathscr{E}, \mathscr{X}) \simeq \mathbf{Top}_{\mathscr{S}}(\mathscr{X}, \mathrm{M}(\mathscr{E}))$. □

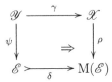

The following results gives an alternative construction of $\mathrm{T}(\mathscr{E})$ in terms of sites. A finite connected limit is one whose diagram is finite, non-empty, and connected. Finite connected limits can be freely adjoined to an arbitrary small category. Let

$$\kappa : \mathbb{C} \longrightarrow \mathbb{C}^{\oplus}$$

denote the *finite connected limit completion* of a small category $\mathbb{C}$.

Lemma 8.2.8 *κ is a final functor.*

Proof. Let $\mathbb{D}_{\oplus}$ denote the finite connected *colimit* completion of a small category $\mathbb{D}$. $\mathbb{D}_{\oplus}$ can be constructed as the full subcategory of $P(\mathbb{D})$ determined by those presheaves that are finite connected colimits of representables. Then the canonical functor $\mathbb{D} \longrightarrow \mathbb{D}_{\oplus}$ is an initial functor, so that, since $\mathbb{C}^{\oplus} = (\mathbb{C}^{\mathrm{op}}{}_{\oplus})^{\mathrm{op}}$, the functor κ is final. □

Proposition 8.2.9 *Let $\mathscr{E} = Sh(\mathbb{C}, J)$, so that $\mathrm{M}(\mathscr{E}) \simeq Sh(\mathbb{C}^{\star}, J^{\star})$, where $\mathbb{C}^{\star}$ is the free lex completion of $\mathbb{C}$. Then $\mathrm{T}(\mathscr{E})$ can be constructed as the pullback*

$$\begin{array}{ccc} \mathrm{T}(\mathscr{E}) & \rightarrowtail & \mathrm{M}(\mathscr{E}) \\ \downarrow & & \downarrow \\ P(\mathbb{C}^{\oplus}) & \rightarrowtail & P(\mathbb{C}^{\star}) \end{array}$$

in $\boldsymbol{Top}_{\mathscr{S}}$, where the bottom geometric morphism is induced by the unique factorization of δ through κ.

$$\begin{array}{ccc} \mathbb{C} & \xrightarrow{\kappa} & \mathbb{C}^{\oplus} \\ & \overset{\delta}{\searrow} & \downarrow \\ & & \mathbb{C}^{\star} \end{array}$$

Proof. We must show that for an arbitrary topos $\mathscr{X}$, the category of $\mathscr{X}$-valued probability distributions on $\mathscr{E}$ is equivalent to the category of cones

$$\begin{array}{ccc} \mathscr{X} & \xrightarrow{h} & \mathrm{M}(\mathscr{E}) \\ {\scriptstyle k}\downarrow & & \downarrow \\ P(\mathbb{C}^{\oplus}) & \rightarrowtail & P(\mathbb{C}^{\star}) \end{array}$$

by an equivalence that is natural in $\mathscr{X}$. Intuitively, this is clear because such a cone is simply an h for which $\mathscr{X} \xrightarrow{h} \mathrm{M}(\mathscr{E}) \rightarrowtail P(\mathbb{C}^{\star})$ factors through $P(\mathbb{C}^{\oplus})$. In any case, some explanation is necessary. Suppose we are given a $\mathscr{X}$-valued probability distribution with corresponding cosheaf $G : \mathbb{C} \longrightarrow \mathscr{X}$, and geometric morphism $\mathscr{X} \xrightarrow{h} \mathrm{M}(\mathscr{E})$. G satisfies $\varinjlim(G) \cong 1$. Then G lifts to a functor $G^{\oplus} : \mathbb{C}^{\oplus} \longrightarrow \mathscr{X}$ that preserves finite connected limits, and which furthermore, since κ is final (Lemma 8.2.8), satisfies $\varinjlim(G^{\oplus}) \cong 1$. It follows that the left extension $k^* : P(\mathbb{C}^{\oplus}) \longrightarrow \mathscr{X}$ of $G^{\oplus}$ preserves finite connected limits and also 1. Therefore, k^* is left exact, so that we have a geometric morphism k and a cone as above.

Conversely, a cone such as above gives a cosheaf $G : \mathbb{C} \longrightarrow \mathscr{X}$ corresponding to h, and at the same time a flat functor $K : \mathbb{C}^{\oplus} \longrightarrow \mathscr{X}$ corresponding to k. K satisfies $\varinjlim K \cong 1$, so that, again since κ is final, we have $\varinjlim (K \cdot \kappa) \cong 1$ also. But since the cone commutes, we have $G \cong K \cdot \kappa$, so that the corresponding distribution is a probability distribution. □

The following diagram depicts the two canonical factorizations of the unit $\delta : \mathscr{E} \rightarrowtail \mathrm{M}(\mathscr{E})$, one through the bag-domain $\mathrm{B}_L(\mathscr{E})$, and the other through the probability distribution classifier $\mathrm{T}(\mathscr{E})$.

$$\begin{array}{ccc}
\mathscr{E} & \rightarrowtail & \mathrm{B}_L(\mathscr{E}) \\
\downarrow & & \downarrow \\
\mathrm{T}(\mathscr{E}) & \rightarrowtail & \mathrm{M}(\mathscr{E})
\end{array}$$

In terms of freely adjoining finite limits to a site for the topos, the above corresponds to the diagram below.

$$\begin{array}{ccc}
\mathbb{C} & \xrightarrow{\text{fp}} & \mathbb{C}_{\text{fp}} \\
{\scriptstyle \text{fc}}\downarrow & & \downarrow{\scriptstyle \text{eq}} \\
\mathbb{C}^{\oplus} & \xrightarrow[\text{fp}]{} & \mathbb{C}^{\star}
\end{array}$$

For example, the functor labeled 'eq' is the unit for freely adjoining equalizers.

8.3 Discrete Complete Spread Structure

Our discussion in this section focuses on a certain notion of discreteness that implies that the lower bagdomain and symmetric topos agree (Remark 8.2.4). It turns out that this notion also provides a suitable 'single universe' for local homeomorphisms and complete spreads.

Definition 8.3.1 *We shall say that a topos $\mathscr{X}$ over a topos $\mathscr{S}$ has* discrete $\mathscr{S}$-complete spread structure *if in any commutative diagram*

$$\begin{array}{ccc}
\mathscr{F} & \xrightarrow{\eta} & \mathscr{X} \\
{\scriptstyle f}\searrow & & \swarrow \\
& \mathscr{S} &
\end{array} \qquad (8.5)$$

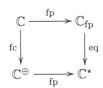

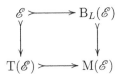

with f locally connected, η is an $\mathscr{S}$-complete spread iff f is discrete. We say a locale has discrete complete spread structure just when its topos of sheaves has. Sometimes we omit the prefix $\mathscr{S}$ when it is clear what is the base topos.

Proposition 8.3.2 *If a topos has discrete complete spread structure, then its bagdomain and symmetric toposes are equivalent.*

Proof. A distribution on a topos with discrete complete spread structure must preserves pullbacks, because the domain of the corresponding complete spread has the form $\mathscr{S}/I \longrightarrow \mathscr{S}$. □

For locales, discrete complete structure may be described with other equivalent conditions.

Proposition 8.3.3 *The following are equivalent for any locale X in a topos $\mathscr{S}$:*

1. *X has discrete complete spread structure;*
2. *any locale morphism $Y \longrightarrow X$ with locally connected domain is a complete spread iff Y is discrete;*
3. *The counit $|X| \longrightarrow X$ is the Gleason core* (locally connected coreflection) *of X, where $|X|$ denotes the object of points of X;*
4. *the canonical functor*

$$\mathscr{S}/|X| \longrightarrow \mathbf{E}(X) = \mathbf{Dist}_{\mathscr{S}}(\mathscr{S}, Sh(X))$$

is an equivalence.

Proof. The first two conditions are equivalent because the localic reflection of a locally connected topos is locally connected. The second and third conditions are obviously equivalent. The third and fourth are equivalent because the complete spread corresponding to the terminal distribution is precisely the Gleason core of the locale. □

A proof of the following may be found in the literature.

Lemma 8.3.4 *The quasi-components of an open set of a topological space are open in its Gleason core.*

By definition, a zero-dimensional space is one in which the clopen sets generate the topology.

Proposition 8.3.5 *A zero-dimensional T_0 topological space has discrete Set-complete spread structure.*

Proof. The Gleason core of a space has the same underlying set as the given space. The Gleason core of a zero-dimensional T_0 space is discrete. Indeed, the quasi-components of a zero-dimensional T_0 space are its singletons. By Lemma 8.3.4, singletons are open in the Gleason core. □

Proposition 8.3.6 *The product of two locales with discrete complete spread structure has complete spread structure.*

Proof. This follows because $|X \times Y| \cong |X| \times |Y|$. □

Proposition 8.3.7 *Let $\mathscr{E}$ denote a locally connected topos over $\mathscr{S}$. Then an $\mathscr{S}$-complete spread has discrete $\mathscr{E}$-complete spread structure.*

Proof. This is Proposition 9.4.1, which we prove in that section. □

Remark 8.3.8 *A local homeomorphism $\mathscr{E}/X \longrightarrow \mathscr{E}$ also has discrete $\mathscr{E}$-complete spread structure (Exercise 5). Thus, the locales in $\mathscr{E}$ with discrete $\mathscr{E}$-complete spread structure contain as full subcategories both the local homeomorphisms and the $\mathscr{S}$-complete spreads. This notion of single universe is largely unexplored. For example, consider Exercise 9.*

Exercises 8.3.9

1. *Show that $\mathrm{M}(\mathscr{E}) \simeq \mathrm{B}_L(\mathscr{E})$ iff every complete spread over $\mathscr{E}$ has totally connected components.*
2. *Show that a zero-dimensional sober topological space has discrete complete spread structure.*
3. *Construct the probability distribution classifier $\mathrm{T}(\mathscr{E})$ by a coinverter argument using $\mathbb{C}^{\oplus}$ instead of $\mathbb{C}^{\star}$.*
4. *It is well known that for any small category $\mathbb{C}$, the (finite) product completion $\mathbb{C}_{\mathrm{fp}}$ is finitely complete iff $\mathbb{C}$ has all (finite) small connected limits (Diers). Moreover, the universal functor $\mathbb{C} \longrightarrow \mathbb{C}_{\mathrm{fp}}$ preserves any connected limits that might exist in $\mathbb{C}$. Deduce from this that*

$$\mathbb{C}^{\star} \simeq (\mathbb{C}^{\oplus})_{\mathrm{fp}}.$$

   In turn, conclude that $\mathrm{M}(\mathscr{E}) \simeq \mathrm{B}_L(\mathrm{T}(\mathscr{E}))$. This gives an alternative proof, in terms of sites, of Theorem 8.2.7.
5. *Show that a locale is locally connected and has discrete complete spread structure iff it is discrete.*
6. *Show that the intersection of a (non-empty) family of locally connected topologies is locally connected.*
7. *Show directly that a topological space has a Gleason core: it has the same underlying set, but retopologized with the smallest locally connected topology larger than the given one.*
8. *Prove Lemma 8.3.4.*
9. *Prove or refute: a discrete Conduché fibration over a small category $\mathbb{C}$ has discrete $\mathrm{P}(\mathbb{C})$-complete spread structure, regarded as a locale in $\mathrm{P}(\mathbb{C})$.*
10. *Give a direct description of the algebras ('convex toposes') for the probability distributions classifier KZ-monad T in $\boldsymbol{Top}_{\mathscr{S}}$.*

8.4 Algebraic Geometry: Coschemes

In this section we discuss an example for the purpose of fixing ideas. We provide more details than an informed reader may need; however, we feel this is worthwhile as it illustrates in detail, and in a special case - the one associated with the Zariski topos (classifier of local rings) - the notions of distribution, distribution algebra, and complete spread geometric morphism. In the process we answer two questions. The first is how is the topos classifier of local rings with a given residue field constructed? Second, what is the nature of a distribution on the Zariski topos?

Throughout, the term 'ring' means a commutative ring with unit. Let *Ring* denote the category of such rings. Let *FP* denote the category of finitely presented rings $A = Z[x_1, \ldots, x_n]/I$. The ring $Z[x]$ is a coring object in *FP* for the tensor product $Z[x] \otimes Z[x] = Z[x, y]$. The comultiplication and coaddition $Z[x] \longrightarrow Z[x, y]$ are given by $x \mapsto xy$ and $x \mapsto x + y$.

We denote the topos of set-valued functors on *FP* by Set^{FP}. As always, we have the Yoneda functor

$$h : FP^{\mathrm{op}} \longrightarrow \mathit{Set}^{FP} .$$

Let $U = h(Z[x])$ denote the covariant representable

$$U(A) = \mathit{Ring}(Z[x], A) = A .$$

U is a universal ring object, or 'affine line,' making Set^{FP} a ringed topos. Set^{FP} classifies rings in the sense that there is a canonical equivalence of categories

$$\mathbf{Top}(\mathit{Set}, \mathit{Set}^{FP}) \simeq \mathbf{Lex}(FP^{\mathrm{op}}, \mathit{Set}) \simeq \mathit{Ring} . \tag{8.6}$$

The equivalence is given on the one hand by associating with a left exact functor F the ring $F(Z[x])$, and on the other hand by associating with a ring R the contravariant representable functor

$$\overline{R}(A) = \mathit{Ring}(A, R) . \tag{8.7}$$

Notice that $F(Z[x])$ is a ring because F is left exact and $Z[x]$ is a coring. The functor $\overline{R}$ is left exact because it is representable. The above equivalence can also be regarded in terms of points of Set^{FP}: $p \mapsto p^*(U)$.

The so-called (gros) Zariski topos denoted $\mathscr{Z}$, may be defined as the topos of sheaves on the site FP^{op} for the Grothendieck topology whose cocovers in *FP* are finite families of localizations $\{A \longrightarrow A[a_i^{-1}]\}$, such that $a_1 + \cdots + a_n = 1$. We call this topology the Zariski topology. The covering sieves are connected, so that $\mathscr{Z}$ is a locally connected topos.

Let *LRing* denote the category of commutative local rings with unit. (The trivial ring in which $0 = 1$ is not considered to be a local ring.) The equivalence (8.6) restricts to one

$$\mathbf{Top}(\mathit{Set}, \mathscr{Z}) \simeq \mathit{LRing} ,$$

so that the Zariski topos classifies local rings. The representables $h(A)$ are sheaves for the Zariski topology. In particular, the universal ring $U = h(Z[x])$ is a sheaf. Moreover, U is a local ring object in $\mathscr{Z}$. Local rings are preserved under inverse image of a geometric morphism. In terms of points of $\mathscr{Z}$, the above equivalence is given by $p \mapsto p^*(U)$. More generally, for any Grothendieck topos $\mathscr{E}$, there is a natural equivalence of categories

$$\mathbf{Top}(\mathscr{E}, \mathscr{Z}) \simeq \mathit{LRing}(\mathscr{E}) .$$

If R is a (local) ring in a topos $\mathscr{E}$, let $\mathit{Ring}_{\mathscr{E}}(R^{\Delta A})$ denote the object of ring homomorphisms from the constant ring ΔA to R in $\mathscr{E}$. If c is an object of a site $\mathbb{C}$ for $\mathscr{E}$, then

$$\mathit{Ring}_{\mathscr{E}}(R^{\Delta A})(c) = \mathit{Ring}(A, R(c)) .$$

This describes the geometric morphism $p : \mathscr{E} \longrightarrow \mathscr{Z}$ corresponding to a local ring object R:

$$p^*(A) = \mathit{Ring}_{\mathscr{E}}(R^{\Delta A}) \rightarrowtail R^n .$$

We sometimes refer to $p^*(A)$ as the R-variety defined by A in $\mathscr{E}$. We have $p^*(Z[x]) = p^*(U) = R$.

Example 8.4.1 *The frame of the Zariski spectrum* $\mathit{Spec}(R)$ *of a ring* R *is the lattice of radical ideals of* R*, ordered by inclusion:*

$$\mathscr{O}(\mathit{Spec}(R)) = \{\text{radical ideals of } R\} .$$

This frame is generated by the basic radical ideals:

$$D(r) = \{a \in R \mid \exists n, t \; a^n = tr\} .$$

We have

$$\mathscr{O}(\mathit{Spec}(R[r^{-1}]) \cong \{\text{radical } I \mid I \subseteq D(r)\} .$$

Thus, it makes sense to denote the locale $\mathit{Spec}(R[r^{-1}])$ *by* $D(r)$*. This is an open sublocale of* $\mathit{Spec}(R)$*.*

We define the structure sheaf on $\mathit{Spec}(R)$*:*

$$\mathscr{O}_R(D(r)) = R[r^{-1}] .$$

For any $r, s \in R$, $D(r) \subseteq D(s)$ *iff* $r \in D(s)$ *iff* $r^n = ts$*, for some natural number* n *and some* $t \in R$*. Therefore, if* $D(r) \subseteq D(s)$*, then the ring homomorphism* $R \longrightarrow R[r^{-1}]$ *inverts* s*, so that it factors through* $R \longrightarrow R[s^{-1}]$*.*

$$\begin{array}{ccc} R & \longrightarrow & R[s^{-1}] \\ & \searrow & \downarrow \\ & & R[r^{-1}] \end{array}$$

The structure sheaf $\mathcal{O}_R$ *is a local ring object in* $\mathcal{E} = Sh(Spec(R))$. *As above, we may explicitly describe the geometric morphism* $\mathcal{E} \xrightarrow{p} \mathcal{Z}$ *that corresponds to* $\mathcal{O}_R$ *as a left exact cosheaf* $FP^{op} \xrightarrow{p^*} \mathcal{E}$:

$$p^*(A)(D(r)) = Ring_{\mathcal{E}}({\mathcal{O}_R}^{\Delta A})(D(r)) = Ring(A, R[r^{-1}]) \ .$$

In particular, we have $p^*(U) \cong \mathcal{O}_R$.

Definition 8.4.2 *A ring homomorphism* with local domain *is a ring homomorphism* $L \longrightarrow R$ *for which* L *is a local ring.*

Example 8.4.3 *Let* k *denote any field, and let* R *denote the local ring* $k[x]_{(x)}$ *with maximal ideal* (x). *The quotient ring* $R/(x)$ *is isomorphic to* k. *A ring homomorphism* $L \xrightarrow{\varphi} R$ *with local domain amounts to a commutative square*

$$\begin{array}{ccc} L/P & \rightarrowtail & R \\ \downarrow & & \downarrow \pi \\ L/Q & \rightarrowtail & k \end{array}$$

where $P = \ker(\varphi)$ *and* $Q = \ker(\pi\varphi) =$ *maximal ideal of* L.

Our goal is to produce the topos classifier of ring homomorphisms with local domain to a given ring R:

$$\mathbf{Top}(\mathcal{E}, \mathcal{Z}_R) \ \simeq \ LRing(\mathcal{E})/\Delta_{\mathcal{E}}(R) \ .$$

Theorem 8.4.4 *Let* R *be a commutative ring with* 1. *Then there is a topos* $\mathcal{Z}_R$ *that classifies ring homomorphisms to* R *with local domain. Furthermore, there is a geometric morphism*

$$\psi_R : \mathcal{Z}_R \longrightarrow \mathcal{Z} \ .$$

If R *is a local ring, then the left exact* $\overline{R}$ (8.7) *is a cosheaf* (for the Zariski topology), *and* ψ_R *is a complete spread geometric morphism:* $\mathcal{Z}_R$ *occurs as the middle topos in the comprehensive factorization of the point Set* $\longrightarrow \mathcal{Z}$ *corresponding to* R. $\mathcal{Z}_R$ *is totally connected, and has a pure point Set* $\longrightarrow \mathcal{Z}_R$.

Proof. We first establish the following equivalences.

$$\mathbf{Top}(Set, Set^{FP/R}) \ \simeq \ \mathbf{Lex}(FP/R^{op}, Set) \ \simeq \ Ring/R \ .$$

FP/R has finite colimits: they are created in FP. Associated with a ring homomorphism $f : T \longrightarrow R$ we have the representable functor

$$\overline{f} : FP/R^{op} \longrightarrow Set \ ,$$

which is left exact.

Conversely, suppose we have a left exact functor

$$F : FP/R^{\mathrm{op}} \longrightarrow Set\,.$$

We regard elements $r \in R$ as homomorphisms $Z[x] \xrightarrow{r} R$. Let

$$f_F : T = \coprod_{r \in R} F(r) \longrightarrow R$$

be the evident projection. We denote elements of T as pairs (t,r), where $t \in F(r)$. We have $f_F(t,r) = r$.

The set T has a commutative ring structure such that:

1. f_F is a ring homomorphism,
2. $\overline{f_F} = F$,
3. $f_{\overline{f}} = f$.

Let (t,r) and (t',r') be elements of T. The object $Z[x,y] \xrightarrow{r,r'} R$ is the coproduct in FP/R of $Z[x] \xrightarrow{r} R$ and $Z[x] \xrightarrow{r'} R$. We use the map

$$F(x+y) : F(r,r') = F(r) \times F(r') \longrightarrow F(r+r')$$

to add elements of T:

$$(t,r) + (t',r') = (F(x+y)(t,t'), r+r')\,.$$

To see what is the 0-element of T observe that we have

$$\begin{array}{ccc} Z[x] & \xrightarrow{\quad 0 \quad} & Z \\ & {\scriptstyle 0_R}\searrow \quad \swarrow{\scriptstyle !} & \\ & R & \end{array}$$

in FP, where $Z \xrightarrow{!} R$ is the initial object of FP/R. Therefore $F(!) \cong 1$ and we have

$$F(0) : 1 \longrightarrow F(0_R)\,.$$

The 0-element of T is then $(F(0), 0_R)$. The remaining details are routinely verified.

We next turn to a construction of $\mathscr{X}_R$. Consider the topos pullback of the essential geometric morphism associated with the discrete fibration $FP/R \longrightarrow FP$ for $\overline{R}$ (8.7).

$$\begin{array}{ccc} \mathscr{X}_R & \rightarrowtail & Set^{FP/R} \\ {\scriptstyle \psi_R}\downarrow & & \downarrow \\ \mathscr{X} & \rightarrowtail & Set^{FP} \end{array}$$

Then a point $Set \longrightarrow \mathscr{X}_R$ corresponds to a commutative square of geometric morphisms (up to isomorphism) as follows.

$$\begin{array}{ccc} Set & \longrightarrow & Set^{FP/R} \\ \downarrow & & \downarrow \\ \mathscr{Z} & \rightarrowtail & Set^{FP} \end{array}$$

This says that a point of $\mathscr{Z}_R$ amounts to a local ring L that as a 'ring' $Set \longrightarrow Set^{FP}$ factors through $Set^{FP/R}$. Thus, a point of $\mathscr{Z}_R$ is equivalently given by a local ring L that is equipped with a ring homomorphism $L \longrightarrow R$.

Finally, if R is a local ring, then $\overline{R}$ preserves Zariski-covers. I.e., the left exact $\overline{R}$ is a cosheaf, and the above construction of ψ_R is just the usual construction of the complete spread factor of the comprehensive factorization of the point $Set \longrightarrow \mathscr{Z}$ corresponding to R. In this case, $\mathscr{Z}_R$ is locally connected, in fact totally connected, and the pure factor is a point $Set \longrightarrow \mathscr{Z}_R$. □

Remark 8.4.5 *The universal ring homomorphism in* $Set^{FP/R}$ *is* $U \xrightarrow{\pi} V$, *where*

$$U(A \xrightarrow{f} R) = A \,,$$

and V *is the constant presheaf* $\Delta(R)$. *The component* $\pi_f : U(f) \longrightarrow V(f)$ *of the natural transformation* π *is given by* $\pi_f = f$. *Also, we have*

$$U = \coprod_{r \in R} h_R(r) \,,$$

where

$$h_R : FP/R^{\mathrm{op}} \longrightarrow Set^{FP/R}$$

denotes Yoneda. As always, we regard an element $r \in R$ *as the object* $Z[x] \xrightarrow{r} R$ *of* FP/R. *We have* $h_R(r)(A \xrightarrow{f} R) = f^{-1}(r)$.

Example 8.4.6 *Consider the finite field* $R = Z/pZ = F_p$, *which is a local ring. If* L *is any local ring, then a ring homomorphism* $L \longrightarrow F_p$ *amounts to an isomorphism of the residue field* L/M *with* F_p. *Thus, the totally connected topos* $\mathscr{Z}_{F_p}$ *classifies local rings paired with an isomorphism of its residue field with* F_p. *The geometric morphism* $\psi_{F_p} : \mathscr{Z}_{F_p} \longrightarrow \mathscr{Z}$ *is a complete spread geometric morphism.*

Example 8.4.6 generalizes as follows.

Corollary 8.4.7 *For any field* k, *there is a topos classifier of local rings with residue field* k. *This topos is a subtopos of* $\mathscr{Z}_k$.

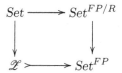

Proof. Let j denote the least topology in $\mathscr{Z}_k$ that forces the universal morphism $U \xrightarrow{\pi} V$ in $\mathscr{Z}_k$ to be an epimorphism. Then $Sh_j(\mathscr{Z}_k)$ classifies local rings whose residue field is isomorphic to k. Indeed, the kernel of a ring epimorphism $L \longrightarrow k$ with local domain must be equal to the single maximal ideal of L, so that the residue field of L is k. □

Corollary 8.4.7 answers our first question. For the second question, we introduce the following terminology.

Definition 8.4.8 *A* coscheme *is a distribution on* $\mathscr{Z}$. *Equivalently, a coscheme is a cosheaf on the Zariski site* FP^{op}.

Left exact coschemes correspond precisely to local rings: if R is a local ring, then $\overline{R} = Ring(\_\,, R)$ is a left exact coscheme. More generally, we may associate with a local ring object R in a locally connected topos $\mathscr{E}$ a coscheme (possibly not left exact):

$$G_{(\mathscr{E},R)}(A) = \pi_0(Ring_{\mathscr{E}}(R^{\Delta A})) \,, \tag{8.8}$$

where $\pi_0 \dashv \Delta$ is the connected components functor for $\mathscr{E}$.

The canonical equivalence of topos distributions with complete spread geometric morphisms implies the following.

Theorem 8.4.9 *The functor defined in* (8.8) *is a coscheme, and every coscheme has this form for some ringed topos* $(\mathscr{E}, R)$, *where* $\mathscr{E}$ *is locally connected, and* R *is local.*

Proof. Let R be a local ring object in a locally connected topos $\mathscr{E}$, corresponding to geometric morphism $\mathscr{E} \xrightarrow{p} \mathscr{Z}$. Then $G_{(\mathscr{E},R)}$ is the functor

$$FP^{\mathrm{op}} \xrightarrow{h} \mathscr{Z} \xrightarrow{p^*} \mathscr{E} \xrightarrow{\pi_0} Set \,,$$

which is a Zariski cosheaf. Conversely, for any coscheme G there is a complete spread geometric morphism

$$\psi : \mathscr{E} \longrightarrow \mathscr{Z}$$

for which $\mathscr{E}$ is locally connected, satisfying $G \cong \pi_0 \cdot \psi^* \cdot h$. But $R = \psi^* U$ is a local ring and $\psi^*(h(A)) \cong Ring_{\mathscr{E}}(R^{\Delta A})$. □

The function $D : R \longrightarrow \mathscr{O}(Spec(R))$ provides a bijection between the set $E(R)$ of idempotents of R and the definable subobjects of 1 in $Sh(Spec(R))$.

Definition 8.4.10 *A ring* R *is* idempotent finite *if every localization* $R[r^{-1}]$ *has only finitely many idempotents. A* minimal idempotent *is an idempotent* e *for which* $D(e)$ *is connected. We write* $E_m(R)$ *for the set of minimal idempotents of a ring* R.

For example, an integral domain is idempotent finite because every localization has no idempotents other than $0, 1$.

We have already mentioned the spatial version of the following result (Exercise 1.1.2).

Lemma 8.4.11 *A ring is idempotent finite iff its Zariski spectrum is a locally connected locale. In this case the terminal cosheaf on Spec(R) is*

$$\pi_0(D(r)) = \{\text{connected components of } D(r)\} \cong E_m(R[r^{-1}]) .$$

Proof. Any locale with only finitely many definable subobjects (= clopens) may be covered with finitely many connected clopens. Thus, if a ring has only finitely many idempotents, then its spectrum may covered with finitely many connected clopens. We may repeat this argument for every $D(r) = Spec(R[r^{-1}]$. The upshot is that the connected opens generate the topology of the spectrum.

If $Spec(R)$ is locally connected, then so is every locale $Spec(R[r^{-1}])$. The connected components of the latter form an open cover. However, $Spec(R[r^{-1}])$ is compact so these components must be finite in number. Hence the definable subobjects of any $D(r)$ are finite in number, which coincides with the number of idempotents in $R[r^{-1}]$. □

Corollary 8.4.12 *If a ring R is idempotent finite, then*

$$G_R(A) = G_{(\mathscr{E},\mathscr{O}_R)}(A) = \pi_0(Ring_{\mathscr{E}}({\mathscr{O}_R}^{\Delta A}))$$

is a coscheme, where $\mathscr{E} = Sh(Spec(R))$.

Example 8.4.13 *(Example 8.4.1 continued.) Consider the sheaf of idempotents of $\mathscr{O}_R$:*

$$E_R(D(r)) = E(D(r)) = \{\text{idempotents of } R[r^{-1}]\} .$$

E is a subsheaf of $\mathscr{O}_R$. E is also a Boolean algebra in $Sh(Spec(R))$: $e \vee f = e + f - ef$, $e \wedge f = ef$. E is isomorphic to the Boolean algebra $\Delta(2)$. If R is idempotent finite, then $E \cong \Delta(2)$ is isomorphic to the distribution algebra $\mathbf{U}(\pi_0)$ for the underlying sheaf functor (Def. 7.1.6)

$$\mathbf{U} : \mathbf{E}(\mathscr{E})^{\mathrm{op}} \longrightarrow \mathscr{E} ,$$

for $\mathscr{E} = Sh(Spec(R))$. I.e., E is the initial distribution algebra in $\mathscr{E}$, in the idempotent finite case.

Example 8.4.14 *To sum up, we may associate with any idempotent finite ring R the following:*

1. *the coscheme G_R, such that $G_R(A)$ equals the set of connected components of the $\mathscr{O}_R$-variety defined by A in $Sh(Spec(R))$;*
2. *the complete spread $\mathscr{V}_R \longrightarrow \mathscr{X}$ associated with G_R, which is the complete spread factor of the comprehensive factorization*

$$\begin{array}{ccc} Sh(Spec(R)) & \xrightarrow{\ \eta\ } & \mathscr{V}_R \\ & \searrow^{p} & \downarrow{\scriptstyle \psi} \\ & & \mathscr{X} \end{array}$$

where p corresponds to the local ring $\mathscr{O}_R$ *(*$\mathscr{V}_R$ *is a kind of completion of* Spec(R)*);*

3. *the distribution algebra* $B_R = p_*(E_R) \cong p_*(\Delta 2)$ *in* $\mathscr{X}$*.* B_R *is a sheaf on* $\mathrm{FP}^{\mathrm{op}}$*.* $B_R(A)$ *is equal to set of the definable subobjects of the* $\mathscr{O}_R$*-variety defined by* A *in* $\mathscr{E} = \mathrm{Sh}(\mathrm{Spec}(R))$*. We may equivalently describe* B_R *as the sheaf*
$$B_R(A) = \mathscr{E}(p^*(A), E_R)$$
where
$$p^*(A)(D(r)) = \mathrm{Ring}(A, R[r^{-1}]) \ ,$$
and
$$E_R(D(r)) = E(R[r^{-1}]) \ .$$
Note: B_R *is covariant in* A *because Yoneda:* $\mathrm{FP}^{\mathrm{op}} \longrightarrow \mathscr{X}$ *is contravariant.*

Exercises 8.4.15

1. *Explicitly describe the Grothendieck topology in* $\mathrm{FP}/R^{\mathrm{op}}$ *that defines* $\mathscr{X}_R$*.*
2. *Show that an idempotent* e *is minimal iff* $R[e^{-1}]$ *has no idempotents other than* $0, 1$ *iff* e *is non-0 and for all idempotents* f*, either* $ef = e$ *or* $ef = 0$*. Show that the sum of two distinct minimal idempotents is an idempotent. Show that* 1 *is minimal iff* 0 *and* 1 *are the only idempotents.*
3. *The Zariski spectrum of a Noetherian ring is a Noetherian space. Show that a Noetherian ring is idempotent finite by showing that a Noetherian space is locally connected, and then use Lemma 8.4.11.*
4. *Let* R *be idempotent finite with coscheme* G_R *(Cor. 8.4.12). Show that*
$$G_R(Z[x]) = \pi_0(\mathscr{O}_R) \cong \left(\coprod_{r \in R} R[r^{-1}] \times E_m(R[r^{-1}]) \right) / \sim \ ,$$
where the equivalence relation $\sim$ *is defined by a colimit:*
$$\varinjlim \left(\mathrm{El}(\mathscr{O}_R) \longrightarrow \mathscr{O}(\mathrm{Spec}(R)) \xrightarrow{\pi_0} \mathrm{Set} \right) \ .$$
$\mathrm{El}(\mathscr{O}_R)$ *denotes the category of elements of the sheaf* $\mathscr{O}_R$*.*
5. *What sort of idempotent finite ring* R *is recovered from its coscheme* G_R*? Technically, the question can be phrased in various and related ways. When is the corresponding geometric morphism* $\mathrm{Sh}(\mathrm{Spec}(R)) \xrightarrow{p} \mathscr{X}$ *is a complete spread, i.e., when is the pure factor* η *in Example 8.4.14 an equivalence? For what* R *is the unit*
$$\psi^*(U) \longrightarrow \eta_*\eta^*(\psi^*(U)) \cong \eta_*(\mathscr{O}_R)$$
a ring isomorphism in $\mathscr{V}_R$*, where* U *is the generic ring in* $\mathscr{X}$*? For what* R *is the canonical ring homomorphism* $\Gamma(\psi^* U) \longrightarrow R$ *an isomorphism?*

8.5 Distributions on the Jonsson-Tarski Topos

In this section we shall consider a Grothendieck topos $\mathscr{C}$ for which $\mathbf{E}(\mathscr{C})$, as it happens, is a Grothendieck topos (Prop. 1.4.13). $\mathscr{C}$ is an étendue, meaning a topos $\mathscr{E}$ that has an object F with full support such that $\mathscr{E}/F$ is a localic topos. $\mathscr{C}$ is not locally connected, but we are able to explicitly describe its Gleason core. The topos $\mathscr{C}$ in question is the topos analogue of Cantor space: it has also been called the Jonsson-Tarski topos.

Definition 8.5.1 A Jonsson-Tarski algebra *is a set X equipped with an isomorphism $X \xrightarrow{(f,g)} X \times X$. We refer to such an object as a* JT-algebra. *A morphism of JT-algebras is a morphism $X \longrightarrow Y$ that commutes with the structure isomorphisms. Let $\mathscr{C}$ denote this category.*

We first exhibit a site presentation of $\mathscr{C}$, showing that $\mathscr{C}$ is a Grothendieck topos. Let M_2 denote the free monoid on two generators a, b. We refer to the elements of M_2 as words. We have the topos $P(M_2)$ of right M_2-sets (= presheaves on M_2). We use $\hom(X, Y)$ to denote the set of right M_2-set maps between two right M_2-sets X and Y. Usually we write $\hat{a} : X \longrightarrow X$ for the action of a in X: $\hat{a}(x) = xa$.

Let $\star$ denote the single object of the monoid M_2. We denote the single representative right M_2-set by $h(\star) = M_2$. It carries an action of M_2 by right multiplication. A word V provides a right M_2-set map $h_V : M_2 \longrightarrow M_2$ given by left multiplication: $h_V(W) = VW$.

The slice category $M_2/\star$ is the partially ordered set of all words, such that $V \leq W$ if W is a prefix of V. We denote the empty word by $\top$ since it is the top word in this ordering. We have

$$P(M_2/\star) \simeq P(M_2)/M_2 \,.$$

We may make a JT-algebra $X \xrightarrow{(f,g)} X \times X$ into a right M_2-set by defining $\hat{a} = f$ and $\hat{b} = g$. On the other hand, a right M_2-set X is a JT-algebra if $X \xrightarrow{(\hat{a},\hat{b})} X \times X$ is an isomorphism.

Put another way, a JT-algebra is a right M_2-set that perceives a certain right M_2-set map as an isomorphism. To be precise, an M_2-set X is a JT-algebra if the following unique extension property holds.

$$\begin{array}{ccc} M_2 + M_2 & \xrightarrow{(h_a,h_b)} & M_2 \\ & \underset{\forall}{\searrow} & \downarrow \exists! \\ & & X \end{array}$$

This condition is of course a sheaf condition. The morphism (h_a, h_b) is a monomorphism. Its image is the set of non-empty words.

Thus, $\mathscr{C}$ is a subtopos of $P(M_2)$. A sieve is a subset of M_2 closed under right multiplication, which we typically denote R. The Grothendieck topology

J in M_2 for JT-algebras is thus a collection of certain covering sieves, which we wish to identify. If $\{W_i\}$ is any set of words, let

$$\langle W_i \rangle = \{W \in M_2 \mid \exists i\, W \leq W_i\}$$

denote the sieve generated by the W_i. For instance, the sieve

$$\langle a, b \rangle = \{W \in M_2 \mid W \text{ begins with } a \text{ or } b\}$$

consists of all non-empty words. This sieve generates J. The top sieve is $M_2 = \langle \top \rangle$. A covering sieve is then one of the form $\langle a, b \rangle$, $\langle aa, ab, ba, bb \rangle$, $\langle aa, ab, b \rangle$, and so on.

J is itself a right M_2-set: the action of M_2 is given by pullback along h_a:

$$R \cdot a = \hat{a}(R) = \{W \mid aW \in R\} ,$$

and similarly for b. We have $\hat{a}(R) \rightarrowtail R : W \mapsto aW$.

We may identify the covering sieves with rooted binary trees of finite depth. For instance, we identify $\langle aa, ab, b \rangle$ with the tree

$$\begin{array}{ccccc} & & & \top & \\ & & a & & b \\ & a & & b & \end{array}$$

We may describe the action of M_2 in J in terms of these trees. For example, $\langle aa, ab, b \rangle \cdot a = \langle a, b \rangle$, and $\langle aa, ab, b \rangle \cdot b = \top$.

If R is a covering sieve, and X is any right M_2-set, then we may identify a member of $\hom(R, X)$ with the tree R whose leaves are paired with elements of X. For instance, if $x \in X$, then the tree

$$\top, x \tag{8.9}$$

depicts the morphism $M_2 \xrightarrow{x} X$. For another example,

$$\begin{array}{ccccc} & & & \top & \\ & & a & & b, z \\ & a, x & & b, y & \end{array}$$

depicts the right M_2-map $\langle aa, ab, b \rangle \longrightarrow X$ that sends aa to x, ab to y, and b to z.

The (covering) sieves are ordered by containment. As binary trees, this ordering appears as reverse inclusion. For instance, we have

$$\langle aa, ab, b \rangle \ \leq \ \langle a, b \rangle ,$$

which we depict as follows.

$$\begin{array}{ccccc} & & \top & & \\ & a & & b & \\ a & & b & & \end{array} \leq \begin{array}{ccc} & \top & \\ a & & b \end{array}$$

The induced map $\hom(\langle a, b\rangle, X) \longrightarrow \hom(\langle aa, ab, b\rangle, X)$ appears as

$$\begin{array}{ccc} & \top & \\ a, x & & b, y \end{array} \quad \mapsto \quad \begin{array}{cccc} & & \top & \\ & a & & b, y \\ a, xa & & b, xb & \end{array}$$

The colimit of this system is a right M_2-set that we denote $L(X)$:

$$L(X) = \varinjlim_{J^{op}} \hom(R, X) \ .$$

If we denote members of $L(X)$ by $[\xi]$, where $\xi : R \longrightarrow X$, then the action of M_2 in $L(X)$ is given by

$$[\xi] \cdot a = [R \cdot a \rightarrowtail R \xrightarrow{\xi} X] \ ,$$

and similarly for b. For example, the action of a on

$$\begin{array}{cccc} & & \top & \\ & a & & b, z \\ a, x & & b, y & \end{array}$$

is equal to

$$\begin{array}{ccc} & \top & \\ a, x & & b, y \end{array}$$

and the action of b on this same tree is equal to $\top, z$.

From sheaf theory, if a presheaf has the uniqueness property for every covering sieve, then it is said to be separated. It turns out that if a right M_2-set X has this property for $\langle a, b\rangle \rightarrowtail M_2$, then it is separated. Moreover, if X is separated, then $L(X)$ is a JT-algebra: $L(X)$ is the best approximation of a separated X by a JT-algebra.

The single representable M_2-set M_2 is not a JT-algebra, but it is separated. An element of the JT-algebra $L(M_2)$ is an equivalence class of binary trees whose leaves are paired with elements of M_2. The canonical morphism of JT-algebras

$$(\, L(h_a), L(h_b)\,) : L(M_2) + L(M_2) \longrightarrow L(M_2) \qquad (8.10)$$

is an isomorphism.

Now consider Cantor space $C = 2^N$. This is a product space, where 2 carries the discrete topology. We may regard C as the set of infinite strings $\{abb \cdots\}$. A typical basic open set U_V of Cantor space is then the collection of infinite strings that have the finite word V as a prefix. We have $V \leq W$ iff $U_V \leq U_W$. Thus, the partial order $M_2/\star$ is isomorphic to this base of open sets $\{U_V\}$. This brings us to the following result due to P. Freyd.

Proposition 8.5.2 *$\mathscr{C}$ is an étendue: $L(M_2)$ has full support in $\mathscr{C}$, and we have $\mathscr{C}/L(M_2) \simeq Sh(C)$. The sheaf on C associated with a JT-algebra morphism $X \xrightarrow{f} L(M_2)$ is*

$$F(U_V) = \{x \in X \mid f(x) = [\top, V]\},$$

where $\top, V$ *is defined in (8.9).*

Proof. The pullback of the JT-topology J in $P(M_2)$ to $P(M_2/\star)$ coincides with the canonical topology in $M_2/\star$. The category of sheaves for this topology is precisely the sheaf topos $Sh(C)$. On the other hand, since $P(M_2/\star) \simeq P(M_2)/M_2$, this pullback coincides with $\mathscr{C}/L(M_2)$. □

Definition 8.5.3 A Kennison algebra, or K-algebra, *in a topos is an object* A *of the topos equipped with an isomorphism* $A + A \longrightarrow A$. *Let* $\mathscr{K}$ *denote the category of K-algebras in Set.*

Example 8.5.4 *The set of natural numbers* N *is a K-algebra in Set such that* $n \mapsto 2n$ *or* $n \mapsto 2n + 1$.

Example 8.5.5 *By (8.10),* $L(M_2)$ *is a K-algebra in* $\mathscr{C}$. *In fact, we shall see in Proposition 8.5.6 that* $L(M_2)$ *is the generic K-algebra.*

A few remarks about the structure of $\mathscr{K}$ follow. In particular, we show that $\mathscr{K} \simeq \mathbf{E}(\mathscr{C})$.

Proposition 8.5.6 *The functor* $\mu \mapsto \mu(L(M_2))$ *is an equivalence* $\mathbf{E}(\mathscr{C}) \simeq \mathscr{K}$. *More generally, for any Grothendieck topos* $\mathscr{F}$, $\mathbf{Dist}(\mathscr{F}, \mathscr{C})$ *is equivalent to the category of K-algebras in* $\mathscr{F}$. *The symmetric topos* $\mathrm{M}(\mathscr{C})$ *classifies Kennison algebras.*

Proof. The category of topos distributions $\mathbf{E}(P(M_2))$ is equivalent to the topos of left (= covariant) M_2-sets. If Y is a left M_2-set, then we denote the left action of a by $\bar{a} : Y \longrightarrow Y$. A left M_2-set Y carries the generating dense monomorphism $(h_a, h_b) : M_2 + M_2 \rightarrowtail M_2$ (of right M_2-sets) to the function $(\bar{a}, \bar{b}) : Y + Y \longrightarrow Y$. A K-algebra is thus a left M_2-set that carries the generating dense monomorphism to an isomorphism. This is precisely a cosheaf on M_2 for the JT-topology. □

Definition 8.5.7 *Let* T_C *denote the left* M_2*-set* 2^N, *such that the left action* $\bar{a}$ *prefixes an infinite string with the generator symbol* a, *and likewise for* $\bar{b}$.

Proposition 8.5.8 T_C *is the terminal K-algebra in Set. Equivalently,* T_C *is the terminal distribution on* $\mathscr{C}$.

Proof. T_C is a cosheaf for the JT-topology because T_C carries the generating dense monomorphism $(h_a, h_b) : M_2 + M_2 \rightarrowtail M_2$ to $2^N + 2^N \xrightarrow{(\bar{a}, \bar{b})} 2^N$, which is an isomorphism.

Let A be any K-algebra. We define the unique left M_2-set map $f : A \longrightarrow 2^N$ as follows. Let $x = x_0 \in A$. Then there is a unique $x_1 \in A$ such that $x_0 = c_0 x_1$, where $c_0 \in \{a, b\}$. Similarly, $x_1 = c_1 x_2$ for a unique x_2. We define $f(x) = c_0 c_1 \ldots$. Thus, T_C is terminal. □

Remark 8.5.9 *J. Kennison has shown that the category* $\mathscr{K}$ *of K-algebras in Set is a topos, and M. Barr has remarked that although* $\mathscr{K}$ *is Boolean, it is not well-pointed. In fact,*

$$\mathscr{K} \simeq \mathrm{Sh}_{\neg\neg}(Set^{M_2}/T_C).$$

We mention an application of the Monadicity Theorem (Thm. 7.1.15) and Proposition 8.5.6. Every K-algebra A has an underlying JT-algebra $\mathbf{U}(A) = 2^A$, which is the distribution algebra in $\mathscr{C}$ associated with the K-algebra A. The right action of M_2 in 2^A is as follows: for any subset $S \subseteq A$,

$$S \cdot a = \{x \in A \mid a \cdot x \in S\} ,$$

and similarly for b. The underlying JT-algebra functor $\mathbf{U}$ is contravariant.

Corollary 8.5.10 *The underlying JT-algebra functor*

$$\mathbf{U} : \mathscr{K}^{\mathrm{op}} \longrightarrow \mathscr{C}$$

is monadic.

We conclude with an instance of the construction of the Gleason core of a topos: the Gleason core of $\mathscr{C}$.

$$\gamma : \widehat{\mathscr{C}} \longrightarrow \mathscr{C}$$

Any geometric morphism $\mathscr{F} \longrightarrow \mathscr{C}$ for which $\mathscr{F}$ is locally connected has an essentially unique factorization through the above γ.

Let $\mathbb{A}$ denote the category whose objects are infinite strings $x = aaba\ldots$ in the two generators a, b. A morphism $x \longrightarrow y$ is a word $V \in M_2$ such that $Vx = y$. $\mathbb{A}$ is not a poset: let $x = \overline{ab} = abab\ldots$, and let $V = ab$. Then $Vx = x$, so V and V^2 are distinct endomorphisms of x. There is a functor $\mathbb{A} \longrightarrow M_2$ that carries a morphism W to itself.

We regard $\mathbb{A}$ as a site as follows. We define a single generating covering sieve of a string y as the set of all non-empty words $W \in M_2$ that appear as a prefix of y: so we have a morphism $W : \frac{y}{W} \xrightarrow{W} y$, where $\frac{y}{W}$ denotes the string y after deleting the prefix W. Let $\widehat{\mathscr{C}}$ denote the topos of sheaves for this site. It happens that $\widehat{\mathscr{C}}$ is part of a topos pullback.

$$\begin{array}{ccc} \widehat{\mathscr{C}} & \rightarrowtail & P(\mathbb{A}) \\ {\scriptstyle\gamma}\downarrow & & \downarrow \\ \mathscr{C} & \rightarrowtail & P(M_2) \end{array}$$

The fact that $\widehat{\mathscr{C}}$ is locally connected is a special case of a fact shown already in a more general context (Proposition 2.4.1). The terminal distribution on $\mathscr{C}$ is $\pi_0 \cdot \gamma^*$, and we recover the terminal Kennison algebra as

$$T_C \cong (\pi_0 \cdot \gamma^*)(L(M_2)) .$$

Exercises 8.5.11

1. *What is the underlying JT-algebra of the natural numbers as a K-algebra (Example 8.5.4)?*
2. *Show that* $\mathscr{C}$ *is not locally connected. Show directly (without appealing to Proposition 2.4.1) that* $\widehat{\mathscr{C}}$ *is locally connected, and that* γ *is the Gleason core of* $\mathscr{C}$*.*
3. *Show that* $\widehat{\mathscr{C}}$ *is an étendue.*
4. *A geometric morphism is said to be a surjection if its inverse image functor is faithful. Show that* $\gamma : \widehat{\mathscr{C}} \longrightarrow \mathscr{C}$ *is a surjection.*

Further reading: Barr & Kennison [BK02], Bunge & Carboni [BC95], Bunge & Funk [BF98], Jibladze & Johnstone [JJ91], Johnstone [Joh82b, Joh02, Joh92, Joh89], Joyal & Tierney [JT84], Kock & Reyes [KR99], Vickers [Vic89, Vic92, Vic95].

9

Topological Aspects

Singular covering maps take historical precedence (in the work of Riemann in analysis) over the more recent concept of covering map that occurs in algebraic topology. This chapter emphasizes the point of view that topos complete spreads may be regarded as a class of generalized singular coverings. We deal with aspects of singular covering toposes of interest in topology, and in particular, we focus on a special class of complete spreads that we call branched coverings.

We call the geometric morphism $\mathscr{E}/X \longrightarrow \mathscr{E}$ associated with any object X of a topos $\mathscr{E}$ a local homeomorphism. We shall see that if X is a locally constant object (or locally trivial, or locally split) in a locally connected topos $\mathscr{E}$, then $\mathscr{E}/X \longrightarrow \mathscr{E}$ is a complete spread geometric morphism. Let us call a local homeomorphism that is also a complete spread an unramified covering, or an unramified covering topos. If X is a locally constant object of a locally connected topos $\mathscr{E}$, then the geometric morphism $\mathscr{E}/X \longrightarrow \mathscr{E}$ is thus unramified in this sense.

Our definition (or intrinsic characterization) of branched coverings in the context of toposes employs the notions of complete spread, pure subobject, locally trivial covering, and a newly isolated concept of purely skeletal geometric morphism. We establish the equivalence of this (axiomatic) definition with an alternative notion of branched covering that is almost directly motivated by Fox's topological concept of branched covering, and which was given independently by M. Bunge and S. Niefield [BN00], and by J. Funk [Fun00].

We show that a van Kampen theorem, obtained in joint work by M. Bunge and S. Lack [BL03], holds for what we call fibrations of regular coverings. These include both the locally constant as well as the larger class of unramified coverings, which are shown to enjoy similar topological properties yet do not agree in general.

Finally, we introduce a notion of index of a complete spread, which is related to branched coverings.

9.1 Locally Constant versus Unramified Coverings

Let $\mathscr{E}$ be a topos bounded over $\mathscr{S}$. For the moment, no assumptions on $\mathscr{E}$ will be made. Occasionally, we will need to suppose that $\mathscr{E}$ is locally connected.

We say that an object has *global support* if its unique morphism to the terminal is an epimorphism.

Definition 9.1.1 *An object X of $\mathscr{E}$ is said to be U-*split *by an object U of $\mathscr{E}$ if there is a morphism $\alpha : S \longrightarrow I$ in $\mathscr{E}$, and a morphism $\eta : U \longrightarrow e^*I$, such that there is a morphism $X \times U \longrightarrow e^*S$ for which the square*

$$\begin{array}{ccc} X \times U & \longrightarrow & e^*S \\ {\scriptstyle \pi_2}\downarrow & & \downarrow{\scriptstyle e^*\alpha} \\ U & \xrightarrow[\eta]{} & e^*I \end{array}$$

is a pullback. An object X of $\mathscr{E}$ is said to be locally constant, *or* locally trivial, *if X is U-split by an object U of $\mathscr{E}$ with global support. We call the geometric morphism $\mathscr{E}/X \longrightarrow \mathscr{E}$ associated with a locally constant object X* a locally constant covering.

Remark 9.1.2 *A notion of* constant object *(as U-split where $U = 1$) is implicit in the notion of locally constant object. In view of the central role which definable morphisms play in this book, it is worthwhile to note that an object X of an $\mathscr{S}$-topos $\mathscr{E}$ is 1-split iff it is a definable object, in the sense that $X \longrightarrow 1$ is a definable morphism. Then implicitly and automatically an object is locally constant iff it is locally definable.*

Lemma 9.1.3 *Assume that X and U are objects of a locally connected topos $\mathscr{E}$. Then the following are equivalent:*

1. *X is U-split.*
2. *there exists a morphism $\alpha : S \longrightarrow e_!U$ in $\mathscr{S}$, and a pullback*

$$\begin{array}{ccc} X \times U & \longrightarrow & e^*S \\ {\scriptstyle \pi_2}\downarrow & & \downarrow{\scriptstyle e^*\alpha} \\ U & \xrightarrow[\eta_U]{} & e^*e_!U \end{array}$$

   where η is the unit of the adjunction $e_! \dashv e^*$.
3. *the adjunction square*

$$\begin{array}{ccc} X \times U & \longrightarrow & e^*e_!(X \times U) \\ {\scriptstyle \pi_2}\downarrow & & \downarrow{\scriptstyle e^*e_!\pi_2} \\ U & \xrightarrow[\eta_U]{} & e^*e_!U \end{array}$$

   is a pullback, where η is the unit of $e_! \dashv e^*$.

It follows from the preservation properties of inverse image functors of geometric morphisms that they preserve locally constant objects. We now wonder about the question of inverse image functors reflecting locally constant objects. We are led to consider restrictions on the geometric morphism.

Lemma 9.1.4 *Pullback along a locally connected surjection reflects locally constant objects.*

Proof. Let $\mathscr{G} \xrightarrow{\varphi} \mathscr{E}$ be a locally connected surjection. Assume that φ^*X is split by some object U of global support, with the help of morphisms $\alpha : S \longrightarrow I$ in $\mathscr{S}$, and $\eta : U \longrightarrow \varphi^*e^*I$ in $\mathscr{G}$, so that there is a pullback diagram

$$\begin{array}{ccc} \varphi^*X \times U & \xrightarrow{\varsigma} & \varphi^*e^*S \\ {\scriptstyle \pi_2}\downarrow & & \downarrow{\scriptstyle \varphi^*e^*\alpha} \\ U & \xrightarrow[\eta]{} & \varphi^*e^*I \end{array}$$

in $\mathscr{G}$. Since φ is locally connected, it preserves definable morphisms in the sense that the diagram

$$\begin{array}{ccc} \varphi_!(\varphi^*X \times U) & \xrightarrow{\varsigma'} & e^*S \\ {\scriptstyle \varphi_!\pi_2}\downarrow & & \downarrow{\scriptstyle e^*\alpha} \\ \varphi_!U & \xrightarrow[\eta']{} & e^*I \end{array}$$

is a pullback in $\mathscr{E}$. By the Frobenius condition for a locally connected geometric morphism over $\mathscr{S}$, which in this case says that the canonical map

$$\varphi_!(\varphi^*X \times U) \longrightarrow X \times \varphi_!U$$

is invertible, the above diagram gives a pullback

$$\begin{array}{ccc} X \times \varphi_!U & \xrightarrow{\varsigma'} & e^*S \\ {\scriptstyle \pi_2}\downarrow & & \downarrow{\scriptstyle e^*\alpha} \\ \varphi_!U & \xrightarrow[\eta']{} & e^*I \end{array}$$

which shows that X is $\varphi_!U$-split.

It remains to prove that if U has global support, then so does $\varphi_!U$. Since $U \longrightarrow 1$ is an epimorphism, also $\varphi_!U \longrightarrow \varphi_!1$ is an epimorphism as $\varphi_!$ is a left adjoint. If φ is a surjection (φ^* is faithful), then $\varphi_!1$ has global support since every component of the counit of $\varphi_! \dashv \varphi^*$ is an epimorphism. In particular, $\varphi_!1 \cong \varphi_!\varphi^*1 \longrightarrow 1$ is an epimorphism. This completes the proof. □

We now turn to a consideration of the 'analytic' notion of an unramified covering, which we take to mean a local homeomorphism that is also a complete spread (Def. 9.1.7). We first analyse some further relevant properties of complete spreads, beginning with the following analogue of Lemma 9.1.4 for complete spreads.

Lemmas 9.1.4 and 9.1.5 will prepare us for Theorem 9.1.6, and for the van Kampen theorem for locally constant coverings and unramified coverings.

Lemma 9.1.5 *Pullback along a locally connected surjection reflects complete spreads.*

Proof. Suppose in a pullback

$$\begin{array}{ccc} \mathscr{P} & \xrightarrow{\xi} & \mathscr{G} \\ \downarrow & & \downarrow{\scriptstyle\varphi} \\ \mathscr{F} & \xrightarrow[\psi]{} & \mathscr{E} \end{array}$$

that ξ is a complete spread, where φ is a locally connected surjection. Then ξ is a spread, so that by Lemma 3.1.10, ψ is a spread. Form the spread completion of ψ and its pullback along φ.

$$\begin{array}{ccccc} \mathscr{P} & \rightarrowtail & \mathscr{X} & \longrightarrow & \mathscr{G} \\ \downarrow & & \downarrow & & \downarrow{\scriptstyle\varphi} \\ \mathscr{F} & \underset{\eta}{\rightarrowtail} & \mathscr{Y} & \longrightarrow & \mathscr{E} \end{array}$$

η is a pure spread, hence an inclusion (Lemma 3.1.12). The pullback of η is an equivalence because ξ is a complete spread. Therefore, η is a surjection, hence an equivalence. This proves that ψ is a complete spread. □

Theorem 9.1.6 *If X is a locally constant object of a locally connected topos $\mathscr{E}$, then $\mathscr{E}/X \longrightarrow \mathscr{E}$ is a complete spread.*

Proof. Let X be locally constant. By Exercise 2.4.12, 3, for any morphism $S \xrightarrow{m} I$ of $\mathscr{S}$, $\mathscr{S}/S \longrightarrow \mathscr{S}/I$ is a complete spread. Complete spreads are pullback stable along locally connected (or even essential) geometric morphisms, so

$$\mathscr{E}/e^*S \longrightarrow \mathscr{E}/e^*I\ ,$$

and hence $\mathscr{E}/X \times U \longrightarrow \mathscr{E}/U$, is a complete spread. By Lemma 9.1.5 we are done. □

Definition 9.1.7 *Assume that $\mathscr{E}$ is locally connected over $\mathscr{S}$. We shall refer to an object X of a topos $\mathscr{E}$ over $\mathscr{S}$ for which $\mathscr{E}/X \longrightarrow \mathscr{E}$ is a complete spread as a* complete spread object. *In this case, we call the geometric morphism $\mathscr{E}/X \longrightarrow \mathscr{E}$* an unramified covering.

Theorem 9.1.6 says, in our terminology, that a locally constant covering is an unramified covering. Example 9.1.8 describes an unramified cover that is not locally constant; the class of unramified coverings is strictly larger than the class of locally constant coverings, even over a locally connected space. The domain space in this example is connected, so this map cannot even be a coproduct of locally constant coverings.

Example 9.1.8 *This example describes a local homeomorphism* $Y \xrightarrow{\psi} X$ *(for which* Y *is connected) into a locally path-connected and connected space* X *that is a complete spread (an unramified covering), but is not locally constant. The space* X *is the 'Hawaiian earring:'* X *is the pencil of tangent circles* C_n *of radius* $\frac{1}{n}$, $n = 1, 2, 3 \ldots$, *topologized as a subspace of the Euclidean plane. We have*

$$X = \bigcup_{n=1}^{\infty} C_n$$

with a single tangent point a *such that* $\forall\, m \neq n$, $C_m \cap C_n = \{a\}$. *The Hawaiian earring is not semi-locally simply connected (defined below). The domain space* Y *consists of countably many copies of the real line* $\mathbb{R}$ *and of* X, *topologized as a subset of Euclidean 3-space. To be precise, let*

$$Y = (\bigcup_{n=1}^{\infty} \mathbb{R}_n) \cup (\bigcup_{|z|=1}^{\infty} X_z),$$

where n *is a natural number and* z *is an integer. Let* $Y \xrightarrow{\psi} X$ *be the map such that:*

1. *each* $\mathbb{R}_n$ *is a homeomorphic copy of the real line, and* ψ *restricted to* $\mathbb{R}_n$ *is a universal covering map* $\mathbb{R}_n \longrightarrow C_n$,
2. $\psi^{-1}(a) = \{\ldots, -2, -1, 1, 2, \ldots\}$ *ordered consecutively on* R_1, *and* $\psi^{-1}(a) \cap \mathbb{R}_n = \{\ldots, -n-1, -n, n, n+1, \ldots\}$,
3. ψ *carries* X_z *homeomorphically onto* $\bigcup_{j=|z|+1}^{\infty} C_j$, $|z| = 1, 2, \ldots$,
4. *each* $y \in Y - \psi^{-1}(a)$ *has an open neighbourhood that is homeomorphic to the real line,*
5. $X_z \cap (\bigcup_{n=1}^{\infty} \mathbb{R}_n) = \{z\}$, $|z| = 1, 2, \ldots$.

The space Y *is connected and locally path-connected. We readily see that the map* ψ *is a local homeomorphism, even at the points of the fiber* $\psi^{-1}(a)$. *Furthermore,* ψ *is a spread, and it also holds that the fiber of any point of* X *is in bijection with its cogerms, so that* ψ *is a complete spread. On the other hand,* ψ *is not locally constant because the point* $a \in X$ *does not have an evenly covered neighbourhood. Indeed, any neighbourhood* B *of* a *contains a circle* C_n, *for some* n. *For this* n, *the point* n *of* $\psi^{-1}(a)$ *(according to our naming convention) is a member of* R_n. *The connected component of* $\psi^{-1}(B)$ *that contains this point must contain all of* $\mathbb{R}_n$, *so that* ψ *cannot restrict to a homeomorphism of this component onto* B.

Remark 9.1.9 *The following result provides further evidence beyond Exercise 6 that unramified coverings are locally constant under hypotheses of the locally simply connected kind. A space is said to be* semi-locally simply connected *if it has a cover $\{U_\alpha\}$ of open neighbourhoods such that each U_α has the property that any two paths in U_α with common endpoints are homotopic in the whole space by a homotopy that fixes the endpoints.*

> *Theorem [FT01]:* A non-0 local homeomorphism over a connected, locally path-connected, semi-locally simply connected space that is also a complete spread is a surjective covering space.

This theorem may be proved using a path-lifting argument, but we omit this proof as we shall not use the result and the proof would take us beyond the scope of this book.

Remark 9.1.10 *In some ways, unramified coverings are better behaved than locally constant coverings. Locally constant coverings do not generally compose: it can happen that $\mathscr{E}/X \longrightarrow \mathscr{E}/Y$ and $\mathscr{E}/Y \longrightarrow \mathscr{E}$ are locally constant, but $\mathscr{E}/X \longrightarrow \mathscr{E}$ is not (even in a locally connected topos $\mathscr{E}$). However, unramified coverings do compose. We shall see also that unramified coverings, just like locally constant coverings, satisfy a (coverings) van Kampen theorem with respect to the same class of geometric morphisms of effective descent, namely locally connected surjections. For this purpose, we shall define a notion of regular covering morphism. The locally constant coverings and the unramified coverings are two examples of such regular classes. Of course, we also think of complete spreads as coverings, but of a singular (or ramified), not regular kind. We shall develop this point of view in § 9.3.*

Let us denote by

$$\mathrm{L} : \mathbf{Top}_{\mathscr{S}}{}^{\mathrm{op}} \longrightarrow \mathbf{CAT}$$

the pseudofunctor that assigns to a topos $\mathscr{E}$ the slice category $\mathbf{Top}_{\mathscr{S}}/\mathscr{E}$ and to a geometric morphism $\mathscr{F} \xrightarrow{\varphi} \mathscr{E}$ the functor given by pulling back along φ.

A geometric morphism $\mathscr{F} \xrightarrow{\psi} \mathscr{E}$ is said *to be of effective descent* if any object X of $\mathscr{F}$ equipped with descent data already comes from $\mathscr{E}$ under ψ^*. We are concerned with classes Φ of geometric morphisms of effective descent that are closed under composition and pullbacks. For instance, the class of locally connected surjections is such a class Φ, which we shall meet again below.

Definition 9.1.11 *Let Φ be a class of geometric morphisms of effective descent in $\boldsymbol{Top}_{\mathscr{S}}$ that is closed under composition and pullbacks. We shall say that a subpseudofunctor Γ of L is a Φ-stack, if for any pullback diagram*

$$\begin{array}{ccc}
\mathscr{P} & \longrightarrow & \mathscr{G} \\
{\scriptstyle \varphi^*(\alpha)}\big\downarrow & & \big\downarrow{\scriptstyle \alpha} \\
\mathscr{F} & \underset{\varphi}{\longrightarrow} & \mathscr{E}
\end{array}$$

of geometric morphisms for which φ *is a member of* Φ, *we have*

$$\varphi^*(\alpha) \in \Gamma(\mathscr{F}) \text{ implies } \alpha \in \Gamma(\mathscr{E}) .$$

Let us return to a pseudofunctor we have already encountered in § 1.3:

$$\mathbf{A} : \mathbf{Top}_{\mathscr{S}}{}^{\mathrm{op}} \longrightarrow \mathbf{CAT} ,$$

such that $\mathbf{A}(\mathscr{E})$ is the topos-frame $\mathscr{E}$ itself. $\mathbf{A}$ is a subpseudofunctor of L. This is a basic example of an intensive quantity that Lawvere has emphasized. By the very definition of effective descent, $\mathbf{A}$ is trivially a Φ-stack for any class Φ of effective descent geometric morphisms, and we have already seen in Proposition 1.3.2 that $\mathbf{A}$ preserves binary products. We make the following definition.

Definition 9.1.12 *Let* $\mathscr{K}$ *be an extensive sub-2-category of* $\boldsymbol{Top}_{\mathscr{S}}$, *and* Φ *a class of geometric morphisms of effective descent, closed under composition and pullbacks. A subpseudofunctor* $\mathscr{K}^{\mathrm{op}} \longrightarrow \mathbf{CAT}$ *of* L *is said to be* a fibration of regular covering morphisms with respect to Φ *if it is a* Φ*-stack, and if it preserves binary products.*

Denote by

$$\mathscr{C} : \mathbf{Top}_{\mathscr{S}}{}^{\mathrm{op}} \longrightarrow \mathbf{CAT}$$

the subpseudofunctor of L such that $\mathscr{C}(\mathscr{E})$ is the full subcategory of $\mathrm{L}(\mathscr{E})$ consisting of the local homeomorphisms determined by its locally constant objects. The objects of $\mathscr{C}(\mathscr{E})$ are called *locally constant coverings of* $\mathscr{E}$. We have already observed that this assignment is pseudofunctorial, as inverse images of geometric morphisms preserve locally constant objects in general. In particular, we may consider $\mathscr{C}$ as a contravariant pseudofunctor defined on $\mathbf{LTop}_{\mathscr{S}}$, the full sub 2-category of $\mathbf{Top}_{\mathscr{S}}$ whose objects are locally connected toposes.

For a locally connected topos $\mathscr{E}$, let $\mathscr{U}(\mathscr{E})$ denote the full subcategory of $\mathrm{L}(\mathscr{E})$ determined by the complete spread objects of $\mathscr{E}$, or unramified coverings of $\mathscr{E}$ (Definition 9.1.7). This assignment extends to a pseudofunctor

$$\mathscr{U} : \mathbf{LTop}_{\mathscr{S}}{}^{\mathrm{op}} \longrightarrow \mathbf{CAT}$$

since unramified geometric morphisms are stable under pullback along geometric morphisms with locally connected domain (Exercise 9.1, 7).

Lemma 9.1.13 *The subpseudofunctors* $\mathscr{C}$ *and* $\mathscr{U}$ *of* L *are both fibrations of regular coverings with respect to the class* Φ *of locally connected surjections.*

Proof. It is not difficult to show that $\mathscr{C}$ and $\mathscr{U}$ preserve binary products (Exercise 9.1, 8). The fact that $\mathscr{C}$ is a Φ-stack depends on Lemma 9.1.4. The same fact for $\mathscr{U}$ depends on Lemma 9.1.5. □

Definition 9.1.14 *Let $\mathscr{K}$ be an extensive sub-2-category of $\mathbf{Top}_{\mathscr{S}}$, and Φ a class of geometric morphisms of effective descent, closed under composition and pullbacks. Let Γ be a pseudofunctor on $\mathscr{K}$. We shall say that* the van Kampen theorem holds *for Γ with respect to Φ if whenever*

$$\begin{array}{ccc} \mathscr{E}_0 & \xrightarrow{\beta_1} & \mathscr{E}_1 \\ {\scriptstyle\beta_2}\downarrow & & \downarrow{\scriptstyle\alpha_1} \\ \mathscr{E}_2 & \xrightarrow[\alpha_2]{} & \mathscr{E} \end{array}$$

is a bipushout (in $\mathbf{Top}_{\mathscr{S}}$) of objects in $\mathscr{K}$ in which the induced map $\mathscr{E}_1 + \mathscr{E}_2 \longrightarrow \mathscr{E}$ is a member of Φ, the diagram

$$\begin{array}{ccc} \Gamma(\mathscr{E}_0) & \xleftarrow{\beta_1^*} & \Gamma(\mathscr{E}_1) \\ {\scriptstyle\beta_2^*}\uparrow & & \uparrow{\scriptstyle\alpha_1^*} \\ \Gamma(\mathscr{E}_2) & \xleftarrow[\alpha_2^*]{} & \Gamma(\mathscr{E}) \end{array}$$

is a bipullback in **CAT**.

The conditions of Definition 9.1.12 imply the condition of Definition 9.1.14, which we state as the next theorem.

Theorem 9.1.15 *Let $\mathscr{K}$ be an extensive sub-2-category of $\mathbf{Top}_{\mathscr{S}}$, and let Φ be a class of effective descent morphisms in $\mathscr{K}$, closed under pullback and composition. Let Γ be a fibration of regular covering morphisms in $\mathscr{K}$ with respect to Φ. Then Γ satisfies the van Kampen theorem with respect to Φ.*

Proof. The proof reduces, using the given pushout (testing it with geometric morphisms whose codomain is the object classifier), and since Γ is a sub-pseudofunctor of L, to showing that an object X of $\mathscr{E}$ is in $\Gamma(\mathscr{E})$ if $X_1 = \alpha_1^*(X)$ and $X_2 = \alpha_2^*(X)$ are in $\Gamma(\mathscr{E}_1)$ and $\Gamma(\mathscr{E}_2)$ respectively. Since Γ preserves binary products, $\alpha^*(X) = (\alpha_1^*(X), \alpha_2^*(X))$ is in $\Gamma(\mathscr{E}_1) \times \Gamma(\mathscr{E}_2) \simeq \Gamma(\mathscr{E}_1 + \mathscr{E}_2)$. Since α is of effective descent for Γ, and by our assumption that Γ is a Φ-stack, X is indeed in $\Gamma(\mathscr{E})$. □

Corollary 9.1.16 *The van Kampen theorem holds for both $\mathscr{C}$ and $\mathscr{U}$ regarded as pseudofunctors $\mathbf{LTop}_{\mathscr{S}}^{\mathrm{op}} \longrightarrow$ **CAT**, with respect to the class Φ of locally connected surjections.*

Proof. This follows directly from Theorem 9.1.15 and Lemma 9.1.13. □

Remark 9.1.17 *Assume that the toposes in Theorem 9.1.15 are all locally connected and* locally simply connected, *in the sense that there is a single $U \twoheadrightarrow 1_{\mathscr{E}}$ that splits all locally constant objects. If $\mathscr{E}$ is locally connected and* *locally simply connected, then $\mathscr{C}(\mathscr{E})$ is a (Galois) topos of the form $\mathscr{B}(\pi_1(\mathscr{E}))$, the classifying topos of the fundamental group of $\mathscr{E}$. In this case, the van Kampen theorem takes the form of "a pushout-to-pushout" result, as follows. Let*

$$\begin{array}{ccc} \mathscr{E}_0 & \xrightarrow{\beta_1} & \mathscr{E}_1 \\ {\scriptstyle \beta_2}\downarrow & & \downarrow{\scriptstyle \alpha_1} \\ \mathscr{E}_2 & \xrightarrow[\alpha_2]{} & \mathscr{E} \end{array}$$

*be a pushout diagram in **LTop**$_{\mathscr{S}}$ in which all four toposes are locally simply connected, and where the induced map $\mathscr{E}_1 + \mathscr{E}_2 \longrightarrow \mathscr{E}$ is a locally connected surjection. Then the diagram*

$$\begin{array}{ccc} \mathscr{C}(\mathscr{E}_0) & \xrightarrow{\beta_1} & \mathscr{C}(\mathscr{E}_1) \\ {\scriptstyle \beta_2}\downarrow & & \downarrow{\scriptstyle \alpha_1} \\ \mathscr{C}(\mathscr{E}_2) & \xrightarrow[\alpha_2]{} & \mathscr{C}(\mathscr{E}) \end{array}$$

*is a pushout diagram in **Top**$_{\mathscr{S}}$. The conclusion uses the fact that a bipushout in **Top**$_{\mathscr{S}}$ is calculated as a bipullback in* **Cat** *via the inverse images of the geometric morphisms. We warn the reader that trying to deduce this result from the possible existence of a reflection of the inclusion of Galois toposes into locally simply connected toposes meets with some difficulties.*

Remark 9.1.18 *A pseudofunctor*

$$\Gamma : \mathscr{K}^{\mathrm{op}} \longrightarrow \mathbf{CAT}$$

provides a notion of homotopy. I.e., by definition, a Γ-homotopy $\psi \Rightarrow \varphi$ between two geometric morphisms with the same domain and codomain toposes is a pseudonatural transformation (or isomorphism)

$$\Gamma(\psi) \Rightarrow \Gamma(\varphi)\,.$$

For example, **A**, *$\mathscr{C}$, and $\mathscr{U}$ are all fibrations of regular coverings with respect to locally connected surjections, but $\mathscr{C}$ and $\mathscr{U}$ are distinguished from* **A** *by their homotopies. Indeed, ordinary homotopies (for locally path-connected topological spaces) induce $\mathscr{C}$ and $\mathscr{U}$-homotopies, but they do not induce* **A***-homotopies.*

Exercises 9.1.19

1. *Show that any constant object e^*A (including 0) is a locally constant object.*
2. *Prove Lemma 9.1.3.*

3. *If $\mathscr{E}$ is connected and locally connected, and the base topos is Set, show that an object X of $\mathscr{E}$ is locally constant (as in Definition 9.1.1) iff there is a $U \twoheadrightarrow 1_{\mathscr{E}}$ and an object S of $\mathscr{S}$ such that $X \times U \cong e^*S \times U$ over U. (This condition was taken by Barr and Diaconescu as the definition of locally constant; however, it is only suitable in the connected case.)*
4. *An open set $V \subseteq X$ is* evenly covered by a map $F \xrightarrow{\pi} X$ *if $\pi^{-1}(V)$ has an open partition such that the restriction of π to each member of the partition is a homeomorphism with V. Then π is a (necessarily surjective)* covering space *if X has a cover of open sets each of which is evenly covered by π. Let X be locally connected. Show that a local homeomorphism $F \longrightarrow X$ is a covering space in this sense iff it is a non-0 locally constant object in* $\mathrm{Sh}(X)$.
5. *Show that presheaf on a connected small category is locally constant iff its transition maps are isomorphisms.*
6. *A presheaf (= discrete fibration) is a complete spread object iff it is also a discrete opfibration. Show that a presheaf on a connected category is locally constant iff it is a complete spread object.*
7. *Prove that the unramified geometric morphisms with locally connected codomain (hence also locally connected domain) are stable under pullback along geometric morphisms with locally connected domain.*
8. *Prove that the pseudofunctors $\mathscr{C}$ and $\mathscr{U}$ preserve binary products.*
9. *Let $\mathscr{F} \xrightarrow{\rho} \mathscr{E}$ be a pure geometric morphism between locally connected toposes. Show that the induced functor*

$$\rho^* : \mathscr{C}(\mathscr{E}) \longrightarrow \mathscr{C}(\mathscr{F})$$

is full and faithful.

9.2 Purely Skeletal Geometric Morphisms

In order to prepare for branched coverings in §9.3, we first investigate the class of geometric morphisms that respect pure (mono)morphisms. Throughout this section, $\mathscr{E}$ denotes a locally connected topos.

Definition 9.2.1 *A* pure morphism *of $\mathscr{E}$ is a morphism $X \longrightarrow Y$ for which $\mathscr{E}/X \longrightarrow \mathscr{E}/Y$ is a pure geometric morphism. We say an object X is pure if $X \longrightarrow 1$ is a pure morphism.*

Lemma 9.2.2 *A morphism $X \xrightarrow{a} Y$ of $\mathscr{E}$ is pure iff the induced morphism*

$$Y \times e^*(\Omega_{\mathscr{S}}) \longrightarrow (Y \times e^*(\Omega_{\mathscr{S}}))^a$$

is an isomorphism. In particular, the lattice of definable subobjects of a pure object is isomorphic to the lattice of definable subobjects of $1_{\mathscr{E}}$.

Proof. This follows directly from the definitions. □

Example 9.2.3 *The balloon's shadow map $S^2 \longrightarrow S^2$ (§ 2.1) has the property that every pure open subset remains pure under inverse image. On the other hand, the image part of this map $S^2 \twoheadrightarrow D$, where D is a closed disk, does not have this pure-respecting property (eg., consider the interior of D). Both these maps are complete spreads.*

A map $D \twoheadrightarrow S^2$ that envelopes the sphere by collapsing the boundary of D to a point of S^2 also has the pure-respecting property.

Definition 9.2.4 *We shall say that* a geometric morphism respects (reflects) pure morphisms *if its inverse image functor preserves (reflects) pure morphisms. We shall call* purely skeletal *any geometric morphism that respects* pure monomorphisms *(meaning simply a monomorphism that is a pure).*

Geometric morphisms that respect pure (mono) morphisms are analogous to geometric morphisms that respect double-negation dense monomorphisms, the so-called skeletal geometric morphisms.

Remark 9.2.5 *In the following diagram if p reflects pure morphisms and φ respects pure morphisms, then ψ respects pure morphisms.*

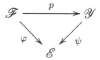

$$\begin{array}{ccc} \mathscr{F} & \xrightarrow{\;p\;} & \mathscr{Y} \\ & {\scriptstyle\varphi}\searrow \quad \swarrow{\scriptstyle\psi} & \\ & \mathscr{E} & \end{array}$$

Proposition 9.2.6 *A locally connected geometric morphism respects pure morphisms. A locally connected surjection reflects pure morphisms.*

Proof. These statements follow from Lemma 2.2.11. □

Proposition 9.2.6 has the following dual statement for pure geometric morphisms.

Proposition 9.2.7 *A pure geometric morphism reflects pure morphisms. A pure inclusion respects pure morphisms.*

Proof. Consider the following diagram for pure p, and object X.

$$\begin{array}{ccc} \mathscr{F}/p^*X & \longrightarrow & \mathscr{E}/X \\ \downarrow & & \downarrow \\ \mathscr{F} & \xrightarrow[p]{} & \mathscr{E} \end{array}$$

The top horizontal is pure. If p^*X is pure, then X is pure by Lemma 2.2.8. We also use Lemma 2.2.8 for the second statement. □

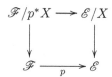

We may always consider the largest topology in a topos for which a given object of the topos is a sheaf. If X denotes an object, then a monomorphism $m : A \rightarrowtail B$ is dense for this largest topology iff $B^*X \longrightarrow B^*X^m$ (transpose of the projection) is an isomorphism. We have the following.

Proposition 9.2.8 *Let $\mathscr{E}$ be a topos bounded over $\mathscr{S}$. Then a monomorphism is dense for the largest topology for which $e^*(\Omega_{\mathscr{S}})$ is a sheaf iff it is a pure morphism. Thus, the pure monomorphisms in $\mathscr{E}$ are the dense monomorphisms for a topology in $\mathscr{E}$.*

Definition 9.2.9 *We refer to the topology of pure monomorphisms as* the pure topology in $\mathscr{E}$, *and to its sheaves as* pure-sheaves. *We refer to monomorphisms that are closed for the pure topology as* pure-closed. *We denote the subtopos of pure-sheaves by $\mathscr{E}_p \rightarrowtail \mathscr{E}$.*

Every topos has a smallest dense subtopos: the subtopos of double-negation sheaves, which is a Boolean topos. Similarly every locally connected topos has a smallest pure subtopos.

Proposition 9.2.10 *If $\mathscr{E}$ is locally connected, then $\mathscr{E}_p$ is locally connected. $\mathscr{E}_p$ is the smallest pure subtopos of $\mathscr{E}$.*

Proof. The unit $e^*(\Omega_{\mathscr{S}}) \longrightarrow i_*i^*e^*(\Omega_{\mathscr{S}})$ for the inclusion $i : \mathscr{E}_p \rightarrowtail \mathscr{E}$ is an isomorphism because $e^*(\Omega_{\mathscr{S}})$ is a pure-sheaf. But this says that i is pure. $\mathscr{E}_p$ is locally connected because a pure subtopos of a locally connected topos is locally connected (2.2.16). Let $Sh_j(\mathscr{E}) \rightarrowtail \mathscr{E}$ be a pure inclusion. By the definition of pure, $e^*(\Omega_{\mathscr{S}})$ is a j-sheaf. Hence, every j-dense monomorphism is pure. Therefore, $\mathscr{E}_p \rightarrowtail Sh_j(\mathscr{E})$. □

Example 9.2.11 *Let R denote the real numbers. Then $\mathrm{Sh}(R)_p = \mathrm{Sh}(R)$ because a pure inclusion of open intervals must be an equality. However, $\mathrm{Sh}(R^2)_p$ is a proper subtopos of $\mathrm{Sh}(R^2)$. For instance, the complement in R^2 of a curve is a pure-closed open subset of R^2. On the other hand, a punctured plane is a pure subset of R^2. The complement of a surface in R^3 is a pure-closed open subset of R^3.*

Example 9.2.12 *The dense topology in a presheaf topos $\mathrm{P}(\mathbb{C})$ is given by the sieves $R \rightarrowtail h_c$ such that for every morphism $d \xrightarrow{f} c$ in $\mathbb{C}$, $f^*(R) \rightarrowtail h_d$ is non-empty. The subtopos of sheaves for this topology is precisely the smallest dense subtopos of $\mathrm{P}(\mathbb{C})$. A sieve $R \rightarrowtail h_c$ is a member of the pure topology in $\mathrm{P}(\mathbb{C})$ if for every f, $f^*(R)$ is connected. Implicitly, a connected sieve is non-empty, so the pure topology is contained in the dense (or double-negation) topology. The smallest pure subtopos $\mathrm{P}(\mathbb{C})_p$ is locally connected.*

Definition 9.2.13 A density object *of an locally connected topos is an object of the form $\mathbf{d}(\mu)$, for some distribution μ, where $\mathbf{d}$ is the density monad, introduced in § 6.2.*

Proposition 9.2.14 *Density objects are pure-sheaves. If $\mathcal{E}$ is locally connected, then $\mathcal{E}_p$ is the smallest subtopos of $\mathcal{E}$ containing the density objects.*

Proof. Let $\mathcal{Y} \xrightarrow{\psi} \mathcal{E}$ denote the complete spread associated with a distribution μ. Let $A \rightarrowtail B$ be a pure monomorphism in a locally connected topos $\mathcal{E}$. We know that morphisms $A \longrightarrow \mathbf{d}(\mu)$ are in bijection with geometric morphisms between the spread completion of $\mathcal{E}/A$ and $\mathcal{Y}$ over $\mathcal{E}$. However, the spread completions of $\mathcal{E}/A$ and $\mathcal{E}/B$ are equivalent because $\mathcal{E}/A \rightarrowtail \mathcal{E}/B$ is a pure geometric morphism. This shows that $\mathbf{d}(\mu)$ is a sheaf for the pure topology. The second statement holds because any constant object is a density object. In particular, the constant object $e^*(\Omega_{\mathcal{S}})$ is a density object. □

Proposition 9.2.15 *For any geometric morphism $\mathcal{F} \xrightarrow{\psi} \mathcal{E}$ over $\mathcal{S}$ with locally connected domain, the following are equivalent:*

1. *ψ is purely skeletal;*
2. *ψ restricts to smallest pure subtoposes;*

$$\begin{array}{ccc} \mathcal{F}_p & \rightarrowtail & \mathcal{F} \\ \downarrow & & \downarrow{\scriptstyle \psi} \\ \mathcal{E}_p & \rightarrowtail & \mathcal{E} \end{array}$$

3. *the distribution algebra $H = \psi_*(f^*\Omega_{\mathcal{S}})$ in $\mathcal{E}$ is a pure-sheaf.*

Proof. These conditions are clearly equivalent once we recall the definitions of pure, and pure-sheaf. □

The spread completion of an object Y of a locally connected topos $\mathcal{E}$ is the complete spread geometric morphism ψ in the diagram

$$\begin{array}{ccc} \mathcal{E}/Y & \xrightarrow{p} & \mathcal{Y} \\ & \searrow \quad \swarrow{\scriptstyle \psi} & \\ & \mathcal{E} & \end{array} \tag{9.1}$$

in which p is pure. These complete spreads correspond to Lawvere's *absolutely continuous distributions*, meaning a distribution of the kind $Y.e_!$.

Proposition 9.2.16 *The spread completion of an object of a locally connected topos is purely skeletal.*

Proof. Use Proposition 9.2.6 and Remark 9.2.5 applied to diagram 9.1. □

Exercises 9.2.17

1. *How are preservation of pure sets under inverse image and change in codimension of singular sets related?*

2. *Show that the inclusion of a single point into the real line is purely skeletal, but not skeletal. Find an example showing that skeletal does not imply purely skeletal.*
3. *Let* $\mathbf{E}_p(\mathscr{E})$ *be the category of distributions on a locally connected topos* $\mathscr{E}$ *that carry pure monomorphisms to isomorphisms. Show that an absolutely continuous distribution* $Y.e_!$ *is a member of* $\mathbf{E}_p(\mathscr{E})$. *Show that* $\mathbf{E}_p(\mathscr{E}) \simeq \mathbf{E}(\mathscr{E}_p)$.
4. *Provide a detailed proof of Proposition 9.2.15.*

9.3 Branched Covering Toposes

Our development of branched coverings stems in part from the ideas of R. H. Fox. We shall define a branched covering of a topos as a special kind of complete spread: we shall define a branched covering as a complete spread that is purely skeletal (Def. 9.2.4), and which is, as we shall say, a purely locally constant covering.

We shall say that an object *has pure support* if its support (which is a subobject of $1_{\mathscr{E}}$) is a pure monomorphism.

Definition 9.3.1 *A geometric morphism* $\mathscr{Y} \xrightarrow{\psi} \mathscr{E}$ *is* a purely locally constant covering *if there is* $U \longrightarrow 1_{\mathscr{E}}$ *with pure support such that* $U^*(\psi)$ *in the pullback*

$$\begin{array}{ccc} \mathscr{Y}/\psi^*(U) & \longrightarrow & \mathscr{Y} \\ {\scriptstyle U^*(\psi)}\downarrow & & \downarrow{\scriptstyle \psi} \\ \mathscr{E}/U & \longrightarrow & \mathscr{E} \end{array}$$

in $\boldsymbol{Top}_{\mathscr{S}}$ *is a local homeomorphism determined by a definable object in* $\mathscr{E}/U$, *in the sense of Remark 9.1.2.*

Note that any locally constant covering of $\mathscr{E}$ is a purely locally constant covering of $\mathscr{E}$.

Definition 9.3.2 *A geometric morphism* $\mathscr{Y} \xrightarrow{\psi} \mathscr{E}$ *is said to be* a branched covering *if:*

1. *it is a complete spread,*
2. *it is a purely locally constant covering, and*
3. *it is purely skeletal.*

We regard branched coverings of $\mathscr{E}$ *as a full subcategory of the category of complete spreads over* $\mathscr{E}$. *Denote by* $\mathscr{B}(\mathscr{E})$ *the corresponding category of branched coverings of* $\mathscr{E}$.

Example 9.3.3 *The balloon's shadow map* $S^2 \longrightarrow S^2$ *(Example 9.2.3) is a complete spread that satisfies the third requirement of branched covering, but* *not the second. The image part* $S^2 \twoheadrightarrow D$ *of this map is a complete spread that satisfies the second requirement, but not the third. Both of these maps exhibit folding, but this folding is detected differently in each case. Intuitively, the last two conditions of a branched covering together rule out folding in a complete spread.*

Any geometric morphism satisfying the last two conditions of Definition 9.3.2, but possibly not the first, can be 'normalized' in the following sense.

Proposition 9.3.4 *The spread completion of a geometric morphism with locally connected domain satisfying the last two conditions of a branched covering is a branched covering.*

Proof. We must verify that the spread completion $\mathscr{Y} \xrightarrow{\psi} \mathscr{E}$ of such a geometric morphism $\mathscr{F} \xrightarrow{\varphi} \mathscr{E}$ satisfies the last two conditions of a branched covering as in Definition 9.3.2. ψ has the third property because pure geometric morphisms reflect pure monomorphisms. The second condition holds because the pure

$$\mathscr{F}/\varphi^*U \longrightarrow \mathscr{Y}/\psi^*U$$

is a complete spread, hence an equivalence, since both $\mathscr{F}/\varphi^*U \longrightarrow \mathscr{E}/U$ and $\mathscr{Y}/\psi^*U \longrightarrow \mathscr{E}/U$ are complete spreads. □

Remark 9.3.5 *The normalization process may have trivial results. The bag map* $D \twoheadrightarrow S^2$ *(collapse the boundary of* D *to a point forming a sphere) satisfies the last two requirements of a branched geometric morphism, but it is not a complete spread. In fact, this map is pure, so its spread completion is the identity on* S^2*. It may be interesting to note that the zipper map* $D \twoheadrightarrow S^2$ *(collapse the boundary of* D *to a closed line segment forming a sphere) is not branched because it is not purely locally constant (but it is purely skeletal). However, it is already normalized, as it is a complete spread.*

The following is immediately clear.

Proposition 9.3.6 *A locally constant covering of a locally connected topos* $\mathscr{E}$ *is a branched covering. For a locally connected topos* $\mathscr{E}$*, the category* $\mathscr{C}(\mathscr{E})$ *of locally constant coverings of* $\mathscr{E}$ *is a full subcategory of the category* $\mathscr{B}(\mathscr{E})$ *of branched coverings of* $\mathscr{E}$*.*

Remark 9.3.7 *If* V *denotes the support of a splitting object* U *for a branched covering* $\mathscr{Y} \xrightarrow{\psi} \mathscr{E}$*, then the support of* $\psi^*(U)$ *is* $\psi^*(V)$*. We refer to the support of* $\psi^*(U)$ *associated with* ψ *as the* non-singular part *of* $\mathscr{Y}$*. By definition, this is a pure subobject of* $1_{\mathscr{Y}}$*. Caution: the non-singular part of* $\mathscr{Y}$ *is not a well-defined subobject of* $1_{\mathscr{Y}}$*. In some cases a largest non-singular part is available, but we have not made this a requirement in Definition 9.3.2. However, it is reasonable to expect that in applications the notion of a branched covering be further specified by including its non-singular part, or its pure splitting object, as part of the data.*

Lemma 9.3.8 *Consider a commutative triangle*

$$\begin{array}{ccc} \mathscr{X} & \xrightarrow{\rho} & \mathscr{Y} \\ & \searrow^{\varphi} \quad {}^{\psi}\swarrow & \\ & \mathscr{E} & \end{array}$$

of localic geometric morphisms. If ρ and φ are local homeomorphisms, and ρ is a surjection, then ψ is a local homeomorphism.

Proof. This fact may be established using the well-known fact that a localic geometric morphism is a local homeomorphism iff it is open and its diagonal is open. □

Proposition 9.3.9 *A branched covering is a locally constant covering iff it has a splitting object with global support.*

Proof. If a branched covering $\mathscr{Y} \xrightarrow{\psi} \mathscr{E}$ has a splitting object with global support, then we have a pullback diagram

$$\begin{array}{ccc} \mathscr{Y}/\psi^*(U) & \longrightarrow & \mathscr{Y} \\ {\scriptstyle U^*(\psi)}\downarrow & & \downarrow{\scriptstyle \psi} \\ \mathscr{E}/U & \longrightarrow & \mathscr{E} \end{array}$$

in which all geometric morphisms are local homeomorphisms except ostensibly ψ. But then by Lemma 9.3.8, ψ must be a local homeomorphism, whence locally constant. □

Lemma 9.3.10 is analogous to the simple fact from topology that in a square

$$\begin{array}{ccc} C & \rightarrowtail & \overline{C} \\ \downarrow & & \downarrow \\ S & \rightarrowtail & X \end{array}$$

where S is a subspace of X, C is a closed subset of S, and $\overline{C}$ is the closure of C in X, we have $C = S \cap \overline{C}$. In other words, the square is a pullback.

Lemma 9.3.10 *Let $\mathscr{F} \rightarrowtail \mathscr{E}$ be a subtopos. Let $\mathscr{X} \xrightarrow{\psi} \mathscr{F}$ be a complete spread. Then the spread completion of ψ over $\mathscr{E}$ forms a topos pullback square.*

$$\begin{array}{ccc} \mathscr{X} & \xrightarrow{\rho} & \mathscr{Y} \\ {\scriptstyle \psi}\downarrow & & \downarrow{\scriptstyle \psi'} \\ \mathscr{F} & \underset{i}{\rightarrowtail} & \mathscr{E} \end{array}$$

Consequently, the pure factor ρ is an inclusion.

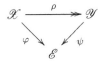

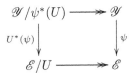

Proof. We may construct the spread completion ψ' as follows. Let λ ($= x_! \cdot \psi^*$) denote the distribution on $\mathscr{F}$ associated with ψ, so that the distribution on $\mathscr{E}$ associated with ψ' is $\lambda \cdot i^*$. We choose any site for $\mathscr{E}$, say with underlying category $\mathbb{C}$. Then ψ' fits in the following pullback.

$$\begin{array}{ccc} \mathscr{Y} & \rightarrowtail & P(\mathbb{Y}) \\ {\scriptstyle \psi'}\downarrow & & \downarrow \\ \mathscr{E} & \rightarrowtail & P(\mathbb{C}) \end{array}$$

$\mathbb{Y}$ denotes the amalgamation site for $\lambda \cdot i^*$. A typical object of $\mathbb{Y}$ is a pair (c, α), where $\alpha \in \lambda i^*(\epsilon_c)$, and where $\mathbb{C} \xrightarrow{\epsilon} \mathscr{E}$ denotes the canonical functor. But we may also regard $\mathbb{C}$ as (the underlying category of) a site for $\mathscr{F}$, where now we have $\mathbb{C} \xrightarrow{\epsilon'} \mathscr{F}$ such that $\epsilon' \cong i^* \cdot \epsilon$. Then $\mathbb{Y}$ is isomorphic to the amalgamation site for λ, since $\lambda \cdot \epsilon' \cong \lambda \cdot i^* \cdot \epsilon$. Hence, we have a commutative diagram

$$\begin{array}{ccccc} \mathscr{X} & \longrightarrow & \mathscr{Y} & \rightarrowtail & P(\mathbb{Y}) \\ {\scriptstyle \psi}\downarrow & & \downarrow{\scriptstyle \psi'} & & \downarrow \\ \mathscr{F} & \underset{i}{\rightarrowtail} & \mathscr{E} & \rightarrowtail & P(\mathbb{C}) \end{array}$$

in which the outer and right squares are pullbacks. The essentially unique factoring morphism $\mathscr{X} \longrightarrow \mathscr{Y}$ must of course be the pure ρ. But then the left square must be a pullback. □

Proposition 9.3.11 *The category of purely skeletal complete spreads over a topos $\mathscr{E}$ is equivalent to the category of complete spreads over $\mathscr{E}_p$. The equivalence is given on the one hand by pullback along $\mathscr{E}_p \rightarrowtail \mathscr{E}$, and on the other by spread completion.*

Proof. Assume first that $\mathscr{X} \xrightarrow{\varphi} \mathscr{E}_p$ is a complete spread. By Lemma 9.3.10, the spread completion $\mathscr{Y} \xrightarrow{\psi} \mathscr{E}$ of the composite of φ with the pure inclusion $\mathscr{E}_p \rightarrowtail \mathscr{E}$ forms a topos pullback with φ. In particular, $\mathscr{X}$ is a pure subtopos of $\mathscr{Y}$. Therefore, $\mathscr{Y}_p$ factors through $\mathscr{X}$, so ψ restricts to smallest pure subtoposes, i.e., ψ respects pure monomorphisms.

On the other hand, assume that a complete spread $\mathscr{Y} \xrightarrow{\psi} \mathscr{E}$ respects pure monomorphisms. We must show that the pullback

$$\begin{array}{ccc} \mathscr{X} & \rightarrowtail & \mathscr{Y} \\ {\scriptstyle \varphi}\downarrow & & \downarrow{\scriptstyle \psi} \\ \mathscr{E}_p & \rightarrowtail & \mathscr{E} \end{array}$$

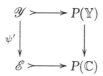

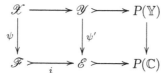

of ψ is a complete spread, and that the inclusion $\mathscr{X} \rightarrowtail \mathscr{Y}$ is pure. By assumption, ψ restricts to $\mathscr{Y}_p \xrightarrow{\psi_p} \mathscr{E}_p$. Since the domain $\mathscr{Y}$ of the complete spread ψ is locally connected, so is $\mathscr{Y}_p$. Form the spread completion of ψ_p, say $\mathscr{X} \xrightarrow{\varphi} \mathscr{E}_p$. But then the given ψ must be the spread completion of $\mathscr{X} \longrightarrow \mathscr{E}$, so again by Lemma 9.3.10, φ is the pullback of ψ. □

Proposition 9.3.11 has the following refinement. The proof technique is essentially the same.

Theorem 9.3.12 *The category $\mathscr{B}(\mathscr{E})$ of branched coverings over $\mathscr{E}$ is canonically equivalent to the category $\mathscr{C}(\mathscr{E}_p)$ of locally constant coverings of its smallest pure subtopos $\mathscr{E}_p$.*

Proof. We pass from a branched covering ψ of $\mathscr{E}$ to a locally constant covering of $\mathscr{E}_p$ by pullback.

$$\begin{array}{ccc} \mathscr{X} & \overset{\rho}{\rightarrowtail} & \mathscr{Y} \\ {\scriptstyle\varphi}\downarrow & & \downarrow{\scriptstyle\psi} \\ \mathscr{E}_p & \underset{i}{\rightarrowtail} & \mathscr{E} \end{array} \tag{9.2}$$

The pullback φ is i^*U-split if ψ is U-split, where U has pure support in $\mathscr{E}$. But then i^*U has global support in $\mathscr{E}_p$. By Lemma 9.3.8, φ must be a local homeomorphism, hence a locally constant covering. We shall show that ρ is pure. We may regard (9.2) as a composite of two pullback squares.

$$\begin{array}{ccccc} \mathscr{X} & \rightarrowtail & \mathscr{Y}/\psi^*V & \rightarrowtail & \mathscr{Y} \\ {\scriptstyle\varphi}\downarrow & & \downarrow & & \downarrow{\scriptstyle\psi} \\ \mathscr{E}_p & \rightarrowtail & \mathscr{E}/V & \rightarrowtail & \mathscr{E} \end{array}$$

Here V denotes the (pure) support of U, which must include the smallest pure subtopos $\mathscr{E}_p$ by a pure inclusion. It follows again by Lemma 9.3.8 that the middle vertical is a local homeomorphism, and in fact it is a locally constant covering. The left top horizontal factor is therefore pure, since it is the pullback of a pure along a locally connected (in fact, a local homeomorphism). The right top horizontal factor is pure by the assumption that ψ is purely skeletal, hence, since $V \rightarrowtail 1$ is pure, so is $\psi^*V \rightarrowtail 1$. This shows that ρ in (9.2) is a composite of pures, hence pure itself. ψ is therefore the spread completion of $i \cdot \varphi$.

If we begin with a locally constant object X of $\mathscr{E}_p$, then we pass to a branched covering of $\mathscr{E}$ by spread completion.

$$\begin{array}{ccc} \mathscr{E}_p/X & \longrightarrow & \mathscr{Y} \\ \downarrow & & \downarrow{\scriptstyle\psi} \\ \mathscr{E}_p & \underset{i}{\rightarrowtail} & \mathscr{E} \end{array}$$

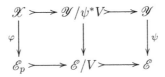

By Lemma 9.3.10, this square is a pullback. We must show that ψ is indeed a branched covering. From Proposition 9.2.16, we get that ψ is purely skeletal. We have a pullback

$$\begin{array}{ccc} \mathscr{E}_p/X & \rightarrowtail & \mathscr{E}/i_*X \\ \downarrow & & \downarrow \\ \mathscr{E}_p & \underset{i}{\rightarrowtail} & \mathscr{E} \end{array}$$

so that the top inclusion is pure. Therefore, ψ is also the spread completion of $\mathscr{E}/i_*X$.

$$\begin{array}{ccc} \mathscr{E}/i_*X & \longrightarrow & \mathscr{Y} \\ & \searrow \quad \swarrow_{\psi} & \\ & \mathscr{E} & \end{array}$$

By composing with the projection $i_*(X \times U) \longrightarrow i_*(X)$ we obtain the following commutative diagram of geometric morphisms.

$$\begin{array}{ccc} \mathscr{E}/i_*(X \times U) & \longrightarrow & \mathscr{Y} \\ \downarrow & & \downarrow{\scriptstyle \psi} \\ \mathscr{E}/i_*U & \longrightarrow & \mathscr{E} \end{array} \tag{9.3}$$

Since the direct image functor of a pure geometric morphism preserves definable objects (Exercise 1.5.7, 2), the left vertical is a definable object. The support $V \rightarrowtail 1_{\mathscr{E}}$ of i_*U is pure because it must be dense for the pure topology. It remains to show that (9.3) is a pullback and that $\psi^*(V)$ is a pure subobject of $1_{\mathscr{Y}}$. Consider the pullbacks

$$\begin{array}{ccccc} i_*(X \times U) & \twoheadrightarrow & Y & \rightarrowtail & i_*X \\ \downarrow & & \downarrow & & \downarrow \\ i_*U & \twoheadrightarrow & V & \rightarrowtail & 1_{\mathscr{E}} \end{array}$$

in $\mathscr{E}$. The top monomorphism is pure so that the top horizontal geometric morphism in the following diagram is pure.

$$\begin{array}{ccc} \mathscr{E}/Y & \longrightarrow & \mathscr{Y} \\ \downarrow & & \downarrow{\scriptstyle \psi} \\ \mathscr{E}/V & \rightarrowtail & \mathscr{E} \end{array}$$

Hence, this is a spread completion diagram. But then $\mathscr{E}/Y \longrightarrow \mathscr{E}/V$ is locally constant, whence a complete spread. By Lemma 9.3.10, this square is a pullback, which shows that (9.3) is a pullback. It also shows that $\psi^*(V) \cong Y$ is a subobject of $1_{\mathscr{Y}}$, which we already know is pure. This completes our argument that ψ is a branched covering, and the proof of the theorem. □

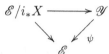

The following result, which relates the definition of a branched covering that we have given with one that is closer in spirit to the topological notion given by R. H. Fox, is implicit in the above proof. Fox does mention in passing that by defining branched cover as a completion of a locally constant (or unramified) map, folds are excluded, at least intuitively. We have seen by means of examples that in our definition of branched covering, folds are also intuitively excluded. A topos-theoretic (or for that matter, a topological) definition of a folded covering has not been given.

Corollary 9.3.13 *A geometric morphism over a topos $\mathscr{E}$ is a branched covering iff it is the spread completion of a locally constant covering of $\mathscr{E}/V$, for some pure subobject $V \rightarrowtail 1_{\mathscr{E}}$. Moreover, if a branched covering $\mathscr{Y} \xrightarrow{\psi} \mathscr{E}$ is the spread completion of a locally constant covering $\mathscr{E}/Y \longrightarrow \mathscr{E}/V$, with pure $V \rightarrowtail 1_{\mathscr{E}}$, then the spread completion diagram*

$$\begin{array}{ccc} \mathscr{E}/Y & \longrightarrow & \mathscr{Y} \\ \downarrow & & \downarrow \psi \\ \mathscr{E}/V & \rightarrowtail & \mathscr{E} \end{array}$$

is a topos pullback.

Remark 9.3.14 *Sometimes we may wish to focus on a particular pure subobject $V \rightarrowtail 1_{\mathscr{E}}$. For instance, in topology V may be the complement of a knot. Given such a V, consider two functors given by completion and by pullback.*

$$\begin{array}{ccccc} & & \mathscr{C}(\mathscr{E}/V) & & \\ & \text{completion} \swarrow & & \searrow \text{pullback} & \\ \mathscr{B}(\mathscr{E}) & & \longrightarrow & & \mathscr{C}(\mathscr{E}_{\mathrm{p}}) \end{array}$$

The bottom arrow is an equivalence, and the other two functors are full and faithful. The pullback functor may be equivalently described just as i^* for $i : \mathscr{E}_{\mathrm{p}} \rightarrowtail \mathscr{E}/V$. By Exercise 9.1.19, 9,

$$i^* : \mathscr{C}(\mathscr{E}/V) \longrightarrow \mathscr{C}(\mathscr{E}_{\mathrm{p}})$$

is full and faithful. Denote by $\mathscr{B}_V(\mathscr{E})$ the full image category of $\mathscr{B}(\mathscr{E})$ under the completion functor: an object of $\mathscr{B}_V(\mathscr{E})$ is thus a branched covering of $\mathscr{E}$ that is the completion of a locally constant covering of $\mathscr{E}/V$. There is an equivalence $\mathscr{B}_V(\mathscr{E}) \simeq \mathscr{C}(\mathscr{E}/V)$. In particular, for a locally connected and locally simply connected topos $\mathscr{E}/V$, $\mathscr{B}_V(\mathscr{E})$ is equivalent to the atomic topos $\mathscr{C}(\mathscr{E}/V)$, also denoted $\Pi_1{}^{\mathrm{c}}(\mathscr{E}/V)$. This is a version of the (coverings) fundamental group of a "knot" in $\mathscr{E}$ with "complement" V.

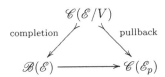

Remark 9.3.15 *Continuing with Remark 9.3.14, if* $V \rightarrowtail 1_{\mathscr{E}}$ *is pure, then we may regard* $\mathscr{C}(\mathscr{E}/V)$ *as a full subcategory of* $\mathscr{C}(\mathscr{E}_p)$ *under the full and faithful* i^*. *For instance, every knot group* $\mathscr{C}(Sh(S^3)/V)$ *can be regarded as a full subcategory of* $\mathscr{C}(Sh(S^3)_p)$. *Hence, any two knot groups have an intersection, by which we mean their overlap in* $\mathscr{C}(Sh(S^3)_p)$.

Proposition 9.3.16 *Branched coverings are stable under pullback along locally connected geometric morphisms.*

Proof. One way to prove this is to work with the formulation of branched cover that Corollary 9.3.13 provides. We invite the reader to complete the details. □

Remark 9.3.17 *One can argue that historically singular coverings precede locally constant coverings in the theory of Riemann surfaces, and that only on account of a desired connection with the fundamental group had additional assumptions been made on the maps, assumptions that in practice have the effect of reducing singular coverings to locally constant coverings. In fact, the familiar concept of locally constant covering is a topological concept formed from the analytical concept of a Riemann surface, or rather, that part of the Riemann surface remaining after the branch points have been deleted.*

Exercises 9.3.18

1. *Establish a version of Theorem 9.3.12 with unramified coverings in place of locally constant ones. Of course the notion of branched covering must be appropriately changed.*
2. *Prove Corollary 9.3.13.*
3. *Prove the following variation of Corollary 9.3.13: a complete spread over* $\mathscr{E}$ *is a branched covering iff it is the spread completion of a locally constant object of* $\mathscr{E}/W$, *for some pure object* $W \longrightarrow 1_{\mathscr{E}}$.
4. *Prove Proposition 9.3.16.*

9.4 The Index of a Complete Spread

In this section we study the category $\mathbf{Dist}_{\mathscr{E}}(\mathscr{E}, \mathscr{Y})$ associated with an $\mathscr{S}$-complete spread $\mathscr{Y} \xrightarrow{\psi} \mathscr{E}$. Note that the base topos is $\mathscr{E}$, not $\mathscr{S}$. We sometimes informally refer to this category as the 'index category' of the $\mathscr{S}$-complete spread ψ for the following reason. Let $W \subseteq Y$ denote the non-singular part of a branched covering $Y \xrightarrow{\psi} X$. The familiar index of branching of ψ is defined in terms of a functor

$$\varsigma : \mathscr{O}(Y) \longrightarrow Sh(X)$$

that associates with an open set $U \subseteq Y$ the locally constant map

$$\psi : U \cap W \longrightarrow \psi(U \cap W) \,,$$

by which we mean the restriction of ψ to $U \cap W$. Indeed, a number $\mathrm{b}(y)$ is then the index of branching of ψ at a point $y \in Y$ if there is a base $\{U_i\}$ of the neighbourhood system of y such that each locally constant map $U_i \cap W \longrightarrow \psi(U_i \cap W)$ has fiber $\mathrm{b}(y)$. Now suppose that $\mathscr{Y} \xrightarrow{\psi} \mathscr{E}$ is a branched covering in our sense. In particular, ψ is the spread completion $\mathscr{E}/Y \xrightarrow{\eta} \mathscr{Y}$ of some $\mathscr{E}/Y \longrightarrow \mathscr{E}$. We may thus interpret the 'index-functor' simply as the functor

$$\varsigma_Y = \Sigma_Y \cdot \eta^* : \mathscr{Y} \longrightarrow \mathscr{E} \,.$$

This functor is an $\mathscr{E}$-distribution that we call *the index-distribution associated with the branched covering*, motivated by the usual notion.

To begin our investigation, we know that for any geometric morphism $\mathscr{Y} \xrightarrow{\psi} \mathscr{E}$, an $\mathscr{E}$-distribution on $\mathscr{Y}$ is isomorphic to $\varphi_! \cdot \eta^*$, for some locally connected φ.

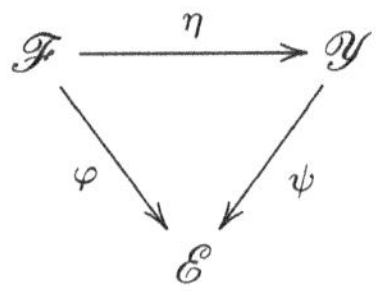

(9.4)

Proposition 9.4.1 *In diagram* (9.4), *assume that ψ is an $\mathscr{S}$-complete spread. Then η is an $\mathscr{E}$-complete spread iff φ is a local homeomorphism. In particular, any $\mathscr{E}$-distribution $\mathscr{Y} \longrightarrow \mathscr{E}$ preserves pullbacks.*

Proof. Assume that φ is a local homeomorphism: $\mathscr{E}/Z \longrightarrow \mathscr{E}$. Consider the $\mathscr{E}$-pure, complete spread factorization of η.

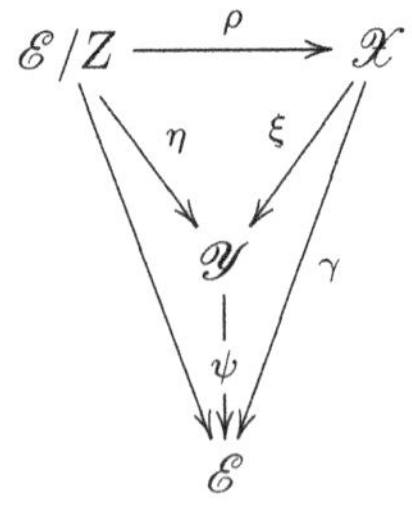

(9.5)

The geometric morphism γ is locally connected, ρ is $\mathscr{E}$-pure so that $\Sigma_Z \cdot \rho^* \cong \gamma_!$, and we have

$$Z \cong \Sigma_Z(\rho^*(\xi^*(1))) \cong \gamma_!(\xi^*(1)) \cong \gamma_!(1) \,.$$

We shall show that ρ is an equivalence. Let $\mathscr{X} \xrightarrow{\bar{\gamma}} \mathscr{E}/Z$ denote the connected part of γ. Then the composite $\bar{\gamma} \cdot \rho$ is uniquely isomorphic to the identity geometric morphism on $\mathscr{E}/Z$.

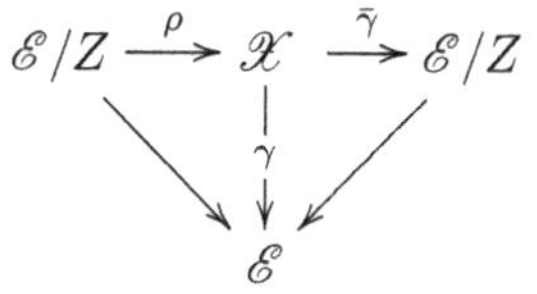

It remains to show that $\rho\cdot\bar{\gamma}$ is isomorphic to the identity on $\mathscr{X}$. Let $\mathscr{Z} \longrightarrow \mathscr{E}$ be the spread completion of $\mathscr{E}/Z \longrightarrow \mathscr{E}$ over $\mathscr{S}$. The $\mathscr{S}$-pure, complete spread factorization of γ is as follows.

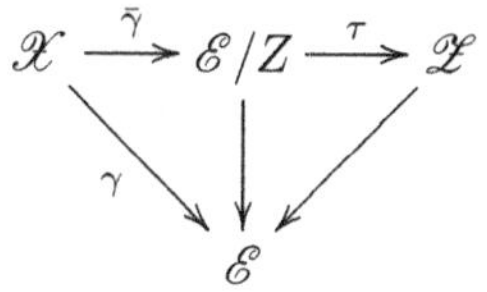

Since the codomain topos of ξ is an $\mathscr{S}$-complete spread, there must be a factorization $\xi \cong \tilde{\xi}\cdot\tau\cdot\bar{\gamma}$, where

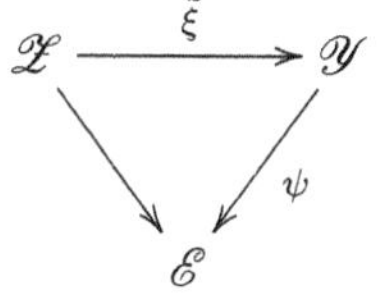

commutes. Thus, $\bar{\gamma}$ is defined over $\mathscr{Y}$ as in the following diagram.

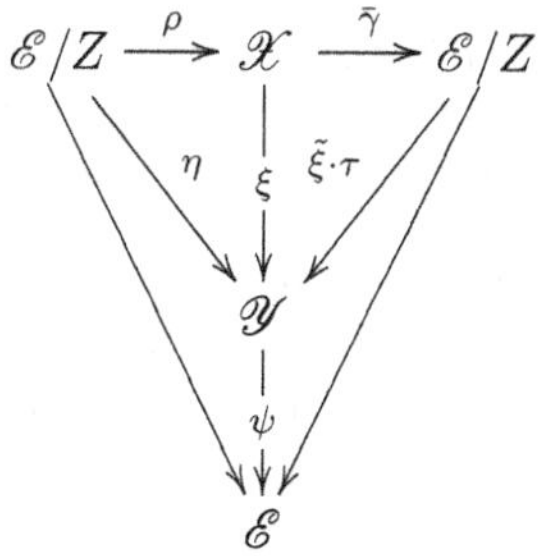

But we also wish to define an isomorphism $\tilde{\xi}\cdot\tau\cdot 1_Z \cong \eta$ such that the isomorphism $\bar{\gamma}\cdot\rho \cong 1_Z$ is over $\mathscr{Y}$. We simply take the composite isomorphism

$$\tilde{\xi}\cdot\tau\cdot 1_Z \cong \tilde{\xi}\cdot\tau\cdot\bar{\gamma}\cdot\rho \cong \xi\cdot\rho \cong \eta\ .$$

The identity geometric morphism on $\mathscr{X}$ and $\rho\cdot\bar{\gamma}$ are two geometric morphisms from the $\mathscr{E}$-complete spread ξ to itself.

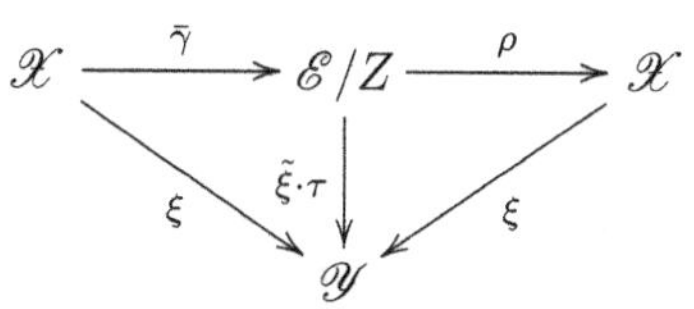

We have $\rho \cdot \bar{\gamma} \cdot \rho \cong \rho$ over $\mathscr{Y}$. Since ρ is $\mathscr{E}$-pure, $\rho \cdot \bar{\gamma}$ must be isomorphic to the identity on $\mathscr{X}$ over $\mathscr{Y}$. This proves that η is an $\mathscr{E}$-complete spread.

Conversely, assume that η is an $\mathscr{E}$-complete spread. We have seen in the previous paragraph that because ψ is an $\mathscr{S}$-complete spread there is a factorization

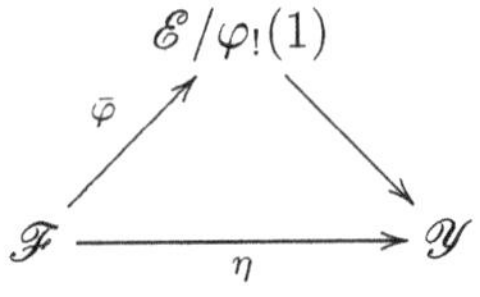

over $\mathscr{E}$. But the connected $\bar{\varphi}$ is $\mathscr{E}$-pure, and a complete spread cannot have a non-trivial first factor that is pure. Thus $\bar{\varphi}$ is an equivalence, so that φ is a local homeomorphism. □

Corollary 9.4.2 *Let Y be any object of a locally connected topos $\mathscr{E}$, with associated complete spread $\mathscr{Y} \xrightarrow{\psi} \mathscr{E}$. Then the $\mathscr{S}$-pure factor $\mathscr{E}/Y \longrightarrow \mathscr{Y}$ is an $\mathscr{E}$-complete spread.*

Intuitively, Corollary 9.4.2 says that an object of a topos $\mathscr{E}$ is $\mathscr{E}$-closed in its $\mathscr{S}$-closure.

Corollary 9.4.3 *A locally connected $\mathscr{S}$-complete spread is a local homeomorphism. Thus, an $\mathscr{S}$-complete spread is an unramified cover iff it is locally connected.*

Proof. If an $\mathscr{S}$-complete spread $\mathscr{Y} \xrightarrow{\psi} \mathscr{E}$ is locally connected, then the identity geometric morphism $\mathscr{Y} \longrightarrow \mathscr{Y}$ is an $\mathscr{E}$-complete spread with locally connected domain. By Proposition 9.4.1, ψ is a local homeomorphism. □

Let $\mathscr{Y} \xrightarrow{\psi} \mathscr{E}$ denote an arbitrary $\mathscr{S}$-complete spread, with interior $\mathscr{E}/X \xrightarrow{\tau} \mathscr{Y}$: $X = \mathbf{d}(\psi)$. ($\mathbf{d}(\psi)$ coincides with the density of the distribution associated with ψ, but in the case of a geometric morphism we sometimes use the term interior.) There is a functor

$$\Phi : \mathscr{E}/X \longrightarrow \mathbf{Dist}_{\mathscr{E}}(\mathscr{E}, \mathscr{Y})$$

that associates with $Z \xrightarrow{m} X$ the distribution $\Sigma_X(m \times \tau^*(\ ))$. Now let λ be an arbitrary $\mathscr{E}$-distribution $\mathscr{Y} \longrightarrow \mathscr{E}$. We know there is a diagram

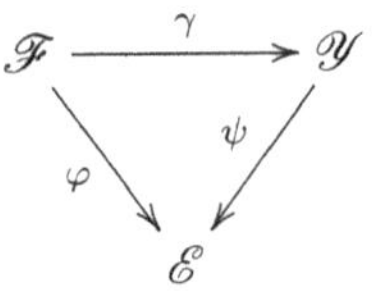

such that φ is locally connected and $\lambda \cong \varphi_! \cdot \gamma^*$. We have $\varphi_!(1) = \lambda(1)$. We know that the $\mathscr{S}$-pure, complete spread factorization of φ is

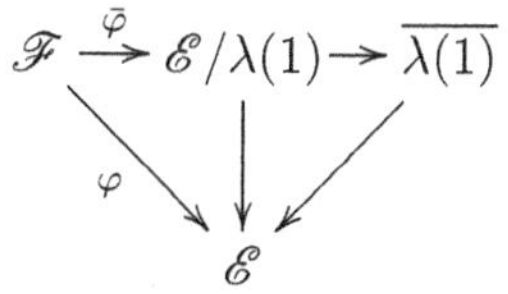

where $\overline{\lambda(1)}$ is the $\mathscr{S}$-spread completion of $\lambda(1)$. Since ψ is an $\mathscr{S}$-complete spread, γ must factor through $\overline{\lambda(1)}$, hence through $\lambda(1)$ over $\mathscr{E}$:

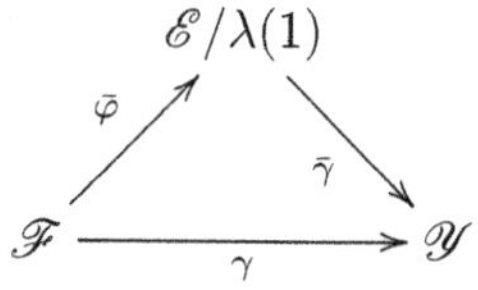

Since $\bar{\varphi}$ is $\mathscr{E}$-pure, we have $\lambda \cong \Sigma_{\lambda(1)} \cdot \bar{\gamma}^*$. Thus, the object $\lambda(1)$ factors through the interior X, say by $\lambda(1) \xrightarrow{m} X$. This gives us a functor

$$\Psi : \mathbf{Dist}_{\mathscr{E}}(\mathscr{E}, \mathscr{Y}) \longrightarrow \mathscr{E}/X$$

that associates with a λ the object $\lambda(1) \xrightarrow{m} X$ just constructed. Moreover, for any λ, we have $\lambda \cong \Sigma_X(\, m \times \tau^*(\ )\,) = \Phi(\Psi(\lambda))$.

Proposition 9.4.4 *Let $\mathscr{Y} \xrightarrow{\psi} \mathscr{E}$ denote an $\mathscr{S}$-complete spread with interior $\mathscr{E}/X \xrightarrow{\tau} \mathscr{Y}$: $X = \mathbf{d}(\psi)$. Then the functors Φ and Ψ establish an equivalence of categories:*

$$\mathscr{E}/X \underset{\Phi}{\overset{\Psi}{\rightleftarrows}} \mathbf{Dist}_{\mathscr{E}}(\mathscr{E}, \mathscr{Y}) \,.$$

Moreover, for any object $Z \xrightarrow{m} X$ of $\mathscr{E}/X$, the $\mathscr{E}$-complete spread associated with the distribution $\Phi(m)$ is $\mathscr{E}/Z \xrightarrow{m} \mathscr{E}/X \xrightarrow{\tau} \mathscr{Y}$.

Proof. For the first statement, we have only to show that for any $Z \xrightarrow{m} X$, we have $\Psi(\Phi(m)) \cong m$. But this is clear because

$$\Phi(m)(1) = \Sigma_X(\, m \times \tau^*(1)\,) = \Sigma_X(m) = Z \,,$$

and the morphism $\Phi(m)(1) \longrightarrow X$ is easily seen to be m. □

Corollary 9.4.5 *Let Y be any object of a locally connected topos $\mathscr{E}$ with $\mathscr{S}$-spread completion $\mathscr{Y} \xrightarrow{\psi} \mathscr{E}$ and interior $\mathscr{E}/X \xrightarrow{\tau} \mathscr{Y}$, as in the following diagram.*

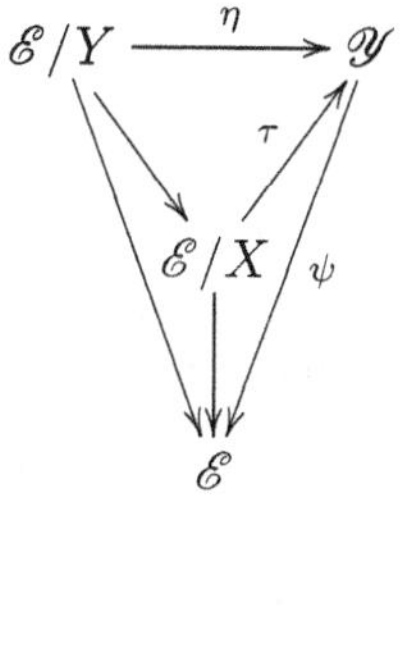

Then the functor $Z \xrightarrow{m} X \mapsto \Sigma_X(m \times \tau^*(\ \ ))$ *is an equivalence*

$$\mathscr{E}/X \simeq \mathbf{Dist}_{\mathscr{E}}(\mathscr{E}, \mathscr{Y})\,.$$

Remark 9.4.6 *Let* $\mathscr{Y} \xrightarrow{\psi} \mathscr{E}$ *be an* $\mathscr{S}$*-complete spread, with interior* $\mathscr{E}/X \xrightarrow{\tau} \mathscr{Y}$*. Then the terminal object of* $\mathbf{Dist}_{\mathscr{E}}(\mathscr{E}, \mathscr{Y})$ *is the* $\mathscr{E}$*-distribution* $\Sigma_X \cdot \tau^*$*, and* τ *is the terminal* $\mathscr{E}$*-complete spread.*

The previous discussion examines $\mathscr{E}$-distributions on a given arbitrary $\mathscr{S}$-complete spread to $\mathscr{E}$. We now begin with a locally connected geometric morphism $\mathscr{F} \xrightarrow{\varphi} \mathscr{E}$ over a topos $\mathscr{S}$, and then consider its $\mathscr{S}$-spread completion.

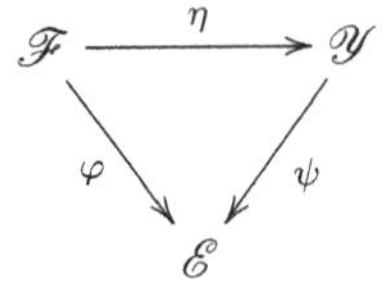

We may also consider the $\mathscr{E}$-pure, complete spread factorization of φ, which is the perimeter of the following diagram.

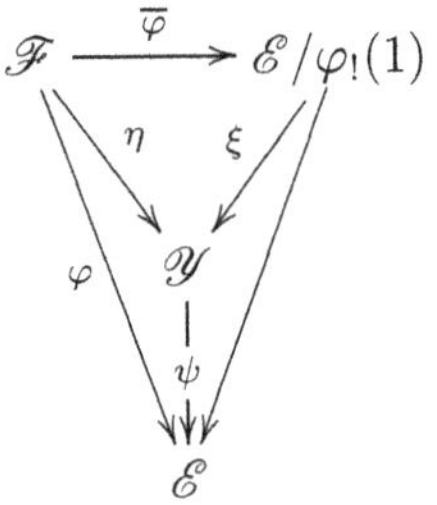

(9.6)

Remark 9.4.7 *By Proposition 9.4.1, the* $\mathscr{S}$*-pure* ξ *in diagram (9.6) is an* $\mathscr{E}$*-complete spread, so that* $\overline{\varphi}$ *and* ξ *give the* $\mathscr{E}$*-pure, complete spread factorization of* η*.*

We conclude with the following result describing the nature of the index-distribution that we had introduced at the beginning of this section.

Corollary 9.4.8 *For any* Y *in a locally connected topos* $\mathscr{E}$*, the functor*

$$\mathscr{E}/Y \longrightarrow \mathbf{Dist}_{\mathscr{E}}(\mathscr{E}, \mathscr{Y})/\varsigma_Y\,; \quad E \xrightarrow{m} Y \mapsto \varsigma_m = \Sigma_Y(m \times \eta^*(\ \ ))$$

is an equivalence, where

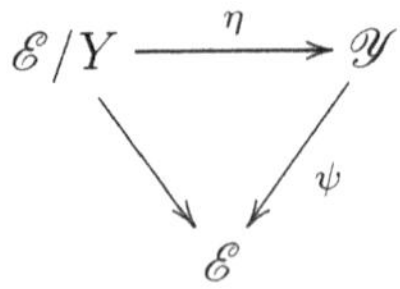

is the $\mathscr{S}$*-spread completion of* Y*, and* ς_Y *denotes the index-distribution* $\Sigma_Y \cdot \eta^*$*. For any* Y *in* $\mathscr{E}$*, if* η *is an inclusion* (as in the case of a branched cover), *then* ς_Y *is a weak terminal object in* $\mathbf{Dist}_{\mathscr{E}}(\mathscr{E}, \mathscr{Y})$.

Proof. We have

$$\mathscr{E}/Y \simeq \mathbf{Dist}_{\mathscr{E}}(\mathscr{E}, \mathscr{E}/Y) \simeq \mathbf{Dist}_{\mathscr{E}}(\mathscr{E}, \mathscr{Y})/\varsigma_Y .$$

Indeed, the first equivalence is by Exercise 1.3.8, 3. The second equivalence is by Proposition 2.5.4, since $\mathscr{E}/Y \xrightarrow{\eta} \mathscr{Y}$ is an $\mathscr{E}$-complete spread (Remark 9.4.7). The second statement of the proposition holds because for any object $E \xrightarrow{m} Y$ there is exactly one natural transformation $m \times Y^* \Rightarrow Y^*$, hence exactly one $Y^* \Rightarrow Y^*(\ )^m$. If η is an inclusion, then there is exactly one natural transformation $\eta_* Y^* \Rightarrow \eta_*(Y^*(\ )^m)$. Now consider left adjoints to see that there is exactly one natural transformation $\varsigma_m \Rightarrow \varsigma_Y$. This shows that ς_Y is a weak terminal because if there is a morphism $\mu \Rightarrow \varsigma_Y$, then $\mu \cong \varsigma_m$ for some m. □

Further reading: Barr & Diaconescu [BD81, BD80], Brown [Bro88], Bunge [Bun04], Bunge & Funk [BF98], Bunge & Lack [BL03], Bunge & Moerdijk [BM97], Bunge & Niefield [BN00], Bunge & Paré [BP79], Fox [Fox57], Funk [Fun00], Funk & Tymchatyn [FT01], Janelidze [Jan90], Johnstone [Joh02], Joyal & Tierney [JT84], Kock & Reyes [KR99], Mulero [Mul98], Springer [Spr57].

Bibliography

[AGV72a] M. Artin, A. Grothendieck, and J. L. Verdier, *Théorie des topos*, SGA4, exposés I-VI, second edition. LNM 420, Springer-Verlag, Berlin, 1972, pp. 1–340.

[AGV72b] ———, *Théorie des topos et cohomologie étale des schémas*, LNM 269, 270, and 305, Springer-Verlag, Berlin, 1972.

[Arn00] V. I. Arnold, *Singularity theory*, Development of Mathematics: 1950 - 2000 (Jean-Paul Pier, ed.), Birkhäuser, Basel-Boston-New York, 2000, pp. 63–95.

[BC95] M. Bunge and A. Carboni, *The symmetric topos*, J. Pure Appl. Alg. **105** (1995), 233–249.

[BD80] M. Barr and R. Diaconescu, *Atomic toposes*, J. Pure Appl. Alg. **17** (1980), 1–24.

[BD81] ———, *On locally simply connected toposes and their fundamental groups*, Cahiers de Top. et Géom. Diff. Catégoriques **22** (1981), no. 3, 301–314.

[Ben75] J. Benabou, *Fibrations petites et localement petites*, C. R. Acad. Sc. Paris **281** (1975), 897–900.

[Ber91] G. M. Bergman, *Co-rectangular bands and cosheaves in categories of algebras*, Algebra Universalis **28** (1991), 188–213.

[BF96a] M. Bunge and J. Funk, *Constructive theory of the lower power locale*, Math. Struct. Comp. Sci. **6** (1996), 1–15.

[BF96b] ———, *Spreads and the symmetric topos*, J. Pure Appl. Alg. **113** (1996), 1–38.

[BF98] ———, *Spreads and the symmetric topos II*, J. Pure Appl. Alg. **130** (1998), no. 1, 49–84.

[BF99] ———, *On a bicomma object condition for KZ-doctrines*, J. Pure Appl. Alg. **143** (1999), 69–105.

[BF00] M. Bunge and M. Fiore, *Unique factorization lifting functors and categories of processes*, Math. Structures in Comp. Sci. **10** (2000), no. 2, 137–163.

[BFJS00] M. Bunge, J. Funk, M. Jibladze, and T. Streicher, *Distribution algebras and duality*, Advances in Mathematics **156** (2000), 133–155.

[BFJS02] ———, *The Michael completion of a topos spread*, J. Pure Appl. Alg. **175** (2002), 63–91.

[BFJS04a] ———, *Definable completeness*, Cahiers de Top. et Géom. Diff. Catégoriques **XLV-4** (2004), 1–24.

[BFJS04b] ———, *Distribution classifiers*, in preparation, 2004.

[Bir84] G. J. Bird, *Limits in 2-categories of locally presentable categories*, Ph.D. thesis, University of Sydney, 1984.

[BK02] M. Barr and J. Kennison, *Interesting topos*, June 2002, Email communication: categories@mta.ca.

[BL03] M. Bunge and S. Lack, *Van kampen theorems for toposes*, Advances in Mathematics **179/2** (2003), 291–317.

[BM60] A. Borel and J. Moore, *Homology theory for locally compact spaces*, Michigan Math. J. **7** (1960), 137–159.

[BM97] M. Bunge and I. Moerdijk, *On the construction of the Grothendieck fundamental group of a topos by paths*, J. Pure Appl. Alg. **116** (1997), 99–113.

[BN00] M. Bunge and S. Niefield, *Exponentiability and single universes*, J. Pure Appl. Alg. **148** (2000), 217–250.

[BP79] M. Bunge and R. Paré, *Stacks and equivalences of categories*, Cahiers de Top. et Géom. Diff. Catégoriques **20** (1979), no. 4, 373–399.

[BP80] M. Barr and R. Paré, *Molecular toposes*, J. Pure Appl. Alg. **17** (1980), 127–152.

[Bre67] G. Bredon, *Sheaf theory*, Mc-Graw-Hill, New York, St. Louis, San Francisco, Toronto, London, Sydney, 1967.

[Bro88] R. Brown, *Topology*, Ellis Horwood Ltd., Chichester, England, 1988.

[Bun74] M. Bunge, *Coherent extensions and relational algebras*, Trans. Amer. Math. Soc. **197** (1974), 355–390.

[Bun95] ———, *Cosheaves and distributions on toposes*, Algebra Universalis **34** (1995), 469–484.

[Bun04] ———, *Galois groupoids and covering morphisms in topos theory*, Proceedings of the Fields Institute: Workshop on Descent, Galois Theory and Hopf algebras, Fields Institute Communications, American Mathematical Society, 2004, pp. 131–162.

[CJ95] A. Carboni and P. T. Johnstone, *Connected limits, familial representability and Artin glueing*, Math. Struct. Comp. Sci. **5** (1995), 441–449.

[Deh94] P. Dehornoy, *Braid groups and left distributive operations*, Trans. Amer. Math. Soc. **345** (1994), no. 1, 115–150.

[Fio95] M. Fiore, *Lifting as a KZ-doctrine*, Category Theory and Computer Science, LNCS 953 (D. Pitt et al., eds.), Springer, 1995, pp. 146–158.

[Fox57] R. H. Fox, *Covering spaces with singularities*, Algebraic Geometry and Topology: A Symposium in Honor of S. Lefschetz (R. H. Fox et al., eds.), Princeton University Press, Princeton, 1957, pp. 243–257.

[FT01] J. Funk and E. D. Tymchatyn, *Unramified maps*, JP Journal of Geometric Topology **1(3)** (2001), 249–280.

[Fun95] J. Funk, *The display locale of a cosheaf*, Cahiers de Top. et Géom. Diff. Catégoriques **36** (1995), no. 1, 53–93.

[Fun99] ———, *The locally connected coclosure of a Grothendieck topos*, J. Pure Appl. Alg. **137** (1999), 17–27.

[Fun00] ———, *On branched covers in topos theory*, Theory and Applications of Categories **7** (2000), no. 1, 1–22.

[Fun01] ———, *The Hurwitz action and braid group orderings*, Theory and Applications of Categories **9** (2001), no. 7, 121–150.

[Gle63] A. M. Gleason, *Universal locally connected refinements*, Illinois. J. Math **7** (1963), 521–531.

[Gro55] A. Grothendieck, *Produits tensorieles topologiques et espaces nucléaires*, Memoirs of the American Mathematical Society, vol. 16, American Mathematical Society, 1955.

[Jan90] G. Janelidze, *Pure galois theory in categories*, J. of Algebra **132** (1990), 249–280.

[JJ82] P. T. Johnstone and A. Joyal, *Continuous categories and exponentiable toposes*, J. Pure Appl. Algebra **25** (1982), 255–296.

[JJ91] M. Jibladze and P. T. Johnstone, *The frame of fibrewise closed nuclei*, Cahiers de Top. et Géom. Diff. Catégoriques **32** (1991), no. 2, 99–112.

[Joh77] P. T. Johnstone, *Topos theory*, Academic Press, Inc., London, 1977.

[Joh82a] ———, *Factorization theorems for geometric morphisms II*, Categorical aspects of topology and analysis, Proc. Carleton University, Ottawa 1980, LNM 915, Springer, Berlin, 1982, pp. 216–233.

[Joh82b] ———, *Stone spaces*, Cambridge University Press, Cambridge, 1982.

[Joh89] ———, *A constructive "closed subgroup theorem" for localic groups and groupoids*, Cahiers de Top. et Géom. Diff. Catégoriques **30** (1989), 3–23.

[Joh92] ———, *Partial products, bagdomains and hyperlocal toposes*, Applications of Categories in Computer Science (M. P. Fourman, P. T. Johnstone, and A. M. Pitts, eds.), London Mathematical Society Lecture Notes Series 177, Cambridge University Press, 1992, pp. 315–339.

[Joh99] ———, *A note of discrete Conduché fibrations*, Theory and Applications of Categories **5** (1999), no. 1, 1–11.

[Joh02] ———, *Sketches of an elephant: A topos theory compendium*, Clarendon Press, Oxford, 2002.

[JT84] A. Joyal and M. Tierney, *An extension of the Galois theory of Grothendieck*, vol. Memoirs of the American Mathematical Society 309, American Mathematical Society, 1984.

[Kel82] G. M. Kelly, *Basic concepts of enriched category theory*, London Mathematical Society Lecture Note Series, vol. 64, Cambridge University Press, Cambridge, 1982.

[Koc75] A. Kock, *Monads for which structure are adjoints to units*, J. Pure Appl. Alg. **104** (1975), 41–59.

[Koc83] ———, *Some problems and results in synthetic functional analysis*, Category Theoretic Methods in Geometry (Aarhus) (A. Kock, ed.), Matematisk Institut, Aarhus Universitet, Various Publications Series 35, 1983, pp. 168–191.

[Koc90] ———, *Relatively Boolean toposes*, Aahus Preprint Series 21, Aahus University, 1989/90.

[KR94] A. Kock and G. E. Reyes, *Relatively Boolean and de Morgan toposes and locales*, Cahiers de Top. et Géom. Diff. Catégoriques **35** (1994), 249–261.

[KR99] ———, *A note on frame distributions*, Cahiers de Top. et Géom. Diff. Catégoriques **40** (1999), no. 2, 127–140.

[Law66] F. W. Lawvere, *Integration on presheaf toposes*, Proc. Meeting Oberwolfach, 1966.

[Law68] ———, *Equality in hyperdoctrines and the comprehension schema as an adjoint functor*, Proc. Symp. Pure Math. 17, Amer. Math. Soc., 1968, pp. 1–14.

[Law83] ———, *Intensive and extensive quantities*, Notes for the lectures given at the workshop on Categorical Methods in Geometry, Aarhus, 1983.

[Law92] ———, *Categories of space and of quantity*, The Space of Mathematics (J. Echeverria et al., eds.), W. de Gruyter, Berlin-New York, 1992, pp. 14–30.

[Law00a] ———, *Comments on the development of topos theory*, Development of Mathematics 1950-2000 (Jean-Paul Pier, ed.), Birkhäuser, Basel-Boston-Berlin, 2000, pp. 715–734.

[Law00b] ———, *Volterra's functionals and covariant cohesion of space*, Supplemento dei Rendiconti del Circolo Matematico di Palermo **64** (2000), 24–204.

[Mic63] E. Michael, *Completing a spread (in the sense of Fox) without local connectedness*, Indag. Math. **25** (1963), 629–633.

[Mik76] C.-J. Mikkelsen, *Lattice theoretic and logical aspects of elementary topoi*, Ph.D. thesis, Aarhus Universiteit, Matematisk Institut, 1976.

[MM92] S. Mac Lane and I. Moerdijk, *Sheaves in geometry and logic*, Springer-Verlag, Berlin-Heidelberg-New York, 1992.

[Moe82] J. -L. Moens, *Charactérisation des topos des faisceaux sur un site interne à un topos*, Ph.D. thesis, Université Catholique de Louvain-la-Neuve, 1982.

[Mul98] M. A. Mulero, *Algebraic characterization of finite (branched) coverings*, Fundamenta Mathematicae **158** (1998), 165–180.

[MV00] I. Moerdijk and J. J. C. Vermeulen, *Proper maps of toposes*, Memoirs of the American Mathematical Society **148** (2000), no. 705, 108.

[Nie27] J. Nielsen, Untersuchungen zur Topologie der geschlossenen zweiseitigen Flachen, Acta Math. **50** (1927), 189–358.

[Nie82] S. B. Niefield, *Cartesianness: topological spaces, uniform spaces, and affine schemes*, J. Pure Appl. Alg. **23** (1982), 147–167.

[Par74] R. Paré, *Colimits in topoi*, Bull. Amer. Math. Soc. **80** (1974), 556–561.

[Par80] ———, *Indexed categories and generated topologies*, J. Pure Appl. Alg. **19** (1980), 385–400.

[Pit85] A. M. Pitts, *On product and change of base for toposes*, Cahiers de Top. et Géom. Diff. Catégoriques **26** (1985), no. 1, 43–61.

[PS78] R. Paré and D. Schumacher, *Abstract families and the adjoint functor theorems*, Indexed Categories and Their Applications, LNM 661, Springer-Verlag, Berlin, 1978, pp. 1–125.

[PS96] V. V. Prasolov and A. B. Sossinsky, *Knots, links, braids and 3-manifolds*, Translations of Mathematical Monographs 154, American Mathematical Society, Providence, Rhode Island, 1996.

[Ros86] G. Rosolini, *Continuity and effectiveness in topoi*, Ph.D. thesis, Oxford University, 1986.

[Sch66] L. Schwartz, *Théorie des distributions*, Hermann, Paris, 1966.

[Spr57] G. Springer, *Introduction to Riemann surfaces*, Addison-Wesley, 1957.

[Sti86] J. Stillwell, *Jakob Nielsen, collected mathematical papers*, Birkhauser, Boston-Basel-Stuttgart, 1986.

[Str] R. Street, *Conspectus of variable categories*, Unpublished.

[Str74] ———, *Fibrations and Yoneda's lemma in a 2-category*, Category Seminar, Sydney 1972/73, LNM 420, Springer-Verlag, Berlin, 1974, pp. 104–133.

[Str03] T. Streicher, *Fibered categories à la Jean Bénabou*, Unpublished, 2003.

[SW73] R. Street and R. F. C. Walters, *The comprehensive factorization of a functor*, Bull. Amer. Math. Soc. **79** (1973), 936–941.

[SW00] H. Short and B. Wiest, *Orderings of mapping class groups after Thurston*, Ens. Math. **46** (2000), 279–312.

[Tie74] M. Tierney, *Forcing topologies and classifying topoi*, in Alex Heller, Myles Tierney (eds.), Algebra, Topology, and Category Theory, Academic Press, New York, 1974, pp. 211–219.

[Ver94] J. J. C. Vermeulen, *Proper maps of locales*, J. Pure Appl. Alg. **92** (1994), 79–107.

[Vic89] S. Vickers, *Topology via logic*, Cambridge Tracts in Theoretical Computer Science, Cambridge University Press, Cambridge, UK, 1989.

[Vic92] ———, *Geometric theories and databases*, Applications of Categories in Computer Science, London Math. Soc. Lecture Note Series (P. T. Johnstone M. P. Fourman and A. M. Pitts, eds.), vol. 177, Cambridge University Press, Cambridge, 1992, pp. 288–314.

[Vic95] ———, *Locales are not pointless*, Theory and Formal Methods 1994: Proceedings of the Second Imperial College, Department of Computing, Workshop on Theory and Formal Methods (Cambridge) (Chris Hankin et al., eds.), Imperial College Press, 1995.

[Wae67] L. Waelbroeck, *Differentiable mappings into b-spaces*, Journal of Functional Analysis **1** (1967), 409–418.

[Wil68] S. Willard, *General topology*, Addison-Wesley Series in Mathematics, Addison-Wesley Publishing Company, Reading, Massachussetts. Menlo Park, California. London. Amsterdam. Don Mills, Ontario. Sydney, 1968.

[Zoe76] V. Zoeberlein, *Doctrines on 2-categories*, Math. Zeit (1976), 267–279.

Index

Lecture Notes in Mathematics

For information about earlier volumes
please contact your bookseller or Springer
LNM Online archive: springerlink.com

Vol. 1691: R. Bezrukavnikov, M. Finkelberg, V. Schechtman, Factorizable Sheaves and Quantum Groups (1998)
Vol. 1692: T. M. W. Eyre, Quantum Stochastic Calculus and Representations of Lie Superalgebras (1998)
Vol. 1694: A. Braides, Approximation of Free-Discontinuity Problems (1998)
Vol. 1695: D. J. Hart el, Markov Set-Chains (1998)
Vol. 1696: E. Bouscaren (Ed.): Model Theory and Algebraic Geometry (1998)
Vol. 1697: B. Cockburn, C. Johnson, C.-W. Shu, E. Tadmor, Advanced Numerical Approximation of Nonlinear Hyperbolic Equations. Cetraro, Italy, 1997. Editor: A. Quarteroni (1998)
Vol. 1698: M. Bhattacharjee, D. Macpherson, R. G. Möller, P. Neumann, Notes on In nite Permutation Groups (1998)
Vol. 1699: A. Inoue,Tomita-Takesaki Theory in Algebras of Unbounded Operators (1998)
Vol. 1700: W. A. Woyczyński, Burgers-KPZ Turbulence (1998)
Vol. 1701: Ti-Jun Xiao, J. Liang, The Cauchy Problem of Higher Order Abstract Differential Equations (1998)
Vol. 1702: J. Ma, J. Yong, Forward-Backward Stochastic Differential Equations and Their Applications (1999)
Vol. 1703: R. M. Dudley, R. Norvaiša, Differentiability of Six Operators on Nonsmooth Functions and p-Variation (1999)
Vol. 1704: H. Tamanoi, Elliptic Genera and Vertex Operator Super-Algebras (1999)
Vol. 1705: I. Nikolaev, E. Zhuzhoma, Flows in 2-dimensional Manifolds (1999)
Vol. 1706: S. Yu. Pilyugin, Shadowing in Dynamical Systems (1999)
Vol. 1707: R. Pytlak, Numerical Methods for Optimal Control Problems with State Constraints (1999)
Vol. 1708: K. Zuo, Representations of Fundamental Groups of Algebraic Varieties (1999)
Vol. 1709: J. Azéma, M. Émery, M. Ledoux, M. Yor (Eds.), Séminaire de Probabilités XXXIII (1999)
Vol. 1710: M. Koecher, The Minnesota Notes on Jordan Algebras and Their Applications (1999)
Vol. 1711: W. Ricker, Operator Algebras Generated by Commuting Projećtions: A Vector Measure Approach (1999)
Vol. 1712: N. Schwartz, J. J. Madden, Semi-algebraic Function Rings and Re ectors of Partially Ordered Rings (1999)
Vol. 1713: F. Bethuel, G. Huisken, S. Müller, K. Steffen, Calculus of Variations and Geometric Evolution Problems. Cetraro, 1996. Editors: S. Hildebrandt, M. Struwe (1999)
Vol. 1714: O. Diekmann, R. Durrett, K. P. Hadeler, P. K. Maini, H. L. Smith, Mathematics Inspired by Biology. Martina Franca, 1997. Editors: V. Capasso, O. Diekmann (1999)
Vol. 1715: N. V. Krylov, M. Röckner, J. Zabczyk, Stochastic PDE's and Kolmogorov Equations in In nite Dimensions. Cetraro, 1998. Editor: G. Da Prato (1999)
Vol. 1716: J. Coates, R. Greenberg, K. A. Ribet, K. Rubin, Arithmetic Theory of Elliptic Curves. Cetraro, 1997. Editor: C. Viola (1999)
Vol. 1717: J. Bertoin, F. Martinelli, Y. Peres, Lectures on Probability Theory and Statistics. Saint-Flour, 1997. Editor: P. Bernard (1999)
Vol. 1718: A. Eberle, Uniqueness and Non-Uniqueness of Semigroups Generated by Singular Diffusion Operators (1999)
Vol. 1719: K. R. Meyer, Periodic Solutions of the N-Body Problem (1999)
Vol. 1720: D. Elworthy, Y. Le Jan, X-M. Li, On the Geometry of Diffusion Operators and Stochastic Flows (1999)
Vol. 1721: A. Iarrobino, V. Kanev, Power Sums, Gorenstein Algebras, and Determinantal Loci (1999)
Vol. 1722: R. McCutcheon, Elemental Methods in Ergodic Ramsey Theory (1999)
Vol. 1723: J. P. Croisille, C. Lebeau, Diffraction by an Immersed Elastic Wedge (1999)
Vol. 1724: V. N. Kolokoltsov, Semiclassical Analysis for Diffusions and Stochastic Processes (2000)
Vol. 1725: D. A. Wolf-Gladrow, Lattice-Gas Cellular Automata and Lattice Boltzmann Models (2000)
Vol. 1726: V. Marić, Regular Variation and Differential Equations (2000)
Vol. 1727: P. Kravanja M. Van Barel, Computing the Zeros of Analytic Functions (2000)
Vol. 1728: K. Gatermann Computer Algebra Methods for Equivariant Dynamical Systems (2000)
Vol. 1729: J. Azéma, M. Émery, M. Ledoux, M. Yor (Eds.) Séminaire de Probabilités XXXIV (2000)
Vol. 1730: S. Graf, H. Luschgy, Foundations of Quantization for Probability Distributions (2000)
Vol. 1731: T. Hsu, Quilts: Central Extensions, Braid Actions, and Finite Groups (2000)
Vol. 1732: K. Keller, Invariant Factors, Julia Equivalences and the (Abstract) Mandelbrot Set (2000)
Vol. 1733: K. Ritter, Average-Case Analysis of Numerical Problems (2000)
Vol. 1734: M. Espedal, A. Fasano, A. Mikelić, Filtration in Porous Media and Industrial Applications. Cetraro 1998. Editor: A. Fasano. 2000.
Vol. 1735: D. Yafaev, Scattering Theory: Some Old and New Problems (2000)
Vol. 1736: B. O. Turesson, Nonlinear Potential Theory and Weighted Sobolev Spaces (2000)
Vol. 1737: S. Wakabayashi, Classical Microlocal Analysis in the Space of Hyperfunctions (2000)
Vol. 1738: M. Émery, A. Nemirovski, D. Voiculescu, Lectures on Probability Theory and Statistics (2000)
Vol. 1739: R. Burkard, P. Deu hard, A. Jameson, J.-L. Lions, G. Strang, Computational Mathematics Driven by In-

dustrial Problems. Martina Franca, 1999. Editors: V. Capasso, H. Engl, J. Periaux (2000)
Vol. 1740: B. Kawohl, O. Pironneau, L. Tartar, J.-P. Zolesio, Optimal Shape Design. Tróia, Portugal 1999. Editors: A. Cellina, A. Ornelas (2000)
Vol. 1741: E. Lombardi, Oscillatory Integrals and Phenomena Beyond all Algebraic Orders (2000)
Vol. 1742: A. Unterberger, Quantization and Nonholomorphic Modular Forms (2000)
Vol. 1743: L. Habermann, Riemannian Metrics of Constant Mass and Moduli Spaces of Conformal Structures (2000)
Vol. 1744: M. Kunze, Non-Smooth Dynamical Systems (2000)
Vol. 1745: V. D. Milman, G. Schechtman (Eds.), Geometric Aspects of Functional Analysis. Israel Seminar 1999-2000 (2000)
Vol. 1746: A. Degtyarev, I. Itenberg, V. Kharlamov, Real Enriques Surfaces (2000)
Vol. 1747: L. W. Christensen, Gorenstein Dimensions (2000)
Vol. 1748: M. Ruzicka, Electrorheological Fluids: Modeling and Mathematical Theory (2001)
Vol. 1749: M. Fuchs, G. Seregin, Variational Methods for Problems from Plasticity Theory and for Generalized Newtonian Fluids (2001)
Vol. 1750: B. Conrad, Grothendieck Duality and Base Change (2001)
Vol. 1751: N. J. Cutland, Loeb Measures in Practice: Recent Advances (2001)
Vol. 1752: Y. V. Nesterenko, P. Philippon, Introduction to Algebraic Independence Theory (2001)
Vol. 1753: A. I. Bobenko, U. Eitner, Painlevé Equations in the Differential Geometry of Surfaces (2001)
Vol. 1754: W. Bertram, The Geometry of Jordan and Lie Structures (2001)
Vol. 1755: J. Azéma, M. Émery, M. Ledoux, M. Yor (Eds.), Séminaire de Probabilités XXXV (2001)
Vol. 1756: P. E. Zhidkov, Korteweg de Vries and Nonlinear Schrödinger Equations: Qualitative Theory (2001)
Vol. 1757: R. R. Phelps, Lectures on Choquet's Theorem (2001)
Vol. 1758: N. Monod, Continuous Bounded Cohomology of Locally Compact Groups (2001)
Vol. 1759: Y. Abe, K. Kopfermann, Toroidal Groups (2001)
Vol. 1760: D. Filipović, Consistency Problems for Heath-Jarrow-Morton Interest Rate Models (2001)
Vol. 1761: C. Adelmann, The Decomposition of Primes in Torsion Point Fields (2001)
Vol. 1762: S. Cerrai, Second Order PDE's in Finite and In nite Dimension (2001)
Vol. 1763: J.-L. Loday, A. Frabetti, F. Chapoton, F. Goichot, Dialgebras and Related Operads (2001)
Vol. 1764: A. Cannas da Silva, Lectures on Symplectic Geometry (2001)
Vol. 1765: T. Kerler, V. V. Lyubashenko, Non-Semisimple Topological Quantum Field Theories for 3-Manifolds with Corners (2001)
Vol. 1766: H. Hennion, L. Hervé, Limit Theorems for Markov Chains and Stochastic Properties of Dynamical Systems by Quasi-Compactness (2001)
Vol. 1767: J. Xiao, Holomorphic Q Classes (2001)
Vol. 1768: M.J. P aum, Analytic and Geometric Study of Strati ed Spaces (2001)
Vol. 1769: M. Alberich-Carramiñana, Geometry of the Plane Cremona Maps (2002)
Vol. 1770: H. Gluesing-Luerssen, Linear Delay-Differential Systems with Commensurate Delays: An Algebraic Approach (2002)
Vol. 1771: M. Émery, M. Yor (Eds.), Séminaire de Probabilités 1967-1980. A Selection in Martingale Theory (2002)
Vol. 1772: F. Burstall, D. Ferus, K. Leschke, F. Pedit, U. Pinkall, Conformal Geometry of Surfaces in S^4 (2002)
Vol. 1773: Z. Arad, M. Muzychuk, Standard Integral Table Algebras Generated by a Non-real Element of Small Degree (2002)
Vol. 1774: V. Runde, Lectures on Amenability (2002)
Vol. 1775: W. H. Meeks, A. Ros, H. Rosenberg, The Global Theory of Minimal Surfaces in Flat Spaces. Martina Franca 1999. Editor: G. P. Pirola (2002)
Vol. 1776: K. Behrend, C. Gomez, V. Tarasov, G. Tian, Quantum Comohology. Cetraro 1997. Editors: P. de Bartolomeis, B. Dubrovin, C. Reina (2002)
Vol. 1777: E. García-Río, D. N. Kupeli, R. Vázquez-Lorenzo, Osserman Manifolds in Semi-Riemannian Geometry (2002)
Vol. 1778: H. Kiechle, Theory of K-Loops (2002)
Vol. 1779: I. Chueshov, Monotone Random Systems (2002)
Vol. 1780: J. H. Bruinier, Borcherds Products on O(2,1) and Chern Classes of Heegner Divisors (2002)
Vol. 1781: E. Bolthausen, E. Perkins, A. van der Vaart, Lectures on Probability Theory and Statistics. Ecole d' Eté de Probabilités de Saint-Flour XXIX-1999. Editor: P. Bernard (2002)
Vol. 1782: C.-H. Chu, A. T.-M. Lau, Harmonic Functions on Groups and Fourier Algebras (2002)
Vol. 1783: L. Grüne, Asymptotic Behavior of Dynamical and Control Systems under Perturbation and Discretization (2002)
Vol. 1784: L.H. Eliasson, S. B. Kuksin, S. Marmi, J.-C. Yoccoz, Dynamical Systems and Small Divisors. Cetraro, Italy 1998. Editors: S. Marmi, J.-C. Yoccoz (2002)
Vol. 1785: J. Arias de Reyna, Pointwise Convergence of Fourier Series (2002)
Vol. 1786: S. D. Cutkosky, Monomialization of Morphisms from 3-Folds to Surfaces (2002)
Vol. 1787: S. Caenepeel, G. Militaru, S. Zhu, Frobenius and Separable Functors for Generalized Module Categories and Nonlinear Equations (2002)
Vol. 1788: A. Vasil'ev, Moduli of Families of Curves for Conformal and Quasiconformal Mappings (2002)
Vol. 1789: Y. Sommerhäuser, Yetter-Drinfel'd Hopf algebras over groups of prime order (2002)
Vol. 1790: X. Zhan, Matrix Inequalities (2002)
Vol. 1791: M. Knebusch, D. Zhang, Manis Valuations and Prüfer Extensions I: A new Chapter in Commutative Algebra (2002)
Vol. 1792: D. D. Ang, R. Goren o, V. K. Le, D. D. Trong, Moment Theory and Some Inverse Problems in Potential Theory and Heat Conduction (2002)
Vol. 1793: J. Cortés Monforte, Geometric, Control and Numerical Aspects of Nonholonomic Systems (2002)
Vol. 1794: N. Pytheas Fogg, Substitution in Dynamics, Arithmetics and Combinatorics. Editors: V. Berthé, S. Ferenczi, C. Mauduit, A. Siegel (2002)
Vol. 1795: H. Li, Filtered-Graded Transfer in Using Noncommutative Gröbner Bases (2002)
Vol. 1796: J.M. Melenk, hp-Finite Element Methods for Singular Perturbations (2002)
Vol. 1797: B. Schmidt, Characters and Cyclotomic Fields in Finite Geometry (2002)

Vol. 1849: Martin L. Brown, Heegner Modules and Elliptic Curves (2004)
Vol. 1850: V. D. Milman, G. Schechtman (Eds.), Geometric Aspects of Functional Analysis. Israel Seminar 2002-2003 (2004)
Vol. 1851: O. Catoni, Statistical Learning Theory and Stochastic Optimization (2004)
Vol. 1852: A.S. Kechris, B.D. Miller, Topics in Orbit Equivalence (2004)
Vol. 1853: Ch. Favre, M. Jonsson, The Valuative Tree (2004)
Vol. 1854: O. Saeki, Topology of Singular Fibers of Differential Maps (2004)
Vol. 1855: G. Da Prato, P.C. Kunstmann, I. Lasiecka, A. Lunardi, R. Schnaubelt, L. Weis, Functional Analytic Methods for Evolution Equations. Editors: M. Iannelli, R. Nagel, S. Piazzera (2004)
Vol. 1856: K. Back, T.R. Bielecki, C. Hipp, S. Peng, W. Schachermayer, Stochastic Methods in Finance, Bressanone/Brixen, Italy, 2003. Editors: M. Fritelli, W. Runggaldier (2004)
Vol. 1857: M. Émery, M. Ledoux, M. Yor (Eds.), Séminaire de Probabilités XXXVIII (2005)
Vol. 1858: A.S. Cherny, H.-J. Engelbert, Singular Stochastic Differential Equations (2005)
Vol. 1859: E. Letellier, Fourier Transforms of Invariant Functions on Finite Reductive Lie Algebras (2005)
Vol. 1860: A. Borisyuk, G.B. Ermentrout, A. Friedman, D. Terman, Tutorials in Mathematical Biosciences I. Mathematical Neurosciences (2005)
Vol. 1861: G. Benettin, J. Henrard, S. Kuksin, Hamiltonian Dynamics – Theory and Applications, Cetraro, Italy, 1999. Editor: A. Giorgilli (2005)
Vol. 1862: B. Helffer, F. Nier, Hypoelliptic Estimates and Spectral Theory for Fokker-Planck Operators and Witten Laplacians (2005)
Vol. 1863: H. Fürh, Abstract Harmonic Analysis of Continuous Wavelet Transforms (2005)
Vol. 1864: K. Efstathiou, Metamorphoses of Hamiltonian Systems with Symmetries (2005)
Vol. 1865: D. Applebaum, B.V. R. Bhat, J. Kustermans, J. M. Lindsay, Quantum Independent Increment Processes I. From Classical Probability to Quantum Stochastic Calculus. Editors: M. Schürmann, U. Franz (2005)
Vol. 1866: O.E. Barndorff-Nielsen, U. Franz, R. Gohm, B. Kümmerer, S. Thorbjønsen, Quantum Independent Increment Processes II. Structure of Quantum Lévy Processes, Classical Probability, and Physics. Editors: M. Schürmann, U. Franz, (2005)
Vol. 1867: J. Sneyd (Ed.), Tutorials in Mathematical Biosciences II. Mathematical Modeling of Calcium Dynamics and Signal Transduction. (2005)
Vol. 1868: J. Jorgenson, S. Lang, $Pos_n(R)$ and Eisenstein Sereies. (2005)
Vol. 1869: A. Dembo, T. Funaki, Lectures on Probability Theory and Statistics. Ecole d'Eté de Probabilités de Saint-Flour XXXIII-2003. Editor: J. Picard (2005)
Vol. 1870: V.I. Gurariy, W. Lusky, Geometry of Müntz Spaces and Related Questions. (2005)
Vol. 1871: P. Constantin, G. Gallavotti, A.V. Kazhikhov, Y. Meyer, S. Ukai, Mathematical Foundation of Turbulent Viscous Flows, Martina Franca, Italy, 2003. Editors: M. Cannone, T. Miyakawa (2006)
Vol. 1872: A. Friedman (Ed.), Tutorials in Mathematical Biosciences III. Cell Cycle, Proliferation, and Cancer (2006)
Vol. 1873: R. Mansuy, M. Yor, Random Times and Enlargements of Filtrations in a Brownian Setting (2006)
Vol. 1874: M. Yor, M. Émery (Eds.), In Memoriam Paul-André Meyer - Séminaire de Probabilités XXXIX (2006)
Vol. 1875: J. Pitman, Combinatorial Stochastic Processes. Ecole d'Eté de Probabilités de Saint-Flour XXXII-2002. Editor: J. Picard (2006)
Vol. 1876: H. Herrlich, Axiom of Choice (2006)
Vol. 1877: J. Steuding, Value Distributions of L-Functions (2006)
Vol. 1878: R. Cerf, The Wulff Crystal in Ising and Percolation Models, Ecole d'Eté de Probabilités de Saint-Flour XXXIV-2004. Editor: Jean Picard (2006)
Vol. 1879: G. Slade, The Lace Expansion and its Applications, Ecole d'Eté de Probabilités de Saint-Flour XXXIV-2004. Editor: Jean Picard (2006)
Vol. 1880: S. Attal, A. Joye, C.-A. Pillet, Open Quantum Systems I, The Hamiltonian Approach (2006)
Vol. 1881: S. Attal, A. Joye, C.-A. Pillet, Open Quantum Systems II, The Markovian Approach (2006)
Vol. 1882: S. Attal, A. Joye, C.-A. Pillet, Open Quantum Systems III, Recent Developments (2006)
Vol. 1883: W. Van Assche, F. Marcellàn (Eds.), Orthogonal Polynomials and Special Functions, Computation and Application (2006)
Vol. 1884: N. Hayashi, E.I. Kaikina, P.I. Naumkin, I.A. Shishmarev, Asymptotics for Dissipative Nonlinear Equations (2006)
Vol. 1885: A. Telcs, The Art of Random Walks (2006)
Vol. 1886: S. Takamura, Splitting Deformations of Degenerations of Complex Curves (2006)
Vol. 1887: K. Habermann, L. Habermann, Introduction to Symplectic Dirac Operators (2006)
Vol. 1888: J. Van der Hoven, Transseries and differential Algebra (2006)
Vol. 1889: G. Osipenko, Dynamical Systems, Graphs, and Algorithms (2006)
Vol. 1890: M. Bunge, J. Funk, Singular Coverings of Toposes (2006)

Recent Reprints and New Editions

Vol. 1618: G. Pisier, Similarity Problems and Completely Bounded Maps. 1995 – Second, Expanded Edition (2001)
Vol. 1629: J.D. Moore, Lectures on Seiberg-Witten Invariants. 1997 – Second Edition (2001)
Vol. 1638: P. Vanhaecke, Integrable Systems in the realm of Algebraic Geometry. 1996 – Second Edition (2001)
Vol. 1702: J. Ma, J. Yong, Forward-Backward Stochastic Differential Equations and their Applications. 1999. – Corrected 3rd printing (2005)

4. Careful preparation of the manuscripts will help keep production time short besides ensuring satisfactory appearance of the finished book in print and online. After acceptance of the manuscript authors will be asked to prepare the final LaTeX source files (and also the corresponding dvi-, pdf- or zipped ps-file) together with the final printout made from these files. The LaTeX source files are essential for producing the full-text online version of the book (see http://www.springerlink.com/openurl.asp?genre=journal&issn=0075-8434 for the existing online volumes of LNM).

   The actual production of a Lecture Notes volume takes approximately 8 weeks.

5. Authors receive a total of 50 free copies of their volume, but no royalties. They are entitled to a discount of 33.3 % on the price of Springer books purchased for their personal use, if ordering directly from Springer.

6. Commitment to publish is made by letter of intent rather than by signing a formal contract. Springer-Verlag secures the copyright for each volume. Authors are free to reuse material contained in their LNM volumes in later publications: A brief written (or e-mail) request for formal permission is sufficient.

Addresses:

Professor J.-M. Morel, CMLA,
École Normale Supérieure de Cachan,
61 Avenue du Président Wilson, 94235 Cachan Cedex, France
E-mail: Jean-Michel.Morel@cmla.ens-cachan.fr

Professor F. Takens, Mathematisch Instituut,
Rijksuniversiteit Groningen, Postbus 800,
9700 AV Groningen, The Netherlands
E-mail: F.Takens@math.rug.nl

Professor B. Teissier, Institut Mathématique de Jussieu,
UMR 7586 du CNRS, Équipe "Géométrie et Dynamique",
175 rue du Chevaleret
75013 Paris, France
E-mail: teissier@math.jussieu.fr

For the "Mathematical Biosciences Subseries" of LNM:

Professor P. K. Maini, Center for Mathematical Biology,
Mathematical Institute, 24-29 St Giles,
Oxford OX1 3LP, UK
E-mail : maini@maths.ox.ac.uk

Springer, Mathematics Editorial, Tiergartenstr. 17,
69121 Heidelberg, Germany,
Tel.: +49 (6221) 487-8410
Fax: +49 (6221) 487-8355
E-mail: lnm@springer-sbm.com